MANAGING BUSINESS INTERFACES

**INTERNATIONAL SERIES
IN QUANTITATIVE MARKETING**

Series Editor:
Jehoshua Eliashberg
The Wharton School
University of Pennsylvania
Philadelphia, Pennsylvania USA

Other books in the series:

Zaccour, G.
Differential Games in Marketing
Wind ,Yoram (Jerry) and Green, Paul E.
Marketing Research and Modeling:
Progress and Prospects
Erickson, Gary M.:
Dynamic Models of Advertising Competition, 2nd Ed
Hanssens, D., Parsons, L., and Schultz, R.:
Market Response Models:
Econometric and Time Series Analysis, 2nd Ed
Mahajan, V., Muller, E. and Wind, Y.:
New-Product Diffusion Models
Wierenga, B. and van Bruggen, G.:
Marketing Management Support Systems:
Principles, Tools, and Implementation
Leeflang, P., Wittink, D., Wedel, M. and Naert, P.:
Building Models for Marketing Decisions
Wedel, M. and Kamakura, W.G.:
Market Segmentation, 2nd Ed
Wedel, M. and Kamakura, W.G.:
Market Segmentation
Nguyen, D.:
Marketing Decisions Under Uncertainty
Laurent, G., Lilien, G.L., Pras, B.:
Research Traditions in Marketing
Erickson, G.:
Dynamic Models of Advertising Competition
McCann, J. and Gallagher, J.:
Expert Systems for Scanner Data Environments
Hanssens, D., Parsons, L., and Schultz, R.:
Market Response Models:
Econometric and Time Series Analysis
Cooper, L. and Nakanishi, M.:
Market Share Analysis

MANAGING BUSINESS INTERFACES

Marketing and Engineering Issues in the Supply Chain and Internet Domains

edited by

Amiya K. Chakravarty
A.B. Freeman School of Business
Tulane University

Jehoshua Eliashberg
The Wharton School
The University of Pennsylvania

Springer

Library of Congress Cataloging-in-Publication Data

Library of Congress Cataloging-in-Publication Data

A C.I.P. Catalogue record for this book is available
from the Library of Congress.

ISBN 1-4020-7614-2 (HC) ISBN 0-387-24378-X (SC) e-ISBN 0-387-25002-6

Printed on acid-free paper

First softcover printing, 2005

Printed in the United States of America.

9 8 7 6 5 4 3 2 1 SPIN 11378341

springeronline.com

To my wife, Indira, my son, Deepak, and my daughter, Nisha, for their love, encouragement, and patience.–Amiya

To my wife, Tova, and my daughter, Keren, for their continuous love and support.–Josh

CONTENTS

INTRODUCTION TO MANAGING BUSINESS INTERFACES: MARKETING, ENGINEERING, AND MANUFACTURING PERSPECTIVES

In the last two decades emerging business technologies have created new opportunities for enhancing both intra and inter organizational effectiveness. Companies are increasingly focusing on subsets of activities, while collaborating with partners in a supply or demand chain. Within an organization there is an increased emphasis on making different functional areas work together seamlessly.

These developments have led to an increased emphasis on research and practice in business that integrate the functional areas within and between business entities. Such initiatives are now evident in developing new products, and managing the associated supply chains. For instance, product development teams, comprising experts from marketing, finance, engineering, and manufacturing, with different backgrounds and perspectives, must coordinate with one another to achieve common goals. Similarly, in a supply chain, independent companies must coordinate their manufacturing and distribution activities to minimize delays and waste in the system.

The research community has recognized the importance of addressing these differing, and often conflicting, business perspectives. This has led to research streams that address issues characterizing the domain of business interfaces. These include: the benefits of coordination, new product development, product portfolio management, supply chain coordination, and partnerships and collaboration in the Internet space.

This book provides state-of-the-art summary as well as new thoughts, in managing business interfaces. Through eleven invited chapters it brings together the latest developments in leading edge research related to new product development, supply chain management, e-business operations, and field studies. The book is organized in to four parts, as follows:

Part 1—**New Products and Existing Product Portfolio Management** includes two contributions. **Chapter 1**: "Investment in Facility Flexibility for Early Market Entry under Competitive Uncertainty" by Franza and Gaimon presents a model that provides insights into the ability of the firm to rapidly changeover its manufacturing facility from one high-volume product to the next generation. This time-based competitive strategy links operations and marketing aspects and it allows the firm to exit and enter new markets with new products effectively. An appropriate investment in the production facility is essential in situations when short product life cycles, product obsolescence, redundancy, and poor performance dictate strategic changes. This theme is also in **Chapter 2**: "Demand Modeling in Product Line Trimming: Substitutability and Variability" by Chong, Ho and Tang. Here, the authors, responding to the need to develop a formal product portfolio trimming procedure, present a model that explicitly captures the product substitution (i.e., redundancy) phenomenon. Such a procedure allows the firm to streamline its marketing, manufacturing, and logistics functions.

Part 2—**Supply Chain Coordination** includes three chapters. **Chapter 3**: "Coordinated Pricing and Production/Procurement Decision: A Review" by Yano and Gilbert provides a review of the literature addressing coordination and simultaneous decision-making in the supply chain. In particular, the authors explore the benefits of coordination under situations such as demand smoothing, demand uncertainty, and set-up cost under different demand conditions. One scenario considered by the authors involves contractual arrangements such as quantity discounts that allow suppliers to extract more profits. These themes are discussed in more depth in the next two chapters. In **Chapter 4**: "Contractual Relationships and Coordination in Distribution Channels" by Iyer and Padmanabhan, the authors examine externalities and moral hazard issues characterizing supply chains consisting of an "upstream" firm that supplies a "downstream" intermediary which in turn supplies the final consumer. **Chapter 5**: "Quantity Discounts for Supply Chain Coordination" by Choi, Lei, and Wang focuses on models addressing supply chain coordination where quantity discount is employed in the transfer price as a mechanism for incentive compatibility. Quantity discount schedules based on transaction efficiency, demand elasticity as well as their combination are reviewed and discussed.

Part 3—**Interfaces in the Internet Space** consists of three chapters. In **Chapter 6**: "Pricing Internet Service" Steinberg argues that if Internet Service Providers want to remain competitive, they need to set their prices carefully because pricing is increasingly seen by engineers as an essential tool for controlling network congestion. This defines an interface domain that links marketing and engineering. He then provides a review of pricing mechanisms such as smart market, Paris Metro pricing, Max-Min fairness, proportional fairness, CPE pricing, and the contract and balancing process. A related topic- usage volume and value segmentation-is analyzed in **Chapter 7**: "Usage Volume and Value Segmentation in Business Information Services" by Bashyam and Karmarkar. Here, the authors identify three options available for users of business information services: package, online (Internet), and self-service. They demonstrate that under certain costs conditions package and online services can co-exist. Under different costs conditions, however, the package service provider may be eliminated from the market. This part of the book concludes integratively with **Chapter 8**: "Collaboration in E-Business: Technology and Strategy" where Chakravarty provides a general framework for collaboration in the Internet space, among members of the supply chain. Activities involving collaboration with *suppliers* such as: information visibility, logistics management, production planning, product development, and resource sharing as well as activities involving collaboration with *customers* such as: revenue enhancement, demand management, order fulfillment, and customer acquisition are reviewed and discussed. The chapter focuses on how collaboration creates value, how online technologies facilitate collaboration, how can the firm invest in an optimal technology portfolio, and how the technology portfolio can be deployed.

Part 4—**Field Studies and Real-World Evidence** consists of three chapters. In **Chapter 9**: "Levers for Improving the NPD Time and Financial Performance" Bajaj, Kekre, and Srinivasan discuss some key managerial drivers, which appear critical, in the design phase of new product development projects—based on research conducted at an avionics and missile guidance manufacturer. The factors identified influence the financial performance and time-to-market metrics. These factors integrate concepts from marketing, operations, and engineering. **Chapter 10**: "Platform Products: Theory and Evidence from International Practice Linking Manufacturing, Marketing, and Product Design" by Kumar and Hadjinicola presents a model as well as various anecdotal evidences for a firm operating in two countries. The objective is to explore the impact of platform products on profitability. Marketing issues considered are: product positioning and pricing. Manufacturing issues considered are: factory location, inventory, economies of scale, levels of product attributes, and postponement. **Chapter 11**: "The Marketing-Manufacturing Interface: Strategic Issues" by Karmarkar and Lele provides an overview, in the context of real-world case examples, of problems that arise in situations where the need for coordination of the marketing and manufacturing functions is not explicitly recognized. Manufacturing issues that receive attention include: capacity deployment, inventory management, and location. Marketing issues covered include: pricing, promotion, forecasting and market intelligence.

Amiya Chakravarty	Jehoshua Eliashberg
New Orleans, Louisiana	Philadelphia, Pennsylvania

I. NEW PRODUCTS AND EXISTING PRODUCT PORTFOLIO MANAGEMENT

1. INVESTMENT IN FACILITY FLEXIBILITY FOR EARLY MARKET ENTRY UNDER COMPETITIVE UNCERTAINTY

RICHARD M. FRANZA and CHERYL GAIMON

ABSTRACT

Time-to-market has become an increasingly important competitive criterion in the global business arena. Pressure for early market entry poses serious challenges to traditional high-volume manufacturers. Such firms rely on dedicated facilities to manufacture standard, low-cost products with long life cycles. The nature of dedicated technology, however, restricts the firms from rapidly exiting obsolete markets and entering new ones.

A model is introduced to analyze the investment in flexible technology for a firm manufacturing a series of high-volume products with short life cycles on a single facility. Here, flexibility measures the firm's ability to rapidly changeover its facility for the manufacture of successive high-volume product generations. The investment in flexibility, however, has implications beyond operations. In particular, the ability to rapidly changeover from the manufacture of one high-volume product to the next allows a firm to pursue an early market entry strategy. Therefore, we characterize an important link between operations and marketing.

The model reflects two key drivers of the competitive environment in which the high-volume manufacturer operates. First, the competitor's market entry time for each new product generation (i.e., the competitor's investment in facility flexibility) is unknown. As a result, we can explore the impact on a firm's investment in changeover flexibility in response to various levels of uncertainty regarding the competitor's market entry strategy. Second, the facility flexibility investment decision is examined in the context of contrasting market conditions. In one competitive environment, the first entrant maintains a sustained market advantage over subsequent entrants. This market phenomenon is often observed for high technology products. In contrast, we also examine the situation where the first entrant's advantage is transient.

Analysis of the model provides important insights linking external market forces to a firm's investment in flexibility. We show that in markets where the pioneer has a sustained competitive advantage, regardless of whether a firm leads or follows, if the competitor delays its market entry then the firm's initial strategy is reinforced. In other words, if the firm is the leader in a market where the pioneer advantage is sustained, then in response to a delay in the competitor's market entry, the firm optimally increases its investment in flexibility to realize an earlier entry strategy. Similarly, in markets where the pioneer has a transient competitive advantage, regardless of whether a firm leads or follows, if the competitor delays its market entry, then the firm's initial strategy is reversed.

Results are obtained contrasting the firm's reaction to economic risk and competitive risk. Specifically, a firm invests in less facility flexibility in response to a greater level of economic risk (discount rate) but more facility flexibility in response to a greater level of uncertainty (risk) associated with the competitor's market entry strategy. Results also demonstrate that regardless of the market condition assumed, as the extent of uncertainty associated with the competitor's market entry strategy increases, the firm optimally invests in more facility flexibility. Lastly, results are obtained demonstrating the competitive advantage realized by the firm capable of pursuing process-based technological innovation or high rates of volume-based learning.

1. INTRODUCTION

Time-to-market has become an increasingly important competitive criterion in the global business arena (Stalk and Hout (1990), Blackburn (1991)). An early entrant in a new product market typically gains market share and revenue advantages over a later entrant (Urban et al. (1986), Kerin et al. (1992)). In addition, shrinking product life cycles, increasing rates of product obsolescence, and intensifying global competitive forces provide further incentives for early market entry (Griffin (1993)).

For high-volume manufacturers, early market entry requires a rapid facility changeover from the production of one generation to the next. The need for high-volume producers to be capable of rapid exit and entry to new markets is a relatively new phenomenon. Traditionally, these firms invest in mass production facilities consisting of dedicated equipment to manufacture a single product with a long life cycle. The dedicated facilities allow these firms to compete favorably in environments that reward high-volume output of consistent quality and low unit cost. Unfortunately, forces such as short product life cycles and rapid product obsolescence render dedicated facilities ineffective due to their inability to exit obsolete markets and enter new markets in a timely and cost effective manner.

Recent technological advances have provided high-volume producers with an alternative to the traditional dedicated facility. For example, Japanese automakers such as Toyota and Nissan have made substantial investments in new facilities capable of rapid changeover between successive models (Johnson and Maskery (1992), Stewart (1992)). With the rapid market entry capability, these firms are able to introduce a greater number of product generations to the marketplace in a given time interval. In response, General Motors and Ford have followed with similar investments.

The concept of manufacturing flexibility has been the subject of a substantial body of research. The primary focus of that research concerns mix flexibility as opposed to changeover flexibility. Mix flexibility is the ability of a manufacturing system to simultaneously produce different product types (Gupta and Somers (1992), Gerwin (1993), Suarez et al. (1995)). Mix flexibility can be measured by the extent of product variety manufactured by a system without incurring major setups (Sethi and Sethi (1990), Gerwin (1993)). In contrast, changeover flexibility refers to the time and cost required to ready a facility for the manufacture of the next high-volume product generation (Gerwin (1993)). A facility with inherently more changeover flexibility completes the high-volume changeover in less time and at a lower cost. While considerable research has explored the tradeoffs of investing in mix flexibility (Hutchinson and Holland (1982), Fine and Li (1988), Fine and Freund (1990)), only limited attention has been given to the investment in changeover flexibility (Gaimon and Singhal (1992) and Franza and Gaimon (1998)). Regardless of the focus on mix or changeover flexibility, the research issues considered are relevant to both marketing and operations. A comprehensive treatment of the interface between marketing and operations is given in Chakravarty (2001).

We examine the investment in changeover flexibility for a high-volume manufacturer. The level of changeover flexibility determines the firm's capability for the early market entry of successive new product generations. Therefore, our analysis provides important insights on the integration of a firm's manufacturing and marketing strategies. We evaluate a firm's internal manufacturing capabilities in relation to various external market forces and competition. Our results demonstrate the value of developing new flexible technology and the importance of reducing the time and cost of changeovers between successive high-volume product generations.

1.1 External market factors

The operations and marketing interface addressed here focuses on flexible technology as an enabler of an early market entry strategy for a firm that manufactures a series of high-volume product generations. The advantages of early market entry have been explored in a significant body of marketing literature (see Kerin et al. (1992) for a review). Numerous empirical studies (e.g., Biggadike (1976), Dillon et al. (1979), Robinson and Fornell (1985), Urban et al. (1986), Robinson (1988), Carpenter and Nakamoto (1989)) have shown that pioneers or first entrants realize market share and revenue advantages over later entrants.

Beyond order of market entry, total revenue is also driven by the amount of time that lapses between the entry of a firm and its competitor (Brown and Lattin (1994), Huff and Robinson (1994)). The earlier a pioneer enters the marketplace relative to the second entrant, the greater the pioneer's competitive advantage. However, in some markets the pioneer's advantage is sustainable while in others the advantage is transient. According to Arthur (1996), in high technology product markets (e.g., pharmaceuticals, computer hardware and software, telecommunications equipment), the pioneer's competitive advantage is sustained or may even increase over the product's life cycle. For example, the pioneer earns increasing returns due to the value of the knowledge required to develop the products and processes, the ability of the first entrant to create the industry standard, and consumer reluctance to switch brands. Conversely, pioneering advantage is transient for firms manufacturing products such as heavy chemicals, foodstuffs, and retail goods where less knowledge is required for development and production and consumer brand loyalty is more limited.

1.2 Internal operations effects

Beyond the marketing-based benefits, high-volume manufacturers must also consider the impact on operations due to investment in changeover flexibility. First, flexible facilities require a higher initial investment than dedicated facilities (Gaimon and Singhal (1992), Stewart (1992)). Second, production efficiency is inferior for flexible as compared to dedicated (task specialized) facilities (Gerwin (1993)). As a result, with a flexible facility, a higher unit cost is incurred at the start of the manufacturing cycle for each product generation. In contrast, investment in changeover flexibility also provides operations benefits. As flexibility increases, the cost incurred to changeover the facility from the manufacture of one high-volume product to the next declines substantially. This is important for firms operating in short life cycle markets.

1.3 Focus of this research

Analysis of changeover flexibility was overlooked until Gaimon and Singhal (1992). While this research provides a reasonable foundation for the study of changeover flexibility, its results are somewhat limited by assumptions regarding the functional relationships between flexibility, costs, and profit. Franza and Gaimon (1998) further develop the manufacturing and marketing strategy interface in a more general model that includes price as decision variable. In particular, the authors explore the effect on the unit production cost and thereby on price due to manufacturing efficiency versus volume-based learning derived from early market entry.

This paper contributes to the literature in several ways. First, in contrast to prior research where the impact of the competitive environment is only inferred, our formulation explicitly measures time-to-market competition between a firm and its competitor. Also in contrast to the prior deterministic approaches, our model is stochastic to capture uncertainty associated with a competitor's market entry time. Third, characteristics of the model are sufficiently general to permit analysis of markets where the first entrant advantage is sustained throughout each product generation life cycle versus transient.

Analytic results are presented in Section 3 demonstrating that the investment in changeover flexibility is substantially different depending on whether the firm operates in a market where the first entrant advantage is sustained or transient. Analytic results are given in Section 4 demonstrating the impact on a firm's market entry strategy in relation to internal operations capabilities such as the extent that flexibility increases the initial facility investment, decreases the facility changeover cost, and reduces manufacturing efficiency. These results are particularly important since they demonstrate the competitive advantage realized by the firm capable of pursuing process-based technological innovation or high rates of volume-based learning. Results are obtained (Sections 3 and 4) contrasting the firm's reaction to economic risk (discount rate) versus competitive risk (level of uncertainty associated with the competitor's market entry strategy). Numerical examples are presented in Section 5 that both illustrate and broaden the above analytic insights. Lastly, Section 6 contains the conclusions and a discussion of potential extensions of this research.

2. THE MODEL

Broadly speaking, we present a model that characterizes the facility planning decision. In particular, we determine the firm's *level of investment in changeover flexibility* for a newly built manufacturing facility that will produce successive generations of high-volume products with short life cycles. It is important to recognize that the firm's level of investment in facility flexibility is a *one-time decision* occurring at the beginning of the planning horizon. Moreover, the one-time decision establishes the firm's long-term resource capability and impacts the market entry, revenue, production cost, and changeover cost in precisely the same way for each successive product generation. For example, as the investment in facility flexibility increases, the firm realizes an earlier market entry for *each* successive product generation. Similarly, as facility flexibility increases, the firm incurs a lower cost for *each* high-volume changeover.

The purpose of the model is to develop insights regarding the impact of flexibility on the firm's competitive strategy over the life of a new manufacturing facility. Consistent with the strategic focus, total profit is expressed as the *average* obtained from each product generation times the number of generations manufactured over the life of the facility. The use of averages reflects the fact that the initial investment in flexibility impacts the firm's ability to generate profit in *precisely the same way* for each product generation. We do not, however, seek insights at the more tactical decision-making level that depict the impact of flexibility on each individual product generation. Instead, our conceptualization allows us to develop strategic insights without undermining the robustness of the model in terms of its ability to capture the key drivers of the one-time facility flexibility decision.

The use of average values has two other important attributes. First, it simplifies the mathematics involved and thereby permits the straightforward interpretation of important strategic results. Second, by using averages, we avoid the difficulties that would be encountered in estimating the revenue and cost parameters for each product generation manufactured over the life of the facility. In fact, the use of average estimates for these parameters essentially pools and diffuses the effect of individual product specific forecast errors. A similar approach was taken in a related but deterministic analysis of this problem (see Franza and Gaimon 1998).

2.1 Facility and product life cycles

The level of changeover flexibility is determined for a manufacturing facility constructed with an estimated *useful life* denoted by **F**. The useful life represents the length of time the firm expects to operate the facility for the manufacture of a series of high-volume product generations. As an exogenous parameter, the facility's life reflects the natural deterioration of manufacturing systems over time. Operating the facility beyond this time leads to prohibitively high maintenance costs. The value of **F** is estimated based on past experience, forecasts of major process or product innovations, and planned maintenance policies.

The length of the *average life cycle* for each high-volume product to be manufactured on the single facility is given by **L**. The value of **L** may be estimated from forecasts of both product and process innovations and forecasts of competitive forces in the marketplace (Porter et al. (1991)). For example, suppose a firm and its competitor are PC makers. Both firms actively pursue the development of new PC generations in anticipation of innovations in computer chip technology. The history of the computer chip industry has supported Moore's 1965 projection that a new generation chip will be developed approximately every 18 months. Therefore, a reasonable estimate of the average life cycle for PC makers is **L** = 18 months.

Time zero denotes the start of demand for each new product generation. At this time, consumers are cognizant of the new product under development that will be released in the near future. Moreover, consumers will only purchase the previous product generation if it is deeply discounted. This phenomenon is empirically observed in the PC industry. Specifically, when the introduction of a new generation PC containing the latest developments in chip technology is imminent, the previous generation can only be sold at a substantial discount. Therefore, once demand for the next generation product begins, the relative contribution to profit from the previous generation is extremely small (Baker (1989), Forbes Special Report (1991)).

Clearly, a firm will not optimally *enter* a product market *after* the next generation's life cycle has begun. As a consequence, we can focus our analysis of investment in facility flexibility on those values associated with market entry times that occur during the intervals of non-overlapping product life cycles. In other words, to determine a firm's investment in facility flexibility, we may assume the start of one generation's life cycle (time zero) effectively triggers the end of the previous generation's life cycle (time **L**).

Given estimates for the useful life of the facility and the length of the average product life cycle, the *number of product generations* manufactured on the single facility is expressed as **n** = **F/L**. Since **F** and **L** are exogenous, the estimated number of products manufactured on the facility is exogenous, as well.

2.2 Flexibility and market entry time

While consumer demand for each new product generation begins at time zero, the firm's ability to earn revenue starts only after the facility changeover is complete. Therefore, the firm's market entry time occurs at some time after the start of demand (time zero). The firm's initial investment in flexibility dictates the speed with which facility changeovers are realized for the manufacture of each successive high-volume product. It follows that the one-time decision regarding changeover flexibility is a proxy for the firm's market entry strategy.

Let **m** denote the *average market entry time* for each successive product following the start of each life cycle at time zero (decision variable). Therefore, **m** represents the average time required to complete each high-volume facility changeover, and $\mathbf{L} - \mathbf{m}$ is the average length of the production cycle for each of the $\mathbf{n}$ products manufactured. A high (low) level of facility flexibility is associated with an early (late) market entry time and a small (large) value of $\mathbf{m}$. Under reasonable conditions, the profit maximizing solution for $\mathbf{m}$ satisfies $\mathbf{m}\varepsilon(0,\mathbf{L})$ for two reasons. First, the initial investment in a facility capable of instantaneous changeover ($\mathbf{m} = 0$) between successive high-volume product generations is extremely high and possibly infinite. (An infinite cost occurs if instantaneous changeover is technologically impossible.) Also, the firm who enters a short product life cycle market after the next generation has been introduced ($\mathbf{m} \geq \mathbf{L}$) earns an extremely limited (possibly negligible) amount of revenue.

The firm's competitor faces an analogous relationship between facility flexibility and market entry. However, from the perspective of the decision-making firm, the competitor's planned initial investment in facility flexibility is unknown. Let $\boldsymbol{\delta}\varepsilon(0,\mathbf{L})$ denote the random variable representing the *competitor's market entry time* (i.e., the competitor's level of facility flexibility). The probability density function and cumulative distribution of the random variable are denoted by $\mathbf{g(\boldsymbol{\delta})}$ and $\mathbf{G(\boldsymbol{\delta})}$, respectively. Clearly, the time required by the competitor to complete each high-volume product changeover is given by $\boldsymbol{\delta}$. Similarly, the competitor's manufacturing cycle for each product generation is measured by $\mathbf{L} - \boldsymbol{\delta}$.

While driven by the firm's one-time investment in facility flexibility, small differences in the market entry times of successive product generations are possible. For example, some manufacturing changeovers may require less flexibility than others. Also, learning may occur so that the facility changeover time for an early product generation may exceed that of a later generation. These minor variations are reflected by our definition of $\mathbf{m}$ as the *average* market entry time for all product generations manufactured on the single facility.

2.3 Cost relationships

In this section, we introduce the costs incurred by the firm over the life of the facility. Each cost is expressed as a function of the firm's market entry time, $\mathbf{m}$.

2.3.1 Initial facility investment cost

$\mathbf{I(m)}$ represents the *initial facility investment cost* and includes manufacturing, material handling equipment, and tooling costs. Moreover, the investment cost embodies the firm's long-term commitment to a particular volume of productive capacity. Naturally, the investment cost is incurred once, at the start-up of the facility. Based on empirical evidence in Lederer and Singhal (1988) as well as data on expenditures for flexible facilities by Japanese automakers (Johnson and Maskery (1992)), we know the investment cost increases at an increasing rate as the level of facility flexibility increases. Alternatively said, $\mathbf{I(m)}$ decreases at a decreasing rate as $\mathbf{m}$ increases. For example, the maximum investment cost is incurred for a facility capable of instantaneous changeovers between successive high-volume generations. Of course, if $\mathbf{m} = 0$ is technologically infeasible, $\mathbf{I}(0)$ is infinite. This gives us the first and second

Table 1. Cost Function Relationships

Cost Component	First Order Condition		Second Order Condition	
Investment	$\mathbf{I}_m < 0$	(1a)	$\mathbf{I}_{mm} > 0$	(1b)
Changeover	$\mathbf{A}_m > 0$	(2a)	$\mathbf{A}_{mm} > 0$	(2b)
Manufacturing	$\mathbf{C}_m < 0$, $\mathbf{C}_L > 0$	(3a)	$\mathbf{C}_{mm} > 0$	(3b)

order conditions in Equation (1), Table 1. Note that first and second order derivatives are denoted as subscripts.

2.3.2 *Average changeover cost*

A(m) denotes the *average changeover cost* of adapting the facility from the manufacture of one high-volume product generation to the next. The cost is incurred **n** times, at the beginning of each changeover cycle. The high-volume changeover of a less flexible facility is more costly due to the greater expenses incurred for new manufacturing and materials handling equipment, new tooling, increased reorganization of the shop floor, and the disposal of a larger portion of the existing equipment. However, as the level of facility flexibility increases, the reduction in changeover cost occurs at a decreasing rate due to diminishing returns. Mathematically, as **m** increases, **A(m)** increases at an increasing rate as depicted by the first and second order derivatives in Equation (2) of Table 1. Theoretically, the minimum changeover cost incurred by the maximum flexible facility (denoted by **A**(0)) is extremely small.

Lastly, consider the magnitude of **I(m)** in relation to **A(m)**. Clearly, the minimum changeover cost is several orders of magnitude smaller than the cost incurred for investing in the maximum flexible facility (i.e., $\mathbf{I}(0) \gg \mathbf{A}(0)$ holds). Moreover, while the difference between the costs for facility investment and changeover declines as **m** increases, given our context of short life cycle products, it is reasonable to assume that $\mathbf{I(m)} \gg \mathbf{A(m)}$ holds for all $\mathbf{m} \varepsilon (0,\mathbf{L})$.

2.3.3 *Total production cost*

The *average production cost* for each of the **n** products manufactured over the life of the single facility, denoted by **C(m,L)**, is expressed as a function of the level of changeover flexibility and the average product life cycle. First, consider the relationship between the production cost and changeover flexibility. A more flexible facility allows the firm to achieve an earlier market entry and thereby increases total demand for each generation. Due to the greater level of output needed to meet demand, the firm incurs a higher total production cost for each product generation. In addition, relative to a more dedicated environment, as facility flexibility increases, the firm's production costs increase at an increasing rate as a result of the considerable loss in efficiency and the need to manage an increasingly complex manufacturing system (Venkatesan (1990)). This gives us the first and second order conditions with respect to **m** in Equation (3) of Table 1.

Second, consider the relationship between the average product life cycle and the total production cost for each generation. As the average life cycle increases, the duration of the average manufacturing cycle increases. Due to the greater volume of output to be met over the extended manufacturing cycle, the firm incurs a higher total production cost for each generation. This gives us the first order condition with respect to **L** in Equation (3a) of Table 1. However, as the duration of the average life cycle increases, the number of products manufactured over the life of the single facility decreases, **(n = F/L)**. Therefore, as **L** increases, the ultimate impact on the total manufacturing cost earned over the life of the facility, **nC(m,L)**, is not known.

The sign of the second order condition characterizing the impact of **L** on **C(m,L)** is unclear since, in short life cycle environments, demand for each product generation may be truncated during growth or maturity (Meredith (1987), Baker (1989)). If truncated during the growth phase, as the average life cycle increases, both the total volume of output and the total manufacturing cost increase at an increasing rate for each generation. In contrast, if demand is truncated during maturity and the average life cycle increases, the total output and the total manufacturing cost increase at a decreasing rate for each generation.

Two forms of learning affect the total manufacturing cost incurred for each product generation. First, the firm's ability to reduce its production cost through volume-based learning is enhanced by longer manufacturing cycles associated with either an earlier market entry strategy or extended product life cycles. Given our context of short life cycles, however, the potential to derive volume-based learning benefits is limited. We assume, therefore, that if a firm invests in more facility flexibility, the loss in manufacturing efficiency dominates the benefits derived from volume-based learning so that the total production cost for each generation increases. Second, the firm may realize considerable benefits from the transfer of learning between successive product generations. In our model, a firm capable of applying knowledge from the manufacture of one generation to the next realizes a lower average production cost across all **n** products (i.e., the value of **C(m,L)**).

Lastly, we recognize that the competitor's market entry strategy impacts the firm's volume of demand and output, and thereby affects the total production cost incurred for each product generation. This relationship, however, is embodied in our characterization of the firm's net revenue function as described in the next section.

2.4 Net Revenue

The nature of the competitive marketplace in which the firm operates is captured by our characterization of net revenue. Before introducing its mathematical form, we discuss the underlying relationship between the firm's market entry time and net revenue.

Empirical research shows that early entrants have market share and revenue advantages over later entrants. For instance, the first entrant "can gain recognition and establish reputation in the marketplace which will generate word-of-mouth" (Lilien and Yoon (1990)). In addition, after purchasing the first entrant's product, buyers often remain loyal customers even if less costly products are available later (Schmalensee (1982)). Biggadike (1976) examines 40 industrial products and finds that pioneers generally maintain market share advantage throughout the product life cycle. Dillon, Calantone, and Worthing (1979) study 174 industrial products and find that pioneering is one of the primary determinants of long-term success of a new product. Based on a study of 129 successful consumer products, Urban et al. (1986) report that the second entrant obtains, on average, only three quarters of the market share of the pioneer and that later entrants capture progressively smaller shares. Other empirical studies (Robinson and Fornell (1985), Robinson (1988), Carpenter and Nakamoto (1989)) report similar findings.

Beyond the order of market entry, market share and revenue are driven by the amount of time that elapses between the firm's market entry and that of its competitor. With additional data appended to the study by Urban et al. (1986), Brown and Lattin (1994) show that the pioneer advantage over the second entrant is related to length of time the pioneer is alone in the market. However, the authors indicate that at least part of the first entrant's competitive advantage is transient; i.e., a portion of the pioneer's advantage is lost upon competitor market entry.

In a study of 34 categories of consumer goods, Huff and Robinson (1994) show that as the difference between the market entries of the first and second entrant increases, the pioneer's market performance increases. In addition, Huff and Robinson's results indicate that, in some markets, the pioneer's advantage is more sustainable while in others the advantage is transient. Lastly, as the duration of competitive rivalry increases (or as product life cycles increase), it is shown that the later entrant has greater opportunity to chip away at the pioneer's advantage.

Characteristics of markets where the pioneer advantage is likely to be sustained are discussed in both the marketing and innovation literature. According to Schmalensee (1982) and Carpenter and Nakamoto (1989), the pioneer gains consumer trust by being first to demonstrate the validity of the espoused new product features. Therefore, even after the second entrant arrives in the market, the consumer is likely to continue to purchase the pioneer's product at a greater rate. This behavior is particularly prevalent for high technology products where rapidly changing technical advances drive state-of-the art features. For example, in her study of the semiconductor industry, Flaherty (1983) states that when a firm purchasing semiconductor components is satisfied with the product offered by the first entrant, that vendor relationship is maintained through subsequent generations thereby providing the first entrant with a sustained advantage. Similar findings are reported by Spital (1983) in the innovation theory domain.

Von Hippel (1983) offers two explanations of the sustained pioneer advantage in high technology markets. First, rapid technology change requires fast decision making on the part of the consumer. Second, when less time is available to fully evaluate alternative product choices, the initial consumer knowledge of the pioneer is afforded relatively more weight. Moreover, according to Arthur (1996), once a customer learns how to use the first high technology product, he or she is highly reluctant to change brands due to the potential switching costs incurred.

Despite the above empirical results, first market entry does not always guarantee long-term success. For example, if the first entrant rushes to market with inadequate product testing, the duration of any competitive advantage is greatly diminished, (Lilien and Yoon (1990), Golder and Tellis (1993)). Also, the first entrant may not realize a sustained advantage if it fails to sufficiently advertise, sets the wrong price, or incorrectly markets the new product (Lilien and Yoon (1990)). Lastly, according to Huff and Robinson (1994), in less technologically dependent product markets, the extent of the first entrant market advantage erodes rapidly over time.

The above empirical results are captured in our model as follows. To maintain our focus on analysis of investment in facility flexibility, we assume the firm adequately tests and markets its products such that the first entrant realizes a revenue-based competitive advantage. However, we do not preclude the possibility that the first entrant's market advantage is transient. Specifically, we compare the strategic interface between marketing and manufacturing for firms operating in markets where the first entrant's advantage is sustained versus transient.

2.4.1 *First order net revenue relationships*

The *average net revenue* earned by the firm for each of the **n** high-volume products sequentially manufactured on the single facility is denoted by $\mathbf{R}^{ij}(\mathbf{m},\boldsymbol{\delta},\mathbf{L})$. Therefore, net revenue is expressed as a function of the average life cycle, market conditions, the firm's market entry time, and the competitor's market entry time. The subscript i denotes the type of market: i = S refers to a market with sustained pioneer advantage, whereas i = T represents a market with transient pioneer advantage. The subscript j indicates whether the decision making firm is the market leader (j = 1) or follower (j = 2). For example, $\mathbf{R}^{S1}$ is the average net revenue earned by a firm that is the first entrant in a market with sustained pioneer advantage.

First, consider the relationship between net revenue and the length of the average product life cycle. As the average life cycle increases (decreases), the manufacturing cycle is extended (decreases) so that the amount of net revenue amassed by the firm for each generation increases (decreases). This gives us Equation (4a) in Table 1. Also, as the average life cycle increases, the number of products manufactured over the life of the facility decreases, $(\mathbf{n} = \mathbf{F}/\mathbf{L})$. Therefore, as **L** increases, the impact on total revenue over the life of the facility, $\mathbf{nR}^{ij}(\mathbf{m},\boldsymbol{\delta},\mathbf{L})$, is unclear.

Second, consider the impact on net revenue due to the order of market entry of the firm and its competitor. We assume a fixed amount of net revenue is available to both firms over each generation's life cycle. Consistent with the marketing literature, as the difference in the market entry times increases ($\boldsymbol{\delta} - \mathbf{m}$ increases), the net revenue earned by the first entrant increases whereas the net revenue earned by the second entrant decreases for each product generation. Clearly, if both firms enter the market at the same time $(\boldsymbol{\delta} = \mathbf{m})$, then total revenue is shared equally. In addition, for each product generation, the firm earns the maximum potential net revenue if $\mathbf{m} = 0$ and $\boldsymbol{\delta} \geq \mathbf{L}$, (if the firm's market entry occurs at the start of the product life

Table 2. Revenue Function Relationships

First Order Conditions

$$\partial \mathbf{R}^{ij}(\mathbf{m},\boldsymbol{\delta},\mathbf{L})/\partial \mathbf{L} = \mathbf{R}^{ij}_{L} > 0, \text{ for } i = S, T \text{ and } j = 1, 2 \quad (4a)$$

$$\partial \mathbf{R}^{ij}(\mathbf{m},\boldsymbol{\delta},\mathbf{L})/\partial(\boldsymbol{\delta} - \mathbf{m}) = \mathbf{R}^{ij}_{(\delta - m)} > 0, \text{ for } i = S, T \text{ and } j = 1, 2 \quad (4b)$$

Second Order Conditions

$\mathbf{R}^{S1}_{(\delta - m)(\delta - m)} > 0$	(5a)	$\mathbf{R}^{S2}_{(\delta - m)(\delta - m)} < 0$	(5b)
$\mathbf{R}^{T1}_{(\delta - m)(\delta - m)} < 0$	(6a)	$\mathbf{R}^{T2}_{(\delta - m)(\delta - m)} > 0$	(6b)

cycle and the competitor never enters the new market). Similarly, for each product generation, the firm earns the minimum net revenue if $\mathbf{m} \geq \mathbf{L}$ and $\boldsymbol{\delta} = 0$. Naturally, if the firm earns the maximum (minimum) net revenue, its competitor earns the minimum (maximum). This gives us the first order condition in Equation (4b) of Table 2. Note that the condition holds whether $\boldsymbol{\delta} - \mathbf{m}$ is positive or negative.

As a final remark, we explain our reference to$\mathbf{R}^{ij}(\mathbf{m},\boldsymbol{\delta},\mathbf{L})$ as the total *net* revenue earned by the firm for each product generation. Suppose the firm's competitor delays its market entry for each new product generation (the competitor invests in less facility flexibility). As a result, the firm's demand and output volume increase so that both the revenue earned and the manufacturing cost incurred for each product generation increase. From Section 2.3.3, the average manufacturing cost for each product generation is defined independent from competitive forces in the marketplace. Therefore, our characterization of net revenue includes any residual effect on the production cost due to an increase in output volume (demand). In other words, the average net revenue earned for each product generation measures the increase in earnings *less* any increase in manufacturing costs.

2.4.2 Sustained leader advantage: market-type S

In this section, we explore the relationship between net revenue and market entry strategy in an environment where the first entrant realizes a sustained advantage. In the sustained Market-Type S ($i = S$), as the difference in the market entry times increases, the net revenue earned by the pioneer for each product generation increases at an increasing rate. In other words, as the follower seeks an earlier market entry, its net revenue increases at a decreasing rate. The decreasing rate at which the follower's net revenue increases reflects the sustained competitive advantage enjoyed by the pioneer.

The above discussion provides us with second order conditions that drive net revenue in a sustained market. Suppose the firm is the first entrant ($j = 1$) in a Market-Type S. Clearly, $\boldsymbol{\delta} - \mathbf{m} > 0$ holds. As the competitor (follower) delays its market entry or as the firm (pioneer) achieves an earlier entry, the firm's net revenue for each product generation increases at an increasing rate (see Equation (5a) in Table 2). Second, suppose the firm is the follower ($j = 2$) in a Market-Type S. Here, $\boldsymbol{\delta} - \mathbf{m} < 0$ holds. As the firm (follower) realizes an earlier market entry or as the

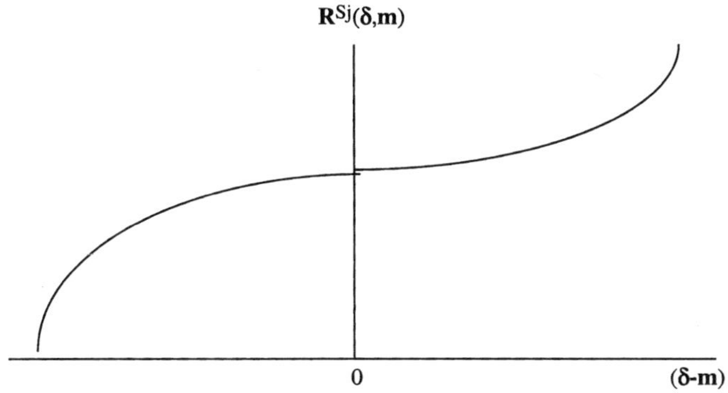

Figure 1. Net revenue for decision-making firm in Market-Type S.

competitor (pioneer) delays its entry, the firm's net revenue for each product generation increases at a decreasing rate, (see Equation (5b) in Table 2). The net revenue earned by the firm for each product generation in a Market-Type S is illustrated in Figure 1.

2.4.3 Transient leader advantage: market-type T

Next, consider a market where the pioneer advantage is transient, i = T. As the difference in market entry times increases in a Market-Type T, the net revenue earned by the pioneer for each product generation increases at a decreasing rate. In other words, as the pioneer pursues an even earlier market entry strategy, the additional contribution to net revenue exhibits diminishing returns. Similarly, the increasing rate at which the follower gains net revenue reflects the transient nature of the pioneer's competitive advantage.

From the above, we obtain the second order conditions on net revenue. First, suppose the firm is the pioneer ($j = 1$) in a Market-Type T so that $\boldsymbol{\delta} - \mathbf{m} > 0$ holds. As the competitor (follower) delays its market entry or as the firm (pioneer) realizes an earlier market entry, the net revenue earned by the firm for each product generation increases at a decreasing rate (see Equation (6a) in Table 2). Second, suppose the firm is the follower in a Market-Type T ($j = 2$) so that $\boldsymbol{\delta} - \mathbf{m} < 0$ holds. As the firm (follower) realizes an earlier market entry or as the competitor (pioneer) delays its market entry, the net revenue amassed by the firm for each product generation increases at an increasing rate (see Equation (6b) in Table 2). Figure 2 illustrates the firm's net revenue for each product generation in a Market-Type T.

2.5 Optimality conditions

The firm's objective is to maximize the expected profit obtained from the manufacture of **n** successive product generations over the life of the facility. Mathematically, expected profit, denoted by $\boldsymbol{\Omega(m,\delta)}$, is expressed as a function of two variables: the firm's optimal market entry time decision, $\mathbf{m}^\star$ (the proxy for facility flexibility), and

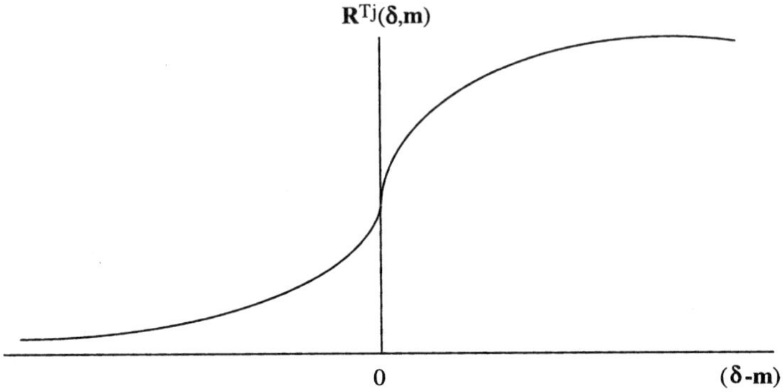

Figure 2. Net revenue for decision-making firm in Market-Type T.

the competitor's unknown market entry time, $\boldsymbol{\delta}$. The objective in Equation (7) captures: (i) the expected net revenue earned from the manufacture of $\mathbf{n}$ high-volume product generations; (ii) the cost associated with $\mathbf{n}$ facility changeovers; (iii) the total production cost of $\mathbf{n}$ generations; and (iv) the initial investment cost for the facility. Note that the cost incurred to setup the facility for the manufacture of the first product is treated as a changeover cost and is not a component of the initial facility investment. In Section 4, we discuss the effect on the optimal solution if the objective were discounted.

$$\Omega(\mathrm{m},\delta) = \underset{(\mathrm{i})}{\mathbf{n}\{\mathrm{E}[\mathbf{R}^{\mathrm{ij}}(\mathbf{m},\boldsymbol{\delta},\mathbf{L})]\}} - \underset{(\mathrm{ii})}{\mathbf{nA}(\mathbf{m})} - \underset{(\mathrm{iii})}{\mathbf{nC}(\mathbf{m},\mathbf{L})} - \underset{(\mathrm{iv})}{\mathbf{I}(\mathbf{m})} \tag{7a}$$

$$\text{with } \mathrm{E}[\mathbf{R}^{\mathrm{ij}}(\mathbf{m},\boldsymbol{\delta},\mathbf{L})] = \int_0^{\mathrm{L}} [\mathbf{R}^{\mathrm{ij}}(\mathbf{m},\boldsymbol{\delta},\mathbf{L})]\mathrm{g}(\delta)\mathrm{d}\delta \tag{7b}$$

The first and second order optimality conditions for the firm's investment in facility flexibility, $\mathbf{m}^\star$, appear in Equations (8) and (9), respectively. For simplicity, functional notation is suppressed wherever possible. The left and right sides of Equation (8) are functions of $\mathbf{m}$ and are denoted by LHS$(\mathbf{m})$ and RHS$(\mathbf{m})$.

$$\underset{\mathrm{LHS}(\mathbf{m})}{\mathbf{n}\{\mathrm{E}[\mathbf{R}^{\mathrm{ij}}(\mathbf{m},\boldsymbol{\delta},\mathbf{L})]\}_{\mathrm{m}}} = \underset{\mathrm{RHS}(\mathbf{m})}{\mathbf{nA}_{\mathbf{m}} + \mathbf{nC}_{\mathbf{m}} + \mathbf{I}_{\mathbf{m}}} \tag{8}$$

$$\underset{\mathrm{LHS}(\mathbf{m})_{\mathrm{m}}}{\mathbf{n}\{\mathrm{E}[\mathbf{R}^{\mathrm{ij}}(\mathbf{m},\boldsymbol{\delta},\mathbf{L})]\}_{\mathrm{mm}}} < \underset{\mathrm{RHS}(\mathbf{m})_{\mathrm{m}}}{\mathbf{nA}_{\mathbf{mm}} + \mathbf{nC}_{\mathbf{mm}} + \mathbf{I}_{\mathbf{mm}}} \tag{9}$$

The interpretation of Equation (8) is straightforward. Corresponding to the optimal level of facility flexibility, $\mathbf{m}^\star$, the marginal expected net revenue equals the marginal total cost over the life of the facility. Since $\mathbf{g}(\boldsymbol{\delta})$ is not a function of $\mathbf{m}$, from Equation (7), any change in $\mathbf{m}$ impacts expected net revenue only through $\mathbf{R}^{\mathrm{ij}}(\mathbf{m},\boldsymbol{\delta},\mathbf{L})$. From Equations (4) and (5), the firm's expected net revenue decreases as $\mathbf{m}$ increases, regardless of whether the first market entrant earns a sustained or transient advantage. Therefore, LHS$(\mathbf{m}) < 0$ holds. For $\mathbf{m}^\star$ to satisfy the first order condition, RHS$(\mathbf{m}^\star) < 0$ must hold. Evaluating each term in RHS$(\mathbf{m})$, we know that $\mathbf{nC}_m$ and $\mathbf{I}_m$ are negative, whereas $\mathbf{nA}_m$ is positive. However, from Section 2.3.2, since $\mathbf{I}_m \gg \mathbf{A}_m$ holds, it is reasonable to assume that a value of $\mathbf{m}^\star$ exists satisfying the first order condition.

Next, consider the second order optimality condition. Since all three terms in RHS$(\mathbf{m})_m$ are positive, we know RHS$(\mathbf{m})$ monotonically increases for all $\mathbf{m}$. Furthermore, from Equation (8), at the optimal level of facility flexibility, $\mathbf{m}^\star$, we obtain $\mathbf{RHS}(\mathbf{m}^\star)_m >$ LHS$(\mathbf{m}^\star)_m$. Therefore, if LHS$(\mathbf{m}^\star)_m \geq 0$ holds, we know that at $\mathbf{m}^\star$, the rate of increase in LHS$(\mathbf{m}^\star)$ is less than the rate of increase in RHS$(\mathbf{m}^\star)$. Alternatively, if LHS$(\mathbf{m}^\star)_m \leq 0$ holds, then LHS$(\mathbf{m}^\star)$ is decreasing at $\mathbf{m}^\star$. Assuming the functions LHS and RHS are well behaved, for $\boldsymbol{\mu} < \mathbf{m}^\star$, we have RHS$(\mu) <$ LHS(μ), whereas for $\mu > \mathbf{m}^\star$ we have RHS$(\mu) >$ LHS(μ). Later, we use these second order relationships to derive analytic insights on the optimal solution.

3. IMPACT OF COMPETITOR'S MARKET ENTRY STRATEGY

We are interested in examining the effect on the firm's investment in facility flexibility due to the stochastic nature of the competitor's market entry strategy. From Equations (8) and (9), we see that the impact of the competitor's market entry time is limited to the term involving expected net revenue. Intuitively, this occurs since the competitor's market entry time has no direct effect on the firm's costs, (RHS$(\mathbf{m})$). Therefore, our analysis focuses on the sensitivity of LHS$(\mathbf{m})$ to characteristics of the competitor's market entry distribution.

We first explore the situation where the probability density of the competitor's market entry strategy is uniformly distributed. Beyond offering mathematical tractability, the uniform distribution reflects a reasonable situation. For example, the uniform distribution is appropriate if a firm is comfortable estimating the time interval of its competitor's market entry, but has limited knowledge regarding the probability of specific entry times during that interval. Later, building upon insights gleaned from the uniform assumption, we develop general results for any symmetric distribution.

Mathematically, let $\mathbf{g}(\boldsymbol{\delta})$ be uniformly distributed over $[\mathbf{a},\mathbf{b}]$ where $\mathbf{a} \geq \mathbf{0}$ and $\mathbf{b} \leq \mathbf{L}$. The competitor's expected market entry time is $[(\mathbf{a} + \mathbf{b})/2]$ and the level of uncertainty in the competitor's market entry strategy is measured by the variance, $[(\mathbf{b} - \mathbf{a})^2/12]$. Substituting into Equation (7), we obtain the firm's expected net revenue for all product generations as shown in Equation (10). Taking the derivative of Equation (10) with respect to $\mathbf{m}$, we have Equation (11), which equals our earlier expression for LHS$(\mathbf{m})$.

$$\mathbf{n}\{E[\mathbf{R}^{ij}(\mathbf{m},\boldsymbol{\delta},\mathbf{L})]\} = \mathbf{n}\int_a^b [\mathbf{R}^{ij}(\mathbf{m},\boldsymbol{\delta},\mathbf{L})][1/(\mathbf{b} - \mathbf{a})]d\delta \tag{10}$$

$$\mathbf{n}\{E[\mathbf{R}^{ij}(\mathbf{m},\boldsymbol{\delta},\mathbf{L})]\}_{\mathbf{m}} = [\mathbf{n}/(\mathbf{b} - \mathbf{a})]\int_a^b [\mathbf{R}^{ij}(\mathbf{m},\boldsymbol{\delta},\mathbf{L})]_{\mathbf{m}}d\delta \tag{11}$$

We are now ready to interpret the effect of the competitor's market entry strategy (i.e., changes in parameters $\mathbf{a}$ and $\mathbf{b}$) on the firm's investment in facility flexibility.

In Section 3.1, we explore the effect of a delay in the competitor's expected market entry strategy (analysis of the mean). In Section 3.2, the firm's investment in flexibility is explored in relation to the extent of uncertainty in the competitor's market entry strategy (analysis of the variance). General results on the firm's investment in facility flexibility when its competitor's market entry strategy obeys any symmetric distribution are given in Section 3.3.

3.1 Effect of a delay in the competitor's expected market entry

To isolate the effect of a delay in the competitor's expected market entry, we need the mean of the density to increase but the variance to remain fixed. Let the competitor's market entry satisfy a uniform distribution over the range $[\mathbf{a} + \varepsilon, \mathbf{b} + \varepsilon]$, for $\varepsilon > 0$. Relative to the situation where $\varepsilon = 0$, the modified distribution has a higher expected value equal to $[(\mathbf{a} + \mathbf{b} + 2\varepsilon)/2]$, but the same variance $[(\mathbf{b} - \mathbf{a})^2/12]$. Next, we compare the values of Equation (11) for $\varepsilon < 0$ versus $\varepsilon = 0$.

Three cases arise, each reflecting a different assumption regarding the relative market entry times of the firm and its competitor. In Scenario 1, the firm enters each new product market prior to the earliest possible entry time of its competitor, $(\mathbf{m} < \mathbf{a})$. This strong pioneering position might reflect the firm's commitment to be the technological leader. In sharp contrast, in Scenario 2, the firm enters each new product market after the time denoting the upper bound of the competitor's market entry distribution, $(\mathbf{m} > \mathbf{b})$. Here, the firm is certain to be the market follower. A firm lacking knowledge of consumer preferences may pursue the Scenario 2 strategy to observe its competitor's actions and reduce the risk of incorrect product positioning. Similarly a firm may commit to be a follower if it lacks the resources necessary to invest in facility flexibility. Lastly, in Scenario 3, the firm's market entry for each product generation lies within the range of the competitor's entry time distribution. Here, it is unclear at the outset which firm will earn the market leader position.

Next, for each scenario, we analyze the investment in facility flexibility in a market where the first entrant realizes a sustained (Section 3.1.1) versus transient (Section 3.1.2) advantage.

3.1.1 Market with sustained first entrant advantage

Recall that in a Market-Type S, the pioneer's net revenue for each product generation increases at an increasing rate as the difference between its market entry time and that of the competitor increases. If the firm is the pioneer in a market with a sustained first entrant advantage, from the relations in Table 2, we obtain the first and second order conditions in Equations (12a) and (12b) of Table 3. Similarly, if the firm is the follower in a Market-Type S, from Table 2, we have Equations (13a) and (13b). Next, we interpret Equations (12c) and (13c). Note that some insights are illustrated by visualizing the change in net revenue that occurs when the firm *delays* its market entry $(\partial \mathbf{R}^{Sj}/\partial \mathbf{m})$ while other insights are illustrated by visualizing the effect on net revenue when the firm pursues an *earlier* market entry $(-\partial \mathbf{R}^{Sj}/\partial \mathbf{m})$.

Suppose the decision making firm is the pioneer in a market where the first entrant advantage is sustained. The additional net revenue earned by the firm due to an earlier

Table 3. Marginal Revenue Analysis

MarkeT-Type S		MarkeT-Type T	
$\mathbf{R}^{S1}{}_{\delta} > 0$, $\mathbf{R}^{S1}{}_{\delta} > 0$	(12a)	$\mathbf{R}^{T1}{}_{\delta} > 0$, $\mathbf{R}^{T1}{}_{\delta} < 0$	(17a)
$\mathbf{R}^{S1}{}_{m} < 0$, $\mathbf{R}^{S1}{}_{mm} > 0$	(12b)	$\mathbf{R}^{T1}{}_{m} < 0$, $\mathbf{R}^{T1}{}_{mm} < 0$	(17b)
$\mathbf{R}^{S1}{}_{m\delta} < 0$, $\mathbf{R}^{S1}{}_{m\delta} < 0$	(12c)	$\mathbf{R}^{T1}{}_{m\delta} > 0$, $\mathbf{R}^{T1}{}_{m\delta} < 0$	(17c)
$\mathbf{R}^{S2}{}_{\delta} > 0$, $\mathbf{R}^{S2}{}_{\delta} < 0$	(13a)	$\mathbf{R}^{T2}{}_{\delta} < 0$, $\mathbf{R}^{T2}{}_{\delta} > 0$	(18a)
$\mathbf{R}^{S2}{}_{m} < 0$, $\mathbf{R}^{S2}{}_{mm} < 0$	(13b)	$\mathbf{R}^{T2}{}_{m} < 0$, $\mathbf{R}^{T2}{}_{mm} > 0$	(18b)
$\mathbf{R}^{S2}{}_{m\delta} > 0$, $\mathbf{R}^{S2}{}_{m\delta} < 0$	(13c)	$\mathbf{R}^{T2}{}_{m\delta} < 0$, $\mathbf{R}^{T2}{}_{m\delta} < 0$	(18c)

market entry strategy ($-\partial\mathbf{R}^{S1}/\partial\mathbf{m} > 0$) increases at an increasing rate as the competitor's expected market entry is delayed (δ increases). This insight is stated mathematically in Equation (12c) and illustrated in Figure 3. In contrast, suppose the firm follows in a Market-Type S. If the competitor (leader) pursues an earlier market entry strategy (δ decreases), then the loss in net revenue realized by a firm that *delays* its market entry ($\partial\mathbf{R}^{S2}/\partial\mathbf{m} < 0$) increases at an increasing rate. This insight is stated mathematically in Equation (13c) and illustrated in Figure 4.

Analysis of Scenarios 1 and 2 are straightforward. In Scenario 1, since we know the decision-making firm is the pioneer ($\mathbf{m} < \mathbf{a}$), Equations (8) and (11) give us Equation (14). From Figure 3, if the competitor's expected market entry time is delayed (i.e., the density shifts from Uniform $[\mathbf{a},\mathbf{b}]$ to Uniform $(\mathbf{a} + \varepsilon, \mathbf{b} + \varepsilon)$ for $\varepsilon > 0$), the area lost ($\mathbf{a} \rightarrow \mathbf{a} + \varepsilon$) is less than the area gained ($\mathbf{b} \rightarrow \mathbf{b} + \varepsilon$). Therefore, the value of LHS($\mathbf{m}$) decreases and the firm optimally enters the market for each product generation earlier ($\mathbf{m}^{\star}$ declines). Proposition 1 summarizes this result.

$$[\mathrm{LHS}(m)] = [\mathbf{n}/(\mathbf{b} - \mathbf{a})]\left\{\int_{a}^{b} [\mathbf{R}^{S1}(\mathbf{m},\delta,\mathbf{L})]_{\mathbf{m}} d\delta\right\} \tag{14}$$

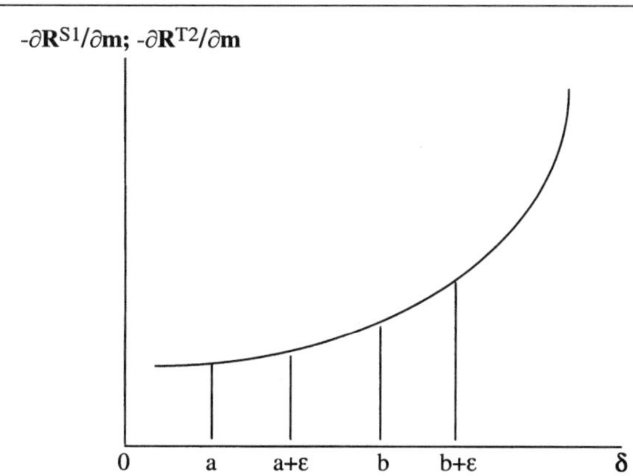

Figure 3. Impact of competitor delay when firm leads in Market-Type S or when firm follows in Market-Type T.

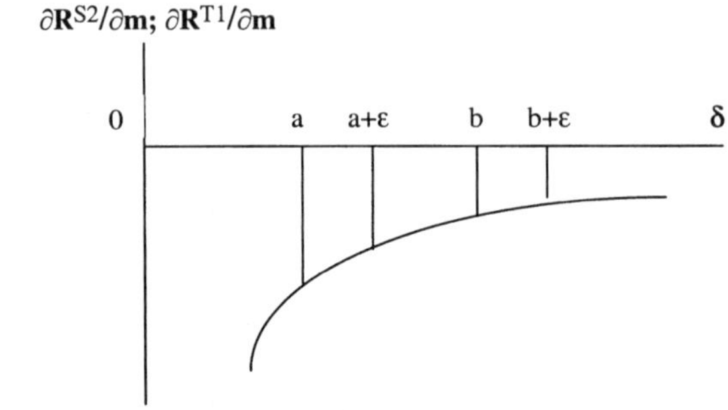

Figure 4. Impact of competitor delay when firm follows in Market-Type S or when firm leads in market-type T.

PROPOSITION 1. Suppose the firm is the leader in a market with sustained first entrant advantage. As the competitor's expected market entry strategy is delayed (occurs earlier), the firm invests in more (less) facility flexibility to achieve earlier (later) market entry for each product generation.

Intuitively, the result in Proposition 1 occurs because, as the competitor's market entry is delayed, the firm's ability to derive sustained pioneering benefits increases. Here, as the leader in a sustained market, the firm's ability to earn net revenue increases as the gap between its market entry and that of its competitor increases. As a result, by pursuing a delayed market entry, the competitor enables the firm to further its commitment to a strategy of market leadership.

Alternatively, suppose the competitor's expected market entry strategy occurs earlier (fast follower) for each product generation. Due to the heightened challenge by its competitor, the firm's ability to reap sustained pioneering benefits from net revenue is reduced (LHS(**m**) increases). The sustained reduction in net revenue drives a lesser investment in facility flexibility so that the firm realizes a more narrow leadership position in the marketplace. Here, fierce time-based competition forces the firm to pursue more balance between the opposing goals of earlier market entry and lower costs. This latter result demonstrates the potential vulnerability of the pioneer's position.

Finally, suppose the firm (pioneer) wants to maintain its market entry strategy despite the greater challenge posed by its competitor (fast follower). From Equation (8), for the value of $\mathbf{m}^*$ to remain fixed while the LHS(**m**) increases, the firm must reduce the cost disadvantage of investing in more facility flexibility (RHS(**m**) must increase). For example, the firm may undertake the development of new technology or production methods that enable a lesser tradeoff between flexibility and efficiency.

Next, consider Scenario 2. Since we know the firm is the market follower $(\mathbf{m} > \mathbf{b})$, Equations (8) and (11) give us Equation (15). From Figure 4, if the competitor's expected market entry time is delayed (the density shifts from Uniform $[\mathbf{a},\mathbf{b}]$ to Uniform $[\mathbf{a} + \varepsilon, \mathbf{b} + \varepsilon]$ for $\varepsilon \geq 0$), the area lost $(\mathbf{a} \rightarrow \mathbf{a} + \varepsilon)$ exceeds the area gained $(\mathbf{b} \rightarrow \mathbf{b} + \varepsilon)$. Therefore, the value of LHS(**m**) increases and the firm optimally pursues a delayed market entry strategy ($\mathbf{m}^*$ increases). Proposition 2 summarizes this result.

$$[\text{LHS}(\mathbf{m})] = [\mathbf{n}/(\mathbf{b} - \mathbf{a})]\left\{\int_{a}^{b} [\mathbf{R}^{S2}(\mathbf{m},\boldsymbol{\delta},\mathbf{L})]_{m} d\delta\right\} \tag{15}$$

PROPOSITION 2. Suppose the firm is the follower in a market with sustained first entrant advantage. As the competitor's expected market entry strategy is delayed (occurs earlier), the firm invests in less (more) facility flexibility and realizes a later (earlier) market entry for each product generation.

Proposition 2 reflects the poor position of the follower in a market with sustained pioneer advantage. If the competitor in a sustained market seeks an even greater leadership position, the firm (follower) optimally invests in more facility flexibility simply to prevent further deterioration of its ability to earn net revenue. Intuitively, this result occurs since the loss in net revenue realized by the firm increases at an increasing rate as the difference between its market entry and that of the competitor increases.

Alternatively, if the competitor (pioneer) delays its market entry strategy, the firm optimally seeks a lower cost solution and thereby reduces its investment in facility flexibility. Here, accepting its position as the follower, the firm sees the pioneer's delayed market entry strategy as an opportunity to reduce its costs. As a result, the firm (follower) furthers its pursuit of a low cost competitive strategy.

Suppose, however, the firm wants to maintain its investment in facility flexibility and forgo the opportunity for cost reduction that is provided by the pioneer's delayed market entry strategy. From Equation (8), for the value of $\mathbf{m}^*$ to remain fixed while the LHS($\mathbf{m}$) increases, the firm must reduce the cost disadvantage of investing in more facility flexibility (i.e., RHS($\mathbf{m}$) must increase). Again, this demonstrates the value to a firm capable of developing new technology or production methods enabling a lesser tradeoff between flexibility and efficiency.

Lastly, in Scenario 3, it is unclear if the decision making firm will be the market leader or follower. From Equations (8) and (11), we find that $\{E[\mathbf{R}^{Sj}(\mathbf{m},\boldsymbol{\delta},\mathbf{L})]\}_m$ or equivalently [LHS($\mathbf{m}$)] satisfies Equation (16). To interpret this equation, we examine the areas under the curves in the two integrals. Since a delay in the firm's market entry causes a reduction in net revenue for each product generation, the values inside the integrals are negative, (see Figures 3 and 4). Therefore, as the area under the curve in each expression increases, the value of Equation (16) grows more negative. From the discussion of net revenue in Section 2.4, we know that the two integrals are symmetric. Specifically, if the firm and its competitor enter each new product market at the same time, total net revenue is equally divided. (If the value of $\mathbf{m}$ is equidistant from $\mathbf{a}$ and $\mathbf{b}$ ($\mathbf{m} - \mathbf{a} = \mathbf{b} - \mathbf{m}$), then the areas under the two curves are equal.) Also, as the difference between both firms' market entry times increases, the leader's revenue increases at the same rate that the follower's revenue declines.

$$[\text{LHS}(\mathbf{m})] = [\mathbf{n}/(\mathbf{b} - \mathbf{a})]\left\{\int_{a}^{m} [\mathbf{R}^{S2}(\mathbf{m},\boldsymbol{\delta},\mathbf{L})]_{m} d\delta + \int_{m}^{b} [\mathbf{R}^{S1}(\mathbf{m},\boldsymbol{\delta},\mathbf{L})]_{m} d\delta\right\} \tag{16}$$

Extending the discussion of Scenario 3, we evaluate the effect on LHS($\mathbf{m}$) if the competitor's market entry distribution changes from Uniform $[\mathbf{a},\mathbf{b}]$ to Uniform

$[\mathbf{a}+\varepsilon,\mathbf{b}+\varepsilon]$ for $\varepsilon > 0$. First, suppose the competitor is more likely to enter earlier than later during each product's life cycle, $(\mathbf{m} - \mathbf{a} > \mathbf{b} - \mathbf{m})$. When the competitor's expected market entry time satisfies Uniform $[\mathbf{a},\mathbf{b}]$, the area under the curve in the first integral exceeds the area under the curve in the second (see Equation (16)). If the competitor's expected market entry time is delayed, ($g(\delta)$ satisfies Uniform $[\mathbf{a} + \varepsilon,\mathbf{b} + \varepsilon]$ for $\varepsilon \geq 0$), then the area under the curve in the first term decreases, whereas the area in the second term increases. However, due to symmetry in the marginal net revenue functions and given the supposition $\mathbf{m} - \mathbf{a} > \mathbf{b} - \mathbf{m}$, the decrease in the area of the first term exceeds the increase in the area of the second. It follows that the value of Equation (16) becomes less negative so that LHS$(\mathbf{m})$ increases. Since we know that as LHS$(\mathbf{m})$ increases the optimal value of $\mathbf{m}^\star$ increases, we have the following result.

Proposition 3. Suppose the firm operates in a market with sustained first entrant advantage. Also, suppose at the time that the firm must determine its investment in facility flexibility, the market leadership position is not definitively known. If the competitor is more likely to enter first, then as the competitor pursues a delayed (an earlier) market entry strategy, the firm invests in less (more) facility flexibility and pursues a delayed (earlier) market entry strategy.

The interpretation of the above result is consistent with Proposition 2. Given its assessment that the competitor is more likely to attain the pioneer position, the firm seeks to improve profitability by lowering its costs as opposed to gaining additional net revenue from early market entry. Here, the firm invests in less facility flexibility and implicitly accepts the follower position. Again, as the competitor delays its market entry strategy, the only way the firm can justify maintaining its entry strategy (and thereby lessen the leader's advantage) is by reducing the cost disadvantage of investing in facility flexibility.

Second, suppose the competitor is more likely to realize later market entry for each new product generation in a sustained market $(\mathbf{m} - \mathbf{a} < \mathbf{b} - \mathbf{m})$. If the competitor's expected market entry time satisfies Uniform $[\mathbf{a},\mathbf{b}]$, the area under the curve in the first integral of Equation (16) is less than the area under the curve in the second term. As before, if the competitor's expected market entry time is delayed ($g(\delta)$ satisfies Uniform $[\mathbf{a} + \varepsilon,\mathbf{b} + \varepsilon]$ for $\varepsilon \geq 0$), the area under the curve in the first term decreases, whereas the area in the second term increases. However, due to the symmetry of the marginal net revenue functions and since $\mathbf{m} - \mathbf{a} < \mathbf{b} - \mathbf{m}$ holds, the decrease in the area of the first term is less than the increase in the area in the second. It follows that the value of Equation (16) becomes more negative so that LHS$(\mathbf{m})$ decreases. As LHS$(\mathbf{m})$ decreases, we know that $\mathbf{m}^\star$ decreases, giving us the following proposition.

Proposition 4. Suppose the firm operates in a market with sustained first entrant advantage. Also, suppose at the time that the firm must determine its investment in facility flexibility, the market leader position is not definitively known. If the competitor is more likely to enter second, then as the competitor delays its market entry strategy (pursues an earlier entry), the firm optimally invests in more (less) facility flexibility to achieve an earlier (a later) market entry.

The interpretation of the above result is consistent with Proposition 1. Given the firm's assessment that it is more likely to be the market leader, as its competitor delays its market entry strategy, the firm's ability to derive sustained pioneering benefits increases. Here, the competitor's behavior enables the firm to enhance its market leadership position.

A synthesis of the above propositions leads to two broad observations. First, in a market with sustained pioneer advantage, regardless of whether a firm leads or follows, if the competitor *delays* its market entry then the firm's initial strategy is reinforced. For instance, if the firm leads, then in response to a delay in the market entry strategy of its competitor, the firm optimally invests in more facility flexibility and thereby furthers its position as the pioneer. In contrast, if the firm follows, then in response to a delayed market entry strategy by its competitor, the firm optimally reduces its investment in flexibility in order to further its position as the low-cost producer. Second, regardless of whether a firm leads or follows, in a market with sustained pioneer advantage, if the competitor pursues an *earlier* market entry strategy, then the firm optimally adjusts its investment in flexibility and seeks more balance between a strategy focused on early entry versus low cost.

3.1.2 Market with transient first entrant advantage

In a Market-Type T, the pioneer's expected net revenue for each product generation increases at a decreasing rate as the difference between its market entry time and that of the competitor increases. If the firm is the pioneer in a market with a transient first entrant advantage, from Table 2, we obtain the first and second order conditions in Equations (17a) and (17b) of Table 3. Similarly, if the firm is the follower in a Market-Type T, from Table 2, we have Equations (18a) and (18b) of Table 3. Next, we interpret Equations (17c) and (18c).

Suppose the decision making firm is the pioneer. Since its first entrant advantage is transient, if the competitor (follower) pursues an earlier entry strategy (δ decreases), then the loss in net revenue realized as the firm *delays* its market entry ($\partial \mathbf{R}^{T1}/\partial \mathbf{m} < 0$) increases at an increasing rate. This insight is stated mathematically in Equation (17c) and is analogous to the situation faced by the follower in the sustained market (see Figure 4). In contrast, suppose the firm enters second in a Market-Type T. As the follower in a transient market, if the firm pursues an earlier entry strategy, its ability to earn net revenue ($-\partial \mathbf{R}^{T2}/\partial \mathbf{m} > 0$) increases at an increasing rate as the competitor's expected entry time is delayed. This insight is stated mathematically in Equation (18c) and is analogous to the situation faced by the pioneer in the sustained market (see Figure 3). Since the mathematical analysis of the transient market is similar to that given for the sustained market, the details are omitted. However, interpretations of the transient market solutions are explored below.

PROPOSITION 5. Suppose the firm is the leader in a market with transient first entrant advantage. As the competitor's expected market entry strategy is delayed (occurs earlier), the firm invests in less (more) facility flexibility and realizes later (earlier) market entry for each product generation.

We can interpret Proposition 5 in two ways. First, in a market with transient pioneer advantage, the firm obtains limited benefits from further increasing its leadership position. Therefore, as the competitor's expected market entry strategy is delayed, the firm seizes the opportunity to increase total profit by reducing its investment in changeover flexibility. Moreover, given the delayed market entry of the competitor, the pioneer can only justify maintaining its market entry strategy if the cost disadvantage of investing in facility flexibility is reduced. Second, suppose the competitor's expected market entry strategy occurs earlier. Here, since its transient pioneering benefits are threatened, the firm optimally invests in more facility flexibility. This demonstrates the vulnerability of the pioneer's position in a transient market.

Proposition 6. Suppose the firm is the follower in a market with transient first entrant advantage. As the competitor's expected market entry strategy is delayed (occurs earlier), the firm invests in more (less) facility flexibility and realizes earlier (later) market entry for each product generation.

Proposition 6 reflects the ability of the follower in a Market-Type T to challenge the leadership position of the first entrant. If the pioneering competitor delays its market entry strategy, then the firm seizes the opportunity to limit the first entrant's advantage and increases its investment in facility flexibility. Therefore, in transient markets, the net revenue advantage of being a "fast follower" offset the additional costs of entering new markets earlier.

Propositions 7 and 8 reflect situations where the market leadership position is not known at the outset. The results are similar to those in Propositions 5 and 6, respectively.

Proposition 7. Suppose the first entrant's market advantage is transient. Moreover, suppose at the time that the firm must decide its investment in facility flexibility, the market leader position is not definitively known. If the competitor is more likely to enter second, then as the competitor delays its market entry (pursues an earlier entry), the firm invests in less (more) facility flexibility and also delays its market entry (pursues an earlier entry).

Proposition 8. Suppose the first entrant's market advantage is transient. Moreover, suppose at the time that the firm must determine its investment in facility flexibility, the market leader position is not definitively known. If the competitor is more likely to enter first, then as the competitor delays its market entry (realizes earlier entry) for each product generation, the firm invests in more (less) facility flexibility and pursues an earlier (delayed) market entry strategy.

Propositions 5 – 8 lead to two broad observations regarding the competitive behavior of a firm operating in a market with a transient pioneer advantage. Moreover, these observations are in direct contrast with the results obtained in a market with a sustained pioneer advantage. First, in a market with transient pioneer advantage, regardless of whether a firm leads or follows, if the competitor *delays* its market entry then the firm optimally adjusts its investment in flexibility and seeks more balance between a strategy focused being a pioneer versus a low-cost producer. For instance, if the firm leads, then in response to a delay in the market entry strategy of its competitor, the firm optimally reduces its investment in facility flexibility in order to reduce production costs. In contrast, if the firm follows, then in response to a delayed market entry strategy by its competitor, the firm optimally increases its investment in flexibility in order to realize earlier market entries. Second, regardless of whether a firm leads or follows, in a market with a transient pioneer advantage, if the competitor pursues an *earlier* market entry strategy, then the firm's initial strategy is reinforced.

So far, we have presented a case-by-case analysis of how the firm's investment in flexibility depends on the market type and the order of market entry. In Section 6.1, the results of Propositions 1–8 are further synthesized and overall competitive strategies are discussed including the impact on profit.

3.2 Impact of uncertainty in the competitor's market entry strategy

In this section, we explore the effect of uncertainty in the competitor's market entry strategy on the firm's investment in flexibility. To isolate the effect of uncertainty, we examine the direction of change in LHS($\mathbf{m}$) (Equations (7b) and (8)) in response to an increase in the variance of the competitor's market entry distribution while holding the mean fixed. With $\mathbf{g(\delta)}$ satisfying a uniform distribution, we analyze the change in LHS($\mathbf{m}$) when $\boldsymbol{\delta}$ varies over the range $\mathbf{[a,b]}$ versus $[\mathbf{a} - \varepsilon, \mathbf{b} + \varepsilon]$ for $\varepsilon > 0$. While the mean of both distributions is $[(\mathbf{a} + \mathbf{b})/2]$, the variance of Uniform $[\mathbf{a} - \varepsilon, \mathbf{b} + \varepsilon]$ is $[(\mathbf{b} - \mathbf{a} + 2\varepsilon)^2/12]$ which exceeds the variance of Uniform $\mathbf{[a,b]}$. Therefore, as ε increases, we observe the effect of an increasing amount of uncertainty in the competitor's market entry around a fixed mean.

As before, we need to consider three scenarios within two market types. Since the derivation of results is similar for all cases, we present a detailed analysis of only the most complex case. Specifically, we analyze Scenario 3 for Market-Type S in which the firm could be either the market leader $(\mathbf{m} < \boldsymbol{\delta})$ or follower $(\mathbf{m} > \boldsymbol{\delta})$. If the firm's market entry time lies in the interval $\mathbf{[a,b]}$, then LHS($\mathbf{m}$) satisfies Equation (14). In contrast, if the firm's market entry lies in the interval $[\mathbf{a} - \varepsilon, \mathbf{b} + \varepsilon]$, then from Equations (8) and (11) we obtain the following expression for $\{E[\mathbf{R}^{Sj}(\mathbf{m},\boldsymbol{\delta},\mathbf{L})]\}_m$ or equivalently for [LHS($\mathbf{m}$)].

$$[\text{LHS}(\mathbf{m})] = [\mathbf{n}/(\mathbf{b} - \mathbf{a} + 2\varepsilon)]\left\{\int_{a-\varepsilon}^{m} [\mathbf{R}^{S2}(\mathbf{m},\boldsymbol{\delta},\mathbf{L})]_m d\delta + \int_{m}^{b+\varepsilon} [\mathbf{R}^{S1}(\mathbf{m},\boldsymbol{\delta},\mathbf{L})]_m d\delta\right\} \quad (19)$$

By comparing Equations (14) and (19), we can observe the effect on the firm's investment in facility flexibility due to a greater level of uncertainty in the competitor's market entry strategy (see Figures 3 and 4). Given the extended limits of integration, and since $[\mathbf{R}^{S2}(\mathbf{m},\boldsymbol{\delta},\mathbf{L})]_m$ and $[\mathbf{R}^{S1}(\mathbf{m},\boldsymbol{\delta},\mathbf{L})]_m$ are negative, we know that the sum of the integrals in Equation (19) is smaller (larger negative value) than the corresponding sum in Equation (14). Furthermore, as $\varepsilon > 0$ increases, the sum of the integrals in Equation (19) becomes smaller relative to the fixed value of the integral sum in Equation (14). Next, consider the terms that are multiplied to the sums of the integrals. The multiplier in Equation (14) is positive and proportional to $1/(\mathbf{b} - \mathbf{a})$, whereas the multiplier in Equation (19) is positive and proportional to $1/(\mathbf{b} - \mathbf{a} + 2\varepsilon)$ for $\varepsilon > 0$. Therefore, as $\varepsilon > 0$ increases, the multiplier in Equation (19) decreases, whereas the multiplier in Equation (14) is fixed.

From the above analysis, to assess whether the magnitude of Equation (19) increases or decreases as $\varepsilon > 0$ increases, we apply l 'Hopital's rule. The numerator is the sum of the integrals; the denominator is $(\mathbf{b} - \mathbf{a} + 2\varepsilon)$. As $\varepsilon > 0$ increases, the value of $\mathbf{b} - \mathbf{a} + 2\varepsilon$ increases linearly, whereas the (negative) magnitude of the numerator increases exponentially. Clearly, the numerator dominates so that as ε increases, the value of Equation (22) grows more negative and LHS$(\mathbf{m})$ decreases. In conclusion, we have shown that, as ε increases, the firm's investment in facility flexibility increases, (i.e., the firm pursues an earlier market entry strategy).

The above analysis has been extended to the two other scenarios for Market-Type S, and to all three scenarios for Market-Type T. In every instance, the same results hold. This important insight is summarized in Proposition 9, below. It states that, regardless of the market type encountered and regardless of whether the firm leads or follows, the firm optimally invests in additional facility flexibility as a hedge against increased competitive uncertainty.

PROPOSITION 9. As the level of uncertainty in the competitor's market entry strategy around a fixed mean increases, the firm optimally invests in more facility flexibility thereby pursuing an earlier market entry for each product generation. Furthermore, these results hold regardless of whether the pioneer's market advantage is sustained or transient and regardless of whether the firm is the market leader or follower.

3.3 Generalization of results for symmetric distributions

To generalize Propositions 1–9 for any symmetric distribution, we need only consider how the probability density impacts the marginal expected net revenue (LHS($\mathbf{m}$)). In Equation (20), the marginal expected net revenue is expressed in a way that is equivalent to Equations (7b) and (8) where the integral is defined over the range of $\boldsymbol{\delta}$ for which $\mathbf{g}(\boldsymbol{\delta})$ is non-zero.

$$\{E[\mathbf{R}^{ij}(\mathbf{m},\boldsymbol{\delta},\mathbf{L})]_{\mathbf{m}} = \int [\mathbf{R}^{S2}(\mathbf{m},\boldsymbol{\delta},\mathbf{L})]_{\mathbf{m}} g(\delta)d\delta \tag{20}$$

Propositions 1–9 are based on analysis of Equation (20) for $\mathbf{g}(\boldsymbol{\delta})$ satisfying a uniform distribution. For $\mathbf{g}(\boldsymbol{\delta})$ satisfying any symmetric distribution, we need to show that, under the same suppositions as those that drive Propositions 1–9, the same directions of change are triggered in Equation (20). Therefore, we must prove that the signs (positive or negative) of the expressions given by Equations (21) and (22) are the same for any symmetric probability density.

$$[\partial/\partial\ ]\{ \int [\mathbf{R}^{ij}(\mathbf{m},\boldsymbol{\delta},\mathbf{L})]_{\mathbf{m}} g(\delta)d\delta\} \tag{21}$$

$$\int [\partial/\partial\ ]\ \{[\mathbf{R}^{ij}(\mathbf{m},\boldsymbol{\delta},\mathbf{L})]_{\mathbf{m}} g(\delta)d\delta\} \tag{22}$$

From the product rule for differentiation, the signs of Equations (21) and (22) could only differ if the signs of $\{[\mathbf{R}^{ij}(\mathbf{m},\boldsymbol{\delta},\mathbf{L})]\}_{m\delta}$ and $[g(\boldsymbol{\delta})]_\delta$ are different. Clearly, since $g(\delta)\varepsilon[0,1]$, the magnitude of $[g(\delta)]_\delta$ is extremely small relative to $[\mathbf{R}^{ij}(\mathbf{m},\boldsymbol{\delta},\mathbf{L})]_{m}$.

Therefore, changes in the terms involving marginal net revenue dominate changes in the terms involving the probability density. As a result, by employing l'Hopital's rule, it can be shown that Equations (21) and (22) have the same sign. In conclusion, we have shown that Propositions 1–9 hold for any symmetric probability density function. In Section 5.2, we numerically explore the situation where the competitor's market entry time distribution is skewed (asymmetric).

4. ANALYTIC SENSITIVITY OF THE OPTIMAL SOLUTION

In this section, we complete our presentation of analytic results. First, we explore the impact of various cost parameters on the firm's investment in flexibility. These analytic results follow from the fact that a parameter change that increases (decreases) RHS($\mathbf{m}$) drives a decrease (increase) in the value of $\mathbf{m}^*$ so that a greater (lesser) investment in facility flexibility is advocated (see Section 2.5). Second, we explore the impact of various revenue parameters on the firm's investment in flexibility. These analytic results follow from the fact that a parameter change that decreases (increases) LHS($\mathbf{m}$) drives a decrease (increase) in the value of $\mathbf{m}^*$ so that the firm pursues an earlier (delayed) market entry strategy.

4.1 Analysis of the cost parameters

First, we develop analytic sensitivity results for the facility investment cost. Recall $\mathbf{I}_m < 0$ represents the marginal savings associated with reducing the level of investment in facility flexibility. Larger absolute values of $\mathbf{I}_m$ indicate greater investment cost savings as the firm delays its market entry strategy. Mathematically, as the magnitude of $\mathbf{I}_m$ increases, the value of RHS($\mathbf{m}$) decreases and the firm's optimal investment in facility flexibility is reduced.

This result demonstrates the value to a firm that, either alone or with the help of an original equipment maker (OEM), develops flexible process technology requiring a lesser investment cost. In particular, if process innovation reduces the cost of investing in facility flexibility, then the firm optimally realizes earlier market entries for successive generations of new products. For example, improvements in process technology used in the auto industry have led to substantial investments in facility flexibility by Japanese and US firms seeking reductions in the time to market for new models (Johnson and Maskery (1992)).

Second, consider the impact of a marginal increase in the facility changeover cost as the firm's market entry strategy is delayed ($\mathbf{A}_m > 0$). As $\mathbf{A}_m$ increases, the firm has a greater opportunity to reduce the cost incurred for the facility changeover from one high-volume product generation to the next. Mathematically, as $\mathbf{A}_m$ increases, the value of RHS($\mathbf{m}$) increases and we obtain an earlier market entry strategy. Again, this result demonstrates the value to a firm that develops process technology capable of more rapid and cost effective facility changeovers.

Next, we consider the firm's market entry strategy in relation to the total production cost incurred for each high-volume product generation. $\mathbf{C}_m < 0$ denotes the marginal savings in the production cost due to a reduction in the level of facility flexibility (i.e., an increase in manufacturing efficiency). As the magnitude of $\mathbf{C}_m$ increases, greater savings in the production cost are obtained from a delayed market entry strategy. Mathematically, as $\mathbf{C}_m$ increases, the value of RHS(**m**) decreases, and the firm optimally pursues a delayed market entry strategy. Therefore, as the potential for cost savings obtained from manufacturing efficiency increases (decreases), the optimal investment in changeover flexibility decreases (increases).

The above result supports the competitive advantage realized by a firm capable of a high rate of volume-based learning. Suppose a firm hones its ability to obtain learning benefits over the longer production cycles associated with an earlier market entry strategy. The cost savings derived from learning partially offset the higher total production cost incurred to operate the more flexible (less efficient) facility. Here, a lesser increase in the total production cost occurs if the firm seeks an earlier market entry strategy by investing in more changeover flexibility. Beyond the benefits obtained from learning, technological improvement may also lead to the development of more efficient flexible technology. If this occurs, a firm's future investment in changeover flexibility may entail a lesser trade-off between the benefits of early market entry and manufacturing efficiency.

The above analytic results are summarized in Proposition 10. The strategic insights gleaned from this proposition are stated in Propositions 11a and 11b.

PROPOSITION 10. The firm optimally invests in less (more) changeover flexibility and enters successive high-volume new product markets later (earlier) if: (i) the marginal cost of investing in changeover flexibility increases (decreases); (ii) the marginal cost of performing facility changeovers decreases (increases); or (iii) the marginal savings in the production cost from manufacturing efficiency increases (decreases).

PROPOSITION 11A. As technological innovation reduces the cost of investing in facility flexibility, a firm optimally invests in more flexibility and pursues an earlier market entry strategy.

PROPOSITION 11B. A firm capable of a high rate of volume-based learning mitigates the increase in the production cost associated with less efficient but more flexible technology. Moreover, the firm optimally invests in more facility flexibility and realizes an earlier market entry strategy.

4.2 Analysis of the revenue parameters

We now consider the impact of various revenue drivers on the firm's investment in facility flexibility. $\mathbf{R}_m < 0$ is the reduction in net revenue due to a marginal delay in the firm's market entry strategy. As the magnitude of $\mathbf{R}_m$ increases, the firm incurs greater losses in net revenue due to the same delay in market entry. Mathematically, when the absolute value of $\mathbf{R}_m$ increases, LHS(**m**) decreases and the optimal market entry for each product generation occurs earlier (investment in flexibility increases). In other words, the additional net revenue earned plus the savings in the changeover cost associated with more flexibility offset the increase in the initial facility investment and the production cost. This result is summarized below.

PROPOSITION 12. As the loss in marginal net revenue due to a delayed market entry strategy increases (decreases), the firm optimally invests in more (less) flexibility to realize faster changeovers for the manufacture of successive high-volume product generations.

Table 4. Functions Used in Numerical Analysis

$\mathbf{I(m)} = i_0[\exp(-\, i_1\mathbf{m})]$	$\mathbf{A(m)} = a_0[\exp(a_1\mathbf{m})]$
$\mathbf{C(m,L)} = c_0\{[\exp(c_1(\mathbf{L} - \mathbf{m})) - 1]\}$	$\rho(\mathbf{L}) = 1 + k\mathbf{L}$
$\mathbf{R}^{S1}(\mathbf{m},\boldsymbol{\delta},\mathbf{L}) = \rho(\mathbf{L})\{r_1[\exp(r_2(\boldsymbol{\delta} - \mathbf{m}))]\}$	$\mathbf{R}^{T1}(\mathbf{m},\boldsymbol{\delta},\mathbf{L}) = \rho(\mathbf{L})\{2r_3 - r_3[\exp(r_4(\boldsymbol{\delta} - \mathbf{m}))]\}$
$\mathbf{R}^{S2}(\mathbf{m},\boldsymbol{\delta},\mathbf{L}) = \rho(\mathbf{L})\{2r_1 - r_1[\exp(-\, r_2(\boldsymbol{\delta} - \mathbf{m}))]\}$	$\mathbf{R}^{T2}(\mathbf{m},\boldsymbol{\delta},\mathbf{L}) = \rho(\mathbf{L})\{r_3[\exp(r_4(\boldsymbol{\delta} - \mathbf{m}))]\}$

Lastly, note that Proposition 12 holds in markets where the pioneer benefits are sustained and transient and whether the firm leads or follows. However, the actual magnitude of the change in the firm's investment in flexibility depends on both the market type and the order of entry.

4.3 Impact of discounting on optimal solution

Although our model does not explicitly consider the effect of discounting, we can easily infer its impact. The investment cost, $\mathbf{I}_{(m)}$, is incurred at the initial time of the planning horizon and is unaffected by discounting. In contrast, discounting reduces the future streams of net revenue, production cost, and changeover cost. Therefore, in a discounted version of the model, the impact of the initial investment cost is intensified since the values of the other terms in the objective are relatively smaller. Mathematically, discounting reduces the value of RHS$(\mathbf{m})$ more than the value of LHS$(\mathbf{m})$ so that the firm optimally pursues a later market entry strategy.

This result has important implications for firms competing in global markets. A firm operating in an economy with a higher (lower) discount rate optimally invests in less (more) facility flexibility. In other words, the firm operating in an economic environment characterized by more risk, optimally foregoes the pursuit of an early market entry strategy in favor of a low cost (low price) strategy. Proposition 13 summarizes this result.

PROPOSITION 13. As the discount rate increases (decreases), the firm optimally invests in less (more) changeover flexibility and realizes a delayed (earlier) market entry strategy.

5. NUMERICAL ANALYSIS

Managerial insights are presented in this section that are based on extensive numerical experimentation. In Section 5.1, the interpretations of our analytic results are enriched with new insights on profit. Results are presented in Section 5.2 based on the assumption that the probability density depicting the competitor's market entry strategy satisfies a triangular distribution. This characterization allows us to extend our earlier results on symmetric distributions to situations where the competitor's market entry distribution is skewed.

The functions used in our numerical experiments appear in Table 4, where exp($\star$) is used to denote $e^{(\star)}$. Since a_0, a_1, c_0, c_1, i_0, i_1, r_1, r_2, r_3, r_4, and $k > 0$, the first and second order conditions in Section 2 hold. Similar functions appear in Franza and Gaimon (1998) and Gaimon and Singhal (1992). The interpretation of parameters is straightforward, as described below.

The parameter i_0 is extremely large since it represents the initial investment cost

Table 5. Functions Used in Numerical Analysis

Parameter	Base Value	Range of Values
Facility Life (**F**)	120 months	Not Applicable
Product Life Cycle (**L**)	30 months	Not Applicable
Number of Products (**n**)	4	Not Applicable
Life Cycle Factor (k)	0.01	Not Applicable
Minimum Changeover Cost (a_0)	$2M	($100K, $50M)
Changeover Cost Sensitivity (a_1)	0.20	(0.05, 0.50)
Maximum Initial Investment Cost(i_0)	$1.5B	($250M, $5B)
Investment Cost Sensitivity (i_1)	0.15	(0.05, 0.50)
Operating Cost Scaling Factor (c_0)	$10M	($2M, $25M)
Operating Cost Sensitivity (c_1)	0.025	(0.01, 0.10)
Market-Type S Neutral Revenue (r_1)	$250M	($50M, $500M)
Market-Type S Revenue Sensitivity (r_2)	0.035	(0.01, 0.10)
Market-Type T Neutral Revenue (r_3)	$250M	($50M, $500M)
Market-T Revenue Sensitivity (r_4)	0.035	(0.01, 0.10)

for the maximum flexible facility ($\mathbf{m} = 0$). The parameter i_1 reflects the sensitivity of the investment cost to the level of facility flexibility. The minimum facility changeover cost is denoted by a_0 ($\mathbf{m} = 0$), whereas a_1 measures the sensitivity of the changeover cost to the level of facility flexibility. The parameter c_0 is a scaling factor of the production cost, and c_1 denotes the sensitivity of the production cost to the lack of efficiency associated with flexible facilities.

The total net revenue earned for each product generation is proportional to $\rho(\mathbf{L})$. The parameter k measures the rate of increase in net revenue as the average product life cycle (**L**) increases. The parameters r_1 and r_3 denote the net revenue earned in Market-Types S and T, respectively, if the firm and its competitor pursue the same market entry strategies ("neutral" net revenue). In Market-Types S and T, the parameters r_2 and r_4, respectively, indicate the sensitivity of the firm's net revenue to the difference between its market entry and that of the competitor.

The baseline input parameter values used in the numerical experiments appear in Table 5 along with the ranges of the sensitivity analysis. The inputs for the facility life cycle, average product life cycle, and the investment and changeover cost parameters reflect estimates of industry data from the automotive industry (also see Franza and Gaimon (1998)). Detailed results of these numerical analyses are contained in a technical appendix available upon request.

5.1 Profit implications

Regardless of the market type and regardless of whether the firm leads or follows, analytic results state that, as uncertainty in the competitor's market entry strategy increases, the firm optimally invests in more facility flexibility to achieve an earlier market entry strategy. Beyond this, numerical results show that, as a consequence of the greater investment in flexibility advocated as a hedge against uncertainty, the firm realizes lower expected profit over the facility life. In other words, the firm optimally sacrifices profit in response to competitive uncertainty. Proposition 14 summarizes this numerical result.

PROPOSITION 14. As the level of uncertainty in the competitor's market entry strategy increases, the expected profit earned by the firm over the life of the facility decreases. This result holds regardless of whether the first market entrant realizes a sustained or transient advantage and regardless of whether the firm is the pioneer or follower.

5.2 Analysis for triangular competitor market entry distribution

The triangular distribution is defined by three parameters. The lower and upper bounds of the distribution range are denoted by x and z, respectively, and the mode (peak probability) of the distribution is denoted by y. The probability of a competitor's market entry is zero outside the range [x,z]. The probability of the competitor entering the market linearly increases over $x < \boldsymbol{\delta} < y$ and linearly decreases over $z > \boldsymbol{\delta} > y$. The mean of the triangular distribution is $(x + y + z)/3$; the variance is $(x^2 + y^2 + z^2 - xy - xz - yz)/18$. If the triangular distribution is symmetric, then $y - x = z - y$ holds and the mode (y) equals the mean. Lastly, note that the symmetric triangular distribution has the same mean as the Uniform [x,z], but a slightly smaller variance.

The numerical results obtained when the competitor's market entry obeys a symmetric triangular distribution mirror those of the uniform. However, a key difference occurs since the variance of the triangular [x,y,z] is smaller than that of the Uniform [x,z]. When the competitor's market entry satisfies a symmetric triangular distribution as opposed to a uniform distribution with the same mean, the firm earns higher profit over the facility life with less investment in flexibility. These results are consistent with Propositions 9 and 14.

Next, suppose the competitor's market entry strategy obeys a skewed triangular distribution. If the x and y values are relatively close, the competitor's market entry distribution is skewed early, (the competitor is more likely to enter earlier in the range [x,y]). Therefore, as the mode of the competitor's market entry time (y) increases, we shift from skewed early, to symmetric, to skewed late triangular distribution (competitor's expected market entry is delayed). Proposition 15 summarizes the numerical results. The key insight in Proposition 15 is that, in both market types and regardless of the order of market entry, the firm earns higher profit whenever the competitor delays its market entry strategy.

PROPOSITION 15. As the mode of the competitor's market entry distribution shifts from skewed early to symmetric to skewed late, we have the following. (i) As the follower in a Market-Type S (Market-Type T), the firm optimally invests in less (more) flexibility. (ii) As the leader in a Market-Type S (Market-Type T), the firm optimally invests in more (less) flexibility. (iii) Lastly, the firm's expected profit increases regardless of the market type, regardless of which firm leads or follows, and regardless of whether the firm increases or decreases its investment in flexibility.

6. CONCLUSIONS

A model is introduced to explore the strategic value of changeover flexibility that enables a firm to rapidly and cost effectively transition its facility for the manufacture of successive generations of high-volume products with short life cycles. The model embodies elements from manufacturing and marketing since investment in changeover flexibility not only impacts the firm's cost structure but also serves as a key determinant of the firm's market entry strategy. In other words, the one-time decision regarding a firm's facility changeover flexibility is a proxy for its market entry strategy. Expressing the firm's total net revenue as a function of market forces and a random variable denoting the competitor's market entry strategy captures the effects of competition and uncertainty. Moreover, the firm's changeover flexibility decision is assessed in the context of two different markets. In the Market Type-S, the first entrant realizes a sustained advantage, whereas in the Market Type-T, the leader's advantage is transient. Lastly, each driver of firm profit (i.e., revenue and cost elements) is expressed in general functional form with first and second order derivative conditions supported by an extensive empirical literature.

6.1 Analysis of competitive behavior

In Sections 3.1.1 and 3.1.2, a detailed analysis is given depicting the impact on the firm's investment in facility flexibility due to both the market type and the order of market entry but assuming that the competitor's market entry is uniformly distributed. In Section 3.3, those results are shown to hold more generally for any symmetric probability density. Finally, numerical analysis provides insights in relation to the impact on profit (Section 5). Next, we synthesize these results and explore the overall competitive implications as depicted in Tables 6-S and 6-T.

First, consider the optimal behavior of the pioneer when the competitor (follower) delays its market entry. If the pioneer advantage is sustained, the firm optimally strengthens its resolve to pursue an early market entry strategy and invests in even more facility flexibility. Here, the gain in net revenue compensates for the higher facility investment and production costs so that firm profit increases. In contrast, in a market where the pioneer advantage is transient, the firm seeks a more balanced strategy in response to a delay in the competitor's (follower's) market entry. Specifically, the firm optimally forgoes a portion of the transient benefits to net revenue obtained from its pioneer position and increases profit by reducing its facility investment and production costs. *In conclusion, in response to a delay in the follower's market entry strategy, the leader in a sustained market seeks an even greater pioneer advantage while the leader in a transient market seeks more balance between an early entry and low cost strategy.*

Second, consider the behavior of a pioneering firm when its competitor (follower) pursues an earlier market entry strategy. In a market where the pioneer advantage is sustained, the heightened time-based competition reduces the pioneer's ability to earn net revenue. As a result, the firm limits its loss in profit by reducing the investment in flexibility and enhancing production efficiency. In contrast, in a market where the pioneer advantage is transient, the leader's position is vulnerable. Here, in response to heightened time-based competition and despite the higher costs incurred, the firm optimally limits the loss in profit by investing in more flexibility. *In conclusion, in response to heightened time-to-market competition, the leader in a sustained market seeks more balance between an early entry and low cost strategy whereas the leader in a transient market protects its position by pursuing an even earlier market entry.*

Table 6-S. Strategy in a market with Sustained Pioneer

If the firm___	and the competitor___,	then the firm___.	Focus of Response
leads	delays its entry	enters earlier	Early Entry
leads	enters earlier	delays its entry	Balance
follows	delays its entry	delays its entry	Low Cost
follows	enters earlier	enters earlier	Balance

Table 6-T. Strategy in a Market with Transient Pioneer

If the firm___	and the competitor___,	then the firm___.	Focus of Response
leads	delays its entry	delays its entry	Balance
leads	enters earlier	enters earlier	Early Entry
follows	delays its entry	enters earlier	Balance
follows	enters earlier	delays its entry	Low Cost

Similar interpretations hold if the firm follows. Consider the response by a firm when its competitor (leader) delays its market entry and thereby lowers its investment and production costs. If the pioneer advantage is sustained, the firm optimally reduces its pursuit of flexibility in order to protect its position as the low-cost producer. This solution also reflects the follower's limited opportunity to earn net revenue when the pioneer advantage is sustained. In contrast, in a market where the pioneer advantage is transient, the follower seizes the opportunity provided by the leader's delayed entry and increases its investment in flexibility. Here, the firm benefits from the leader's market vulnerability and maximizes profit by pursuing a fast-follower strategy. *In conclusion, when the leader delays its market entry, the follower in a sustained market protects its position as the low-cost producer whereas the follower in a transient market invests increases the extent of time-based competition.*

Fourth, consider the response by a firm that follows when its competitor (leader) pursues an earlier market entry strategy. If the first entrant advantage is sustained, the firm optimally invests in more flexibility to limit the additional pioneer advantage obtained by the leader. Here, the firm seeks more balance between its initial focus on a low-cost strategy versus an early market entry strategy in order to limit the overall loss in profit. Conversely, in a market where the pioneer advantage is transient, the follower reduces its investment in flexibility in response to the leader's earlier market entry. Here, the follower willingly forgoes the opportunity to close the market entry gap in order to further its position as the low-cost producer. *In conclusion, when the leader pursues an earlier market entry strategy, the follower in a sustained market increases the level of time-based competition whereas the follower in a transient market places more emphasis on being the low-cost producer.*

The above results show an important pattern of managerial decision-making. First, regardless of whether a firm leads or follows, when the pioneer advantage is sustained, if the competitor delays its market entry then the firm's initial strategy is reinforced. For example, if the firm leads (follows), then in response to a delayed market entry by its competitor, the firm furthers its pioneering position (low-cost producer). Second, regardless of whether a firm leads or follows, when the pioneer advantage is sustained, if the competitor pursues an earlier market entry strategy then the firm's initial strategy is reversed and a more balanced approach is pursued. Specifically, in response to increased time-based competition, if the firm is the leader (follower), it optimally delays its entry strategy (increases its investment in flexibility).

Moreover, the exact opposite responses are elicited if the firm operates in a transient market. First, regardless of whether a firm leads or follows, in a market where the pioneer advantage is transient, if the competitor delays its market entry then the firm seeks a more balanced strategy between early market entry and low-cost. Second, regardless of whether a firm leads or follows, in a market where the pioneer advantage is transient, if the competitor pursues an earlier market entry strategy, then the firm's initial strategy is reinforced. In conclusion, these insights demonstrate the importance of properly characterizing the underlying market environment (i.e., whether the pioneer advantage is sustained or transient).

Beyond the above results that hold for any symmetric distribution, consider the situation where the competitor's market entry strategy satisfies a skewed triangular density. Results are obtained that are analogous to those given in Tables 6-S and 6-T. Specifically, suppose the mode of the competitor's market entry distribution shifts from skewed early to symmetric to skewed late (the competitor delays its entry). As the follower (leader) in Market-Type S, the firm invests in less (more) flexibility and realizes a later (earlier) market entry strategy. Alternatively, as the follower (leader) in Market-Type T, the firm invests in more (less) flexibility and attains an earlier (later) market entry strategy. However, regardless of the market type, regardless of which firm leads or follows, and regardless of whether the firm increases or decreases its investment in flexibility, the firm's expected profit over the life of the facility increases.

Lastly, our results relate to Boulding and Christen (2001) who argue that firms often suffer substantial losses in profit when they attempt an early market entry strategy because they overemphasize the revenue benefits but underemphasize the cost implications. Unfortunately, these authors limit their cost considerations to those incurred for R&D and marketing. Therefore, since we demonstrate that, beyond considering the potential increase in revenue, a firm's decision regarding its market entry strategy must recognize the associated impact on manufacturing costs, our results strengthen the views expressed by Boulding and Christen.

6.2 Implications of uncertainty

Two sources of uncertainty are investigated in terms of their effect on the firm's investment in facility flexibility and profit. First, results are given regarding the effect of uncertainty in the context of the competitive marketplace (i.e., the variance of the competitor's market entry strategy around a fixed mean). Second, results are given regarding the effect of uncertainty in the economic environment as inferred by the discount rate.

Analytic results demonstrate that as the level of uncertainty in the competitor's market entry strategy increases, the firm optimally invests in more changeover flexi-

Table 7. Firm Response to Uncertainty

Due to an increase in___,	the firm invests in___flexibility	and profit___.
competitive uncertainty (variance of g(**δ**))	more	declines.
economic uncertainty (discount rate)	less	declines

bility. Furthermore, the reliance on flexibility as a hedge against competitive uncertainty occurs regardless of the market type (sustained or transient pioneer advantage) and regardless of whether the firm leads or follows. Lastly, numerical results show that the firm earns lower profit over the life of the production facility as the level of uncertainty in the competitive marketplace increases. These results illustrate the value to a firm that is capable of properly estimating its competitor's market entry capability (i.e., reducing uncertainty in the competitor's market entry strategy).

The effect on firm behavior due to the level of economic risk (uncertainty) as measured by the discount rate is particularly important for firms operating in global markets. We show that a firm operating in an economy with a high (low) discount rate optimally invests in less (more) facility flexibility. Intuitively, the firm operating in an economic environment characterized by more risk, optimally foregoes the pursuit of an early market entry strategy in favor of a low cost (low price) strategy and earns lower profit over the facility life.

Lastly, it is interesting to note the sharp contrast between the firm's reaction to economic risk and competitive risk. Specifically, a firm invests in less facility flexibility in response to a greater level of economic risk but more facility flexibility in response to a greater level of risk associated with the competitor's market entry strategy. However, in both situations, higher levels of uncertainty reduce firm profit. The contrasting firm behavior is captured below in Table 7.

6.3 Implications of technology improvement and learning

Throughout the paper, results are obtained demonstrating the importance of pursuing process-based technological innovation. For example, a firm that is capable of reducing the cost of investing in facility flexibility optimally invests in more flexibility, realizes an earlier market entry strategy, and increases profit. Similarly, we show that a firm may simultaneously pursue a low-cost and early market entry strategy if it is capable of developing new technology or production methods that reduce the tradeoff between changeover flexibility and efficiency. Results are also given demonstrating the competitive advantage realized by the firm capable of a high rate of volume-based learning. We show that learning mitigates the increase in production cost associated with less efficient but more flexible technology so that the firm optimally invests in more flexibility, realizes an earlier market entry strategy, and increases profit.

6.4 Possible extensions

A variety of extensions are possible. For example, in contrast to our static formulation, a dynamic model would offer insights on changes that occur over a product's life cycle in relation to pricing and the effect of unit cost reductions due to volume based learning. A dynamic model would also permit analysis of the impact of market growth and price competition that increase over time. Preliminary results on the analysis of a dynamic decision making domain are given in (Gaimon and Morton (2001)). In addition, the development of a two-player game formulation would facilitate further analysis of the effect of competition.

REFERENCES

Arthur, W. B., "Increasing Returns and the New World of Business," *Harvard Bus. Rev.*, (July–Aug. 1996) 100–109.

Baker, W. M., "What Traditional Standard Cost Systems Are Not Effective in Today's Manufacturing Environment," *Industrial Mgt.*, Vol. 31, No. 4, pp. 22–24 (July–Aug. 1989).

Biggadike, R. E., *Entry Strategy and Performance*, Harvard Univ. Press, Cambridge, MA, 1976.

Blackburn, J. D., *Time-Based Competition: The Next Battleground in American Manufacturing*, Business One Irwin, Homewood, IL, 1991.

Brown, C. L., J. M. Lattin, "Investigating the Relationship Between Time in Market and Pioneering Advantage," *Mgt. Sci.*, 40 (1994) 1361–1369.

Boulding, W., M. Christen, "First-Mover Disadvantage," *Harvard Bus. Rev.*, October 2001.

Carpenter, G. S. and K. Nakamoto, "Consumer Preference Formation and Pioneering Advantage," *Jl. of Marketing Res.*, 26 (Aug. 1989) 285–298.

Chakravarty, A., *Market Driven Enterprise: Product Development, Supply Chains, and Manufacturing*, John Wiley & Sons, Inc., NY, 2001.

Davis, D. B., "Beating the Clock," *Electronic Bus.*, (May 29, 1989) 21–29.

Dillon, W. R., R. Calantone, P. Worthing, "The New Product Problem: An Approach for Investigating Product Failures," *Mgt. Sci.*, 25 (Dec. 1979) 1184–1196.

Fine, C., R. M. Freund, "Optimal Investment in Product-Flexible Manufacturing Capacity," *Mgt. Sci.*, 36 (1990) 449–466.

Fine, C., L. Li, "Technology Choice, Product Life Cycles, and Flexible Automation," *Jl. of Manuf. and Opns. Mgt.*, 1 (1988) 372–399.

Flaherty, M. T., "Market Share, Technology Leadership, and Competition in International Semiconductor Markets," *Res. on Technological Innovation, Mgt. and Pol.*, 1 (1983) 69–102.

Forbes Special Report to Management, "Speed to Market," *Forbes* (May 30, 1991).Franza, R. M., C. Gaimon, "Flexibility and Pricing Decisions for High-Volume Products with Short Life Cycles," *Intl. Jl. of Flexible Manuf. Sys.*, 10 (1998) 43–71.

Gaimon, C., V. Singhal, "Flexibility and the Choice of Manufacturing Facilities Under Short Product Life Cycles," *Eur. Jl. of Operational Res.*, 60 (1992) 211–223.

Gaimon, C., A. Morton, "Investment in Changeover Flexibility for Early Market Entry," Working Paper, DuPree College of Management, 2001.

Gerwin, D., "Manufacturing Flexibility: A Strategic Perspective," *Mgt. Sci.*, 39 (1993) 395–410.

Golder, P. N., G. J. Tellis, "Pioneer Advantage: Marketing Logic or Marketing Legend," *Jl. of Marketing Res.*, 30 (1993) 158–170.

Griffin, A., "Metrics for Measuring Product Development Cycle Time," *Jl. of Product Innovation Mgt.*, 10 (1993) 112–125.

Gupta, A. K. and D. L. Wilemon, "Accelerating the Development of Technology-Based New Products," *Calif. Mgt. Rev.*, 32 (1990) 24–44.

Gupta, Y, P. and T. M. Somers, "The Measurement of Manufacturing Flexibility," *Eur. Jl. of Operational Res.*, 60 (1992) 166–182.

Huff, L. C., Robinson, W. T., "Note: The Impact of Leadtime Years of Competitive Rivalry on Pioneer Market Share Advantages," *Mgt. Sci.*, 40 (1994) 1370–1377.

Hutchinson, G. K. and J. R. Holland, "The Economic Value of Flexible Automation," *Jl. of Manuf. Sys.*, 1 (1982) 215–228.

Johnson, R., Maskery, M. A., "Factory Fantasia: New Plants Address Japan's Reality: Few Workers Many Models," *Automotive News*, (July 13, 1992).

Kerin, R. A., P. R. Varadarajan, R. A. Peterson, "First-mover Advantage: A Synthesis, Conceptual Framework, and Research Propositions," *Jl. of Marketing*, 56 (1992) 33–52.

Lederer, P. J., V. R. Singhal, "Effect of Cost Structure and Demand Risk in the Justification of New Technologies," *Jl. of Manuf. and Operations Mgt.*, 1 (1988) 339–371.

Lilien, G. L., E. Yoon, "The Timing of Competitive Market Entry: An Exploratory Study of New Industrial Markets," *Mgt. Sci.*, 36 (1990) 568–585.

Meredith, J., "The Strategic Advantages of New Manufacturing Technologies for Small Firms," *Strategic Mgt. Jl.*, 8 (1987), 249–258.

Porter, A. L., A. T. Roper, T. W. Mason, F. A. Rossini, J. Banks, *Forecasting and Management of Technology*, John Wiley & Sons, Inc., 1991.

Robinson, W. T., "Sources of Market Pioneer Advantages: The Case of Industrial Goods Industries," *Jl. of Marketing Res.*, 25 (1988) 87–94.

Robinson, W. T., C. Fornell, "The Sources of Market Pioneer Advantages in Consumer Goods Industries," *Jl. of Marketing Res.*, 22 (1985) 297–304.

Santo, B., K. Wollard, "The World of Silicon: It's Dog Eat Dog," *IEEE Spectrum*, (Sept. 1988) 30–39.

Schmalensee, R., "Product Differentiation Advantages of Pioneering Brands," *Amer. Economic Rev.*, 72 (1982) 349–365.

Sethi, A. K., Sethi, S. P., "Flexibility in Manufacturing: A Survey," *Intl. Jl. of Flexible Manuf. Sys.*, 2 (1990) 289–328.

Smith, P. G., D. G. Reinertsen, *Developing Products in Half the Time*, Van Nostrand Reinhold, Inc., NY, 1991.

Smith, P. G., "Fast-Cycle Product Development," *Engineering Mgt. Jl.*, 2 (1990) 11–18.

Spital, F. C., "Gaining Market Share Advantage in the Semiconductor Industry by Lead Time in Innovation," *Research on Technological Innovation, Mgt. and Pol.*, 1 (1983) 55–67.

Stalk, G. Jr., T. M. Hout, *Competing Against Time: How Time-Based Competition is Reshaping Global Markets*, The Free Press, NY, 1990.

Stewart, T. A., "Bracing for Japan's Hot New Strategy," *Fortune*, (Sept. 21, 1992) 62–74.

Suarez, F. F., M. A. Cusumano, C. H. Fine, "An Empirical Study of Flexibility in Manufacturing," *Sloan Mgt. Rev.*, 37 (1995), 25–32.

Suarez, F. F., M. A. Cusumano, C. H. Fine, "An Empirical Study of Manufacturing Flexibility in Printed Circuit Board Assembly," *Opns. Res.*, 44 (1996) 223–240.

Urban, G. L., T. Carter, S. Gaskin, Z. Mucha, "Market Share Rewards to Pioneering Brands: An Empirical Analysis and Strategic Implications," *Mgt. Sci.*, 32 (1986) 645–659.

Venkatesan, R., "Cummins Engine Flexes Its Factory," *Harvard Bus. Rev.*, March–April 1990.

von Hippel, E., "Increasing Innovators' Returns from Innovation," *Res. on Technological Innovation, Mgt. and Pol.*, JAI Press Inc., 1 (1983) 35–53.

2. DEMAND MODELING IN PRODUCT LINE TRIMMING: SUBSTITUTABILITY AND VARIABILITY

JUIN-KUAN CHONG, TECK-HUA HO, and CHRISTOPHER S. TANG[1]

ABSTRACT

The number of products or stock keeping units (SKUs) in most product categories has been growing at a phenomenal rate. Even though the number of products increases, the average sale of products can decrease. Due to cannibalization, the sales of some products may even drop below a threshold that makes them unprofitable. This has spurred some firms to remove these under-performing products from their product lines. This ad-hoc "trim the lame duck" procedure can have an adverse effect on the firm's profit for two reasons: first, the "lame duck" may not be the most substitutable product within the line, trimming it results in higher lost sales; second, the "lame duck" may be cheaper to keep with lower inventory cost due to less variability in sales. As an initial step to developing a better product elimination procedure, we use a model that explicitly captures product substitution phenomenon to examine various product portfolios. We compare the mean and the variance of the sales associated with two basic strategies: trimming and no trimming. Our results provide insight into when and which products could be trimmed.

1. INTRODUCTION

Companies are constantly seeking revenue growth. One marketing strategy that aims at revenue growth is *line extension*. The basic rationale for the strategy is that companies

1. We would like to thank Professor Josh Eliashberg, Professor Amiya K. Chakravarty and an anonymous reviewer for their constructive comments on an earlier version of this paper. This research is partially supported by the Center for Technology Management at UCLA and NUS Research Grant R-314-000-018-112.

can increase their revenue if they cater to the diverse preferences of the consumers in the product categories. This strategy is perceived to be less risky since line extensions are launched in an existing product category that already has an established customer base. The wide-spread practice of this strategy has led to an unprecedented increase in the number of stock keeping units (SKUs) in most product categories (see for example Hays (1994), Khermouch (1995), Putsis and Bayus (2001)).

There are several criticisms of the line extension strategy. Bayus and Putsis (1999) show that line extension can significantly increase costs and the cost increase dwarfs the benefits derived from the increase in revenue. Quelch and Kenny (1994) suggested that many companies fail to account for the hidden costs of product proliferation associated with line extension. Perhaps, more critically, these authors challenged the wisdom "more products, more revenues" by arguing that "people do not eat more, drink more, brush their teeth more, or wash their hair more just because they have more products from which to choose." In fact, they showed that the total sales of many product categories (such as cookies and shampoo) decreased despite a significant increase in the number of products. This argument seems plausible for certain mature product categories where growth is often minimal or stagnant.

Quelch and Kenny's finding suggests that increasing the number of SKUs in a product category does not necessarily draw new customers to the category. Consequently, sales of new products must come from the cannibalization on the sales of existing products. This happens especially when the existing products are similar or inferior to the new extensions. This has prompted some companies to trim their product lines (e.g. Khermouch (1995)). There are several rationales for product line trimming. Firstly, the sales of some products can drop below a level which makes it unprofitable for firms to continue supplying them. Secondly, intense competition for limited grocery shelf space drives firms to remove the under-performing products, giving 'space' to more promising new products. In most cases, the companies tend to trim their product line in a *reactive* manner. Product obsolescence can be a cause of such reaction. Avlonitis and James (1982) reported, in a survey of 94 firms, that most firms trimmed their product lines because of product obsolescence. Factors contributing to product obsolescence include change in government regulations, change of buyers' specification, and development of new products. In a mail survey of 166 companies, Hart (1988) concluded that firms also trimmed their product lines because of other factors such as resource shortage, poor product quality, rationalization due to merger and acquisition, and problems associated with raw materials.

On the *proactive* side, firms might want to trim their product lines for two reasons. First, a leaner product line enables a firm to regain its product focus. Some reports suggest that an over-crowded product category can lead to brand loyalty weakening (see for example Hays (1994) and Narisetti (1997)). This brand weakening phenomenon makes the product category very competitive because customers are easily swayed by price promotion (c.f. Quelch and Kenny (1994)). Second, a leaner product line allows a firm to streamline its manufacturing and logistics functions. This is particularly useful when product trimming has only a minimal or no impact on the sales of the firm's entire product line. Indeed, a study conducted by the Food Marketing Institute found that the retailers can reduce the number of SKUs by up to 25% without hurting sales or consumers' perception of variety (c.f. Teinowitz and Lawrence (1993)). However, trimming the wrong product can be costly. For instance, the removal of *Classic Coke* by Coca Cola Company in 1985 caused a strong negative reaction by its consumers. Consequently, the company had to re-introduce the product to reduce the loss of customer to its competitors.

Despite its strategic importance, empirical evidence has shown that most companies do not have a formal product elimination program. Hise and McGinnis (1975) reported that only 31.3% out of their sample of 96 firms have formal procedures. Hise *et al.* (1984) observed a similar pattern: 120 out of a sample of 299 firms have product elimination programs. But only 31% of the 120 firms have a documented program. A survey of U.S. and U.K. firms indicated that less than half of the sample ever utilize any formal methodology in making product elimination decision (c.f. Greenley and Bayus (1994)). The formality of a product elimination program appears to correlate with its frequency of use. In a sample size of 94 firms, Avlonitis (1985) observed that the more products a firm trims, the more formal the process gets; and that the longer it takes to trim a product, the less formal the process is. Because of the rapid rise of the number of SKU, we expect product trimming to receive more managerial attention in the near future. The need to formalize product trimming process can not be overemphasized.

A formal procedure for product line trimming must address the following two fundamental questions: (1) Given the current line of products, should the firm trim its product line? (2) Suppose it is desirable to trim the product line, which product(s) should be eliminated? Several approaches were suggested in earlier works (see for example, Berenson (1963), Alexander (1964), Kotler (1965), Hamelman and Mazze (1972), Browne and Kemp (1976)). Most of these approaches use a rating scheme to rate the attractiveness of the products. Berenson (1963) based his model on five decision factors, namely, financial security, financial opportunity, marketing strategy, social responsibility and organized intervention. Management assigns weight to each factor to reflect its relative importance. The summation of five weighted scores becomes the overall rating of a product. Alexander (1964) used a similar procedure with a different set of factors. These are sales trend, price trend, profit trend, substitute products, product effectiveness and demand on executive time. Kotler (1965) did not provide a list of factors to consider. Instead, he suggested a six-step procedure where the first 3 steps were used by management to develop a comprehensive set of criteria and to generate ratings and weights for these criteria. A "product retention index", which is computed in a similar fashion as those of Berenson (1963) and Alexander (1964), is calculated for each product using the ratings and weights. A different index was suggested by Hamelman and Mazze (1972). Their "Selection Index Number (SIN_i)" is a ratio of the square of product i's percentage contribution margin to its percentage resource cost utilization. Browne and Kemp (1976) suggested a general multiple stage sequential review process. First, weak products are detected based on criteria specified by top management. Further analysis and evaluation of these products are done leading to an elimination decision and finally implementation.

These models are "macro" in the sense that they do not capture consumer choice process and product substitution explicitly. But both components are vital to an accurate assessment of the impact due to trimming. Knowing which products consumers will switch to in the absence of a product allows one to quantify precisely the impact of trimming. Our model incorporates the consumer choice process and product substitution phenomenon explicitly.

The intent of this paper is to highlight the importance of product *substitutability* and to examine its impact on product line trimming in terms of *expected demand* and *demand variability*. Urban (1969) studies the impact of product substitutability on the marketing strategy of a product line. Specifically, Urban (1969) models consumer response to different marketing mixes of a product line and use it to determine an optimal allocation of marketing mix for each product in a product line. In this work, we model consumer response to the removal of a product from a product line and use it to evaluate product trimming decisions.

There are several product portfolio models that incorporate, whether explicitly or implicitly, the expected demand (i.e. return) and demand variability (i.e. risk) in their analyses. These models either assimilate the financial portfolio analysis in their approaches (e.g. Mahajan and Wind (1985), Mahajan, Wind and Bradford (1982))or use a mathematical programming formulation in their analyses (e.g. Larréché and Srinivasan (1982), Corstjens and Weinstein (1982)). The prime objective of these models is to determine the right product lines in the optimal product portfolio and to allocate marketing resources to these product lines. Our work is complementary to these models. We focus our analysis on the impact of product trimming on the expected demand and the variance of demand for the remaining products. The result of our analysis can potentially be used as an input to these optimization models.

One might argue that an alternative strategy to product line trimming is to lower the price of under-performing products.[2] However, this strategy may not be desirable (or even viable) for the following reasons. First, the products may differ only in product attributes such as taste (e.g. fruit drops), flavor (e.g. ice cream) or color (e.g. T-shirts, ball pen). The consumers do not usually expect any price difference in such cases. Second, the manufacturer may have limited freedom to charge different prices for similar products because this may not be acceptable to the retailers and consumers. For example, we can not price a product below another product having less feature or of a smaller package size. Third, in industries where the production lead time is relatively long, product line decision must precede pricing decision. Hence, we focus on product line decision.

To analyze the underlying issue whether to trim or not, we consider a situation in which a firm must select one of the following strategies: *Same Line* (i.e., no products should be eliminated from the line) and *Trim Line* (i.e., some products should be

2. Instead of price decrease, one might suggest price increase to realize a higher profit margin. This is a less viable option since sales volume tends to be low for the under-performing products.

eliminated from the line). Specifically, we analyze and compare the expected sales (or demand) and the variance of the sales associated with the two strategies. Besides expected demand, we analyze the variance of the demand, a variable commonly treated as noise by marketing scientists but studied extensively by operations researchers (see Silver and Peterson (1985)), because it is a good surrogate for the manufacturing and logistics costs. We present a simple framework that enables brand managers to better conceptualize the impact of product trimming. In this framework, we specify conditions under which the Same Line strategy is optimal. In the event that the Trim Line strategy may be optimal, our model suggests potential candidates to be trimmed from the line. Driven by its intuitive appeal, many firms tend to adopt the so-called *lame duck* heuristic: trim the product that has the lowest sales (see surveys by Hart (1988), and Avlonitis and James (1982)). We shall show later that this heuristic can be suboptimal because it neglects demand interaction among products within the same category.

This paper is organized as follows. In the next section, we present a model that captures substitutability among products. Section 3 compares the 'Same Line' and 'Trim Line' strategies. In addition, we derive conditions under which the Same Line strategy is optimal. Section 4 deals with the situations when the Trim Line strategy may be optimal. We shows why and when the "lame duck" heuristic can be seriously flawed. In section 5, we provide a numerical example to illustrate some of the insights generated from our analysis. Section 6 concludes the paper with some suggestions for future research.

2. THE MODEL

Consider a firm that manufactures and sells a line of N products in a selling season. The firm has accumulated some experience in selling the N products during previous seasons. At the beginning of the current season, the firm is contemplating if certain products should be removed from the product line. Essentially, the firm can either offer the same N products (i.e., adopt 'Same Line' strategy) or trim the line by removing some products (i.e., adopt 'Trim Line' strategy). In the event that the Trim Line strategy is preferred, the firm would like to determine the product(s) to be eliminated from the product line. In order to evaluate the pros and cons of the Same Line and Trim Line strategies, one needs to examine the impact of each strategy on the expected sales (or demand) and the variance of demand of the products must be determined. This observation motivates us to develop a simple model that captures the consumer purchasing processes and product substitution.

In our model, the product category has $N + 1$ products indexed by i where $i = 0, 1, \ldots, N - 1, N$. For expository purpose, we aggregate all other products other than those of the firm and denote this 'aggregated product bundle' as product 0. Products 1 to N constitute the firm's product line before line trimming. The product category consists of a population of consumers with heterogeneous product preferences. Each consumer is assumed to have a stationary and zero-order probabilistic product choice process. We consider the case in which the consumers are broadly

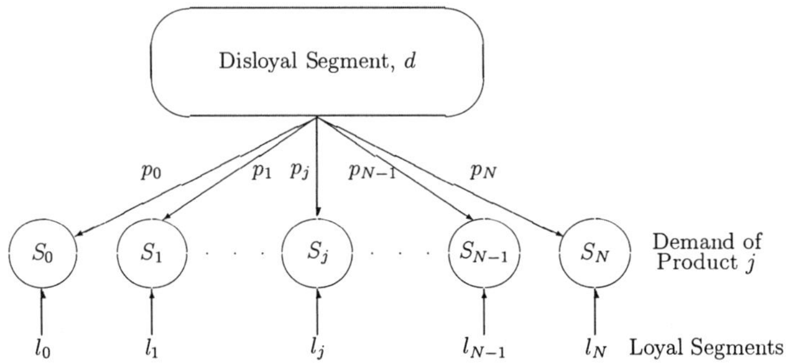

Figure 1. Demand of product j under the same line strategy.

classified into 2 segments: **loyal** and **disloyal**.[3] The loyal segment consists of consumers whose choice of product is pre-specified, while the disloyal segment is comprised of consumers whose product selection is governed by their product choice probabilities. Each product has its own loyal segment but there is an aggregate segment of disloyalists.

We denote the size of loyal segment i by l_i; the size of the disloyal segment by d. Both l_i and d are exogenously given parameters. Each customer in the disloyal segment is assumed to choose product i with probability p_i (see Figure 1). Obviously, $\Sigma_0^N p_i = 1$. Let D_i be the random variable denoting the demand for product i generated from the disloyal segment. Then $(D_0, D_1, \ldots, D_{N-1}, D_N)$ are $N + 1$ multinomial random variables with parameter $(d; p_0, p_1, \ldots, p_{N-1}, p_N)$. It is assumed that the firm has some knowledge about these parameters which can be estimated reliably from the sales data generated during previous selling seasons. The reader is referred to Grover and Srinivasan (1987) for a methodology that estimates these parameters using panel data.

2.1 The same line strategy

Under the Same Line strategy, the firm would sell the same N products as before. Let S_i be the total demand of product i within the selling season. Notice that the demand S_i generates from two pools of consumers: one from the loyal segment and the other from the disloyal segment. Thus, we have

$$S_i = l_i + D_i, \; i = 0, \ldots, N. \tag{2.1}$$

3. Our model is inspired by a model developed by Grover and Srinivasan (1987). Grover and Srinivasan propose a market segment methodology that allows for multiple disloyal segments. For notational convenience, we do not need such a refined partition. We conceptually 'collapse' these disloyal segments into a single aggregate segment and it should be clear later that this will not cause any loss in generalizability of our results. Within each market segment, customers are homogenous in terms of product choice probabilities. Since each product has its own loyal customers, we have altogether $N + 1$ loyal segments.

It follows from (2.1) and the properties of multinomial random variables that the expected value and the variance of S_i can be expressed as:

$$E(S_i) = l_i + dp_i, \qquad i = 0, \ldots, N,$$
$$Var(S_i) = dp_i(1 - p_i), \; i = 0, \ldots, N.$$

In this case, it is easy to check that the mean and variance of the total demand of all products in the firm's product line under the Same Line strategy are given by:

$$E\left(\sum_{1}^{N} S_i\right) = \sum_{j=1}^{N} l_j + d(1 - p_0) \tag{2.2}$$

and,

$$Var\left(\sum_{1}^{N} S_i\right) = d\,p_0(1 - p_0). \tag{2.3}$$

2.2 The trim line strategy

Suppose that the firm has decided to trim product i from its product line. Then there are $N - 1$ remaining products in the line, namely, $1, \ldots, i-1, i+1, \ldots, N$. In this paper, we shall assume that the elimination of product i would have *direct* impact on the purchasing behavior of two specific groups (the loyal segment of product i and the disloyal segment) as follows:

2.2.1 The loyal segment of product i

When we eliminate product i from the product line, the customers in loyal segment of product i have to consider other alternatives. Each of these l_i customers may consider the following alternatives: (1) becomes loyal to other products; (2) joins the disloyal segment and selects the product according to some choice probabilities; or (3) becomes disenchanted and leaves the consumer base. Thus, for each customer who belongs to the loyal segment i, it is assumed that there is a probability q_{ij} that he/she becomes 'loyal' to product j, where $j = 0,1, \ldots, i-1, i+1, \ldots, N$, that there is a probability x_i that he/she joins the disloyal group, and that there is a probability y_i that he/she leaves the consumer base. Clearly, $\Sigma_{j \neq i} q_{ij} + x_i + y_i = 1$, for $i = 1, \ldots, N$.

Notice that the values of q_{ij}, x_i, y_i depend on the nature of product i and the degree of its *substitutability* with other products in the category.[4] When there exists a *highly* substitutable

4. Substitution between products can occur because products are often similar to some degree. Thus, when a product is not available for purchase, some customers may instead buy an alternative product. Intuitively, one would expect product substitution to occur more readily between a pair of similar products than between a pair of dissimilar products. This notion of similarity is implicit in most product positioning models; products that are positioned close to each other in the perceptual space are more similar to each other and are more easily substitutable (see Green and Krieger, 1993). In the classical attraction model (Bell, Keeney and Little, 1975), when a product i is eliminated from the consideration and choice set, the probability for product j ($j \neq i$) is increased from A_j/Σ_k^{Ak} to $A_j/\Sigma_{k \neq i}^{Ak}$ where A_k is the level of attraction of product k. Thus, the attraction model restricts that the ratio of the purchase probabilities of any two remaining products stays constant. This restriction is not required in our model.

product j, it is reasonable to assume that $q_{ij} \approx 1$, and $\{q_{ik}, k \neq j\}$, x_i, $y_i \approx 0$. Notice that the substitutable product j could be one of the remaining products (i.e., $j = 1, \ldots, i-1, i+1, \ldots, N$), or it could be one of the products that belongs to the competitors (i.e., when $j = 0$). However, when all other products are equally substitutable, customers may join the disloyal group (i.e., $x_i \approx 1$). Finally, when the salient features of product i are not captured in other products, it is possible that the i's loyal consumers may leave the product category completely (i.e., $y_i \approx 1$).[5]

Let L_{ij} be the size of the loyal segment j generated from the loyal segment l_i if the firm eliminates product i. Let X_i be the new *addition* to the disloyal segment generated from the loyal segment l_i after trimming product i. Let Y_i be the number of customers who leave the product category. In this case, it is easy to see that the random variables $(L_{i0}, \ldots, L_{i,i-1}, L_{i,i+1}, \ldots, L_{iN}, X_i, Y_i)$ are $N + 2$ multinomially distributed with parameters $(l_i; q_{i0}, \ldots, q_{i,i-1}, q_{i,i+1}, \ldots, q_iN, x_i, y_i)$.

2.2.2 The disloyal segment

There are two disloyal segments that are affected by the elimination of product i. These two disloyal segments are: D_i, the original disloyal segment who intends to buy product i, and X_i, the new *addition* to the disloyal segment generated from the loyal segment l_i after trimming product i. Depending on its substitutability with each of the remaining products j, it is assumed that each consumer who belongs to these two disloyal segments will 'switch' to product j with probability α_{ij}.[6] Notice that $\Sigma_{j=0, j \neq i}^{N} \alpha_{ij} = 1$. Let D_{ij} be the random variable denoting the demand for product j generated from the original disloyal segment after trimming product i from the product line. Then $\{D_{ij} : j = 0, \ldots, i-1, i+1, \ldots, N\}$ are N multinomial random variables with parameters $(D_i; \alpha_{i0}, \alpha_{i1}, \ldots, \alpha_{iN})$. Similarly, let X_{ij} be the random variable denoting the demand for product j generated from the additional disloyal segment coming from the loyal segment of product i. Then $\{X_{ij} : j = 0, \ldots, i-1, i+1, \ldots, N\}$ are N multinomial random variables with parameters $(X_i; \alpha_{i0}, \alpha_{i1}, \ldots, \alpha_{iN})$.

By considering the impact of trimming product i on the loyal segment of product i and the disloyal segment, the effective demand of product j under the Trim Line strategy can be expressed as:

$$T_{ij} = l_j + L_{ij} + D_j + D_{ij} + X_{ij}, j = 0, 1, \ldots, i-1, i+1, \ldots, N.$$

Notice that the total demand T_{ij} for product j is the sum of the various loyal and disloyal segments. The loyal segment consists of the old loyal segment of product j, l_j, and the new addition of loyal segment generated from a migration from the loyal segment of i, L_{ij}. The disloyal segment consists of the existing disloyal segment, D_j, the substitution from the existing disloyal segment of product i, D_{ij}, and the substitu-

5. The elimination of 'Classic Coke' is such a case when many consumers have threatened Coca-Cola company that they will abandon cola drinks and switch to other types of soda.
6. To simplify the exposition of our model, we assume that the 'switching' probabilities are the same for both disloyal segments; however, our analysis can be easily extended to the case when these probabilities are different.

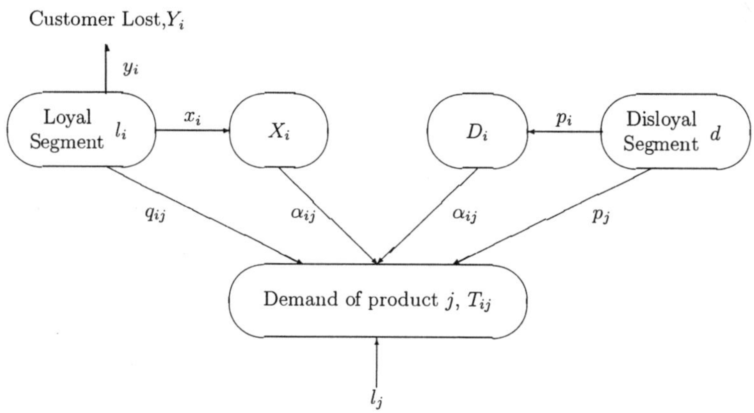

Figure 2. Demand of product j with product i trimmed.

tion from the new disloyal segment generated from a migration of the loyal segment i, X_{ij} (see Figure 2).

Applying the conditional expectation formula, the expected demand for product j after trimming product i can be expressed as:

$$\begin{aligned} E(T_{ij}) &= l_j + l_i q_{ij} + dp_j + \alpha_{ij}\, dp_i + \alpha_{ij}\, l_i\, x_i, \\ &= l_j + d(p_j + \alpha_{ij}\, p_i) + l_i\,(q_{ij} + \alpha_{ij}\, x_i), \\ j &= 0, 1, \ldots, i-1, i+1, \ldots, N. \end{aligned} \tag{2.4}$$

Similarly, we can apply the conditional variance formula to obtain the variance of the demand of product j after trimming product i. In this case, it is easy to show that:

$$\begin{aligned} Var(T_{ij}) &= Var(D_j) + Var(D_{ij}) + Var(L_{ij}) + Var(X_{ij}) \\ &\quad +2Cov(D_j, D_{ij}) + 2Cov(L_{ij}, X_{ij}), \\ &= dp_j\,(1-p_j) + d_{\alpha ij} p_i (1 - \alpha_{ij} p_i) + l_i q_{ij}\,(1-q_{ij}) + l_i \alpha_{ij}\, x_i (1 - \alpha_{ij} x_i) \\ &\quad -2d\alpha_{ij} p_i\, p_j - 2l_i \alpha_{ij}\, x_i\, q_{ij}, \\ &= d(p_j + \alpha_{ij} p_i)(1 - p_j - \alpha_{ij} p_i) + l_i\,(q_{ij} + \alpha_{ij} x_i)(1 - q_{ij} - {}_{\alpha ij} x_i). \end{aligned} \tag{2.5}$$

From Figure 2, we observe that the expressions for the expected demand and the variance of the demand of T_{ij} capture the fact that the demand of product j is generated from three sources: (a) the loyal segment of product j, l_j, (b) the original disloyal segment, d, where each customer will purchase product j with probability $(p_j + \alpha_{ij} p_i)$, and (c) the loyal segment of product i, l_i, where each customer will purchase product j with probability $(q_{ij} + \alpha_{ij} x_i)$.

Observe from equations (2.4) and (2.5) that the expected demand and the variance of T_{ij} depend on the substitutability of product j for product i. This substitutability effect is captured in q_{ij} (in the context of the loyal segment of product i) and α_{ij} (in the context of the disloyal segment). To examine the expectation and the variance of the demand for the remaining product line (i.e., $\Sigma_{j=1, j\neq i}^{N} T_{ij}$), if we trim product i from the line, observe that:

$$\begin{aligned}\sum_{j=1, j\neq i}^{N} T_{ij} &= \sum_{j=1, j\neq i}^{N} \{l_j + D_j + D_{ij} + L_{ij} + X_{ij}\}, \\ &= \sum_{j=1}^{N} l_j + d - D_0 - L_{i0} - D_{i0} - X_{i0} - Y_i.\end{aligned}$$

By applying (2.4) and (2.5), it can be easily shown that:

$$E\left(\sum_{j=1, j\neq i}^{N} T_{ij}\right) = \sum_{j=1}^{N} l_j + d(1 - p_0) - \{\alpha_{i0}p_i + l_i(q_{i0} + \alpha_{i0}x_i + y_i)\}. \tag{2.6}$$

and

$$\begin{aligned}Var\left(\sum_{j=1, j\neq i}^{N} T_{ij}\right) = \; & dp_0(1 - p_0) + d\alpha_{i0}p_i(1 - \alpha_{i0}p_i) \\ & - 2d\alpha_{i0}p_ip_0 + l_iq_{i0}(1 - q_{i0}) \\ & + l_i\alpha_{i0}x_i(1 - \alpha_{i0}x_i) + l_iy_i(1 - y_i) \\ & - 2l_iq_{i0}y_i - 2l_iq_{i0}\alpha_{i0}x_i - 2l_iy_i\alpha_{i0}x_i,\end{aligned}$$

which can be simplified into:

$$\begin{aligned}Var\left(\sum_{j=1, j\neq i}^{N} T_{ij}\right) = \; & d(p_0 - \alpha_{i0}p_i)(1 - p_0 - \alpha_{i0}p_i) \\ & + l_i(y_i + q_{i0} + \alpha_{i0}x_i)(1 - y_i - q_{i0} - \alpha_{i0}x_i).\end{aligned} \tag{2.7}$$

3. SAME LINE VS. TRIM LINE STRATEGIES

In the last section, we derived the expressions for the expectation and the variance of the demand of the entire product line under the same line and trim line strategies (i.e. (2.2), (2.3), (2.6) and (2.7)). In this section, we develop a simple framework for comparing the Same Line and Trim Line strategies. Under the framework, the *pros* and *cons* associated with each of the strategies can be represented as a point on a two-dimensional map. The two dimensions on this map are the expected demand of the entire product line (vertical axis) and the variance of the demand of the entire product line (horizontal axis). The justification of this two-dimensional map is as follows. Observe that the expected revenue increases in expected demand, and that the expected costs (processing cost, production and capacity planning cost, and inventory cost) increases in the variance of the demand (The reader is referred to Silver and Peterson (1984) for an excellent discussion on how costs are affected by the variance

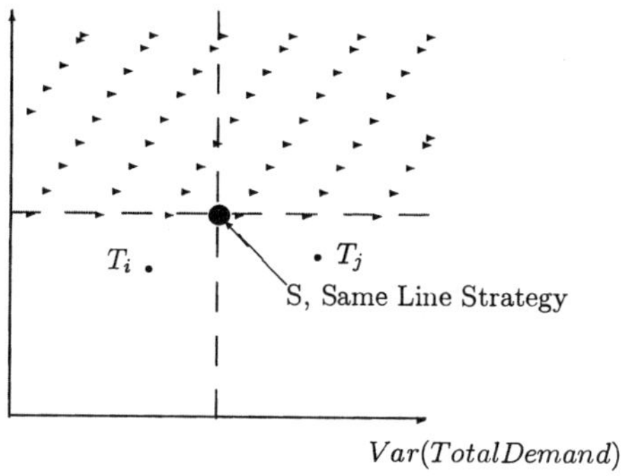

Figure 3. Same line strategy vs. trim line strategies.

of demand.). This observation implies that one can view the expected demand as a surrogate for the revenue and the variance of the demand as a surrogate for cost.

By using (2.2), (2.3), (2.6) and (2.7), we can map each of the strategies on this two dimensional map (see Figure 3). In Figure 3, the point s corresponds to the Same Line strategy that has coordinates $(E(\Sigma_{i=1}^{N} S_i),\ Var(\Sigma_{i=1}^{N} S_i))$. Similarly, the point T_i (or T_j) corresponds to the Trim Line strategy when product i (or product j) is trimmed from the line.[7] In what follows, we shall show that (a) the relative position of each point T_i (or T_j) depends on the product to be trimmed; and (b) T_i and T_j can only be located in the unshaded region (in Figure 3).

In order to compare the Same Line and Trim Line strategy, let us examine the relative position of the point S associated with the Same Line strategy and the point T_i associated with the case when we trim product i under the Trim Line strategy. To do so, we shall first compare the expected demand of the entire product line under the Same Line and the Trim Line strategies, and then compare the variance of the total demand of the entire product line under the two strategies.

Based on equations (2.2) and (2.6), we define $Loss(i)$ as:

$$\begin{aligned} Loss(i) &= E\left(\sum_{j=1}^{N} S_j\right) - E\left(\sum_{j=1, j\neq i}^{N} T_{ij}\right), \qquad (3.1)\\ &= dp_i\alpha_{i0} + l_i(q_{i0} + y_i + x_i\alpha_{i0}) > 0. \end{aligned}$$

The term $Loss(i)$ represents the expected loss in sales as a result of trimming product i. Since the right hand side is always positive, we can conclude that:

7. Although we consider the case of trimming a single product, it is not difficult to see that repeated application of this two dimensional map will allow us to see the impact of trimming more than one product.

Observation 1: *The expected total demand associated with the Trim Line strategy is always lower than that of the Same Line strategy.*[8]

This observation implies that all points T_i will be located in the unshaded region in Figure 3; i.e., either at the south-west quadrant or the south-east quadrant from the point S. Note that the following remarks follow from (3.1):

- *Loss*(i) increases in dp_i and l_i. Thus, products that have a smaller expected demand (i.e., $dp_i + l_i$) tend to result in a smaller loss in the demand for the entire product line. This result may appear to support the 'lame duck' heuristic. However, the expression for *Loss*(i) also suggests that the loss is mediated by other parameters.[9]
- *Loss*(i) increases in y_i. Thus, highly unique products should not be trimmed because they are more likely to lead to customers leaving the product category, i.e., y_i is high.
- *Loss*(i) increases in x_i. In other words, products that exhibit significant 'brand weakening' phenomena, i.e. high probability of joining disloyal segment, should not be trimmed.
- *Loss*(i) increases in α_{i0} and/or q_{i0}. It follows from the definitions of α_{i0} and q_{i0} that α_{i0} and q_{i0} increase when the competitive product 0 is a better substitute for product i than that of other products belonging to the firm. To reduce the expected loss of sales under the Trim Line strategy, it is desirable to trim a product i that has close substitutes offered by the firm but not offered by the competitors. This observation highlights the significance of product substitutability when deciding the product to be trimmed.

Next, based on equations (2.3) and (2.7), we define *Reduction*(i) as:

$$
\begin{aligned}
Reduction(i) &= Var\left(\sum_{j=1}^{N} S_j\right) - Var\left(\sum_{j=1, j\neq i}^{N} T_{ij}\right), \\
&= dp_0(1 - p_0) - \{d\,(p_0 + \alpha_{i0}p_i)(1 - p_0 - \alpha_{i0}p_i) \\
&\quad + l_i(y_i + q_{i0} + \alpha_{i0}x_i)\,(1 - y_i - q_{i0} - \alpha_{i0}x_i)\}, \qquad (3.2)\\
&= d\alpha_{i0}p_i(2p_0 + \alpha_{i0}p_i - 1) \\
&\quad - l_i(y_i + q_{i0} + \alpha_{i0}x_i)(1 - y_i - q_{i0} - \alpha_{i0}x_i). \qquad (3.3)
\end{aligned}
$$

The term *Reduction*(i) represents the reduction of variance in sales as a result of trimming product i. Notice from (3.3) that *Reduction*(i) can be either positive or negative.

8. Note that our model does not incorporate the possibility that a smaller number of products may lead to increased sales because of less customer confusion.
9. Intuitively, $d_{pi} + l_i$ is the maximum potential loss and *Loss*(i) is the expected loss after accounting for consumer switching behaviors.

This implies that the total variance can either decrease or increase as a result of trimming product i. In this case, we can conclude that:

OBSERVATION 2: *Depending on the values of the parameters, the variance of the total demand associated with the Trim Line strategy can be lower or higher than that of the Same Line strategy.*

Based on (3.2), we can make the following remarks:

- When $p_0 < 0.5$ and $(p_{0} + \alpha_{i0} p_i) \leq 0.5$, we have *Reduction* $(i) < 0$.[10] That is, if the firm holds a dominant position in the disloyal segment before line trimming (i.e., $\Sigma_{i=1}^{i=N} p_i = 1 - p_0 > 0.5$), trimming a product always lead to an increase in the variance of the demand of the entire product line. This finding implies that a leader in the disloyal segment is less likely to trim a product. Put differently, a dominant leader may have a wider product line than followers.
- When $p_0 > 0.5$, and $d \gg l_i$, we have *Reduction* $(i) > 0$. In words, if the firm has less than 50% market share in a large disloyal segment, then the firm can reduce its variance of the demand for the entire product line by trimming the product line.

Similarly, the following remarks follow from (3.3):

- *Reduction*(i) increases in p_i when $p_0 \geq 0.5$. This implies that the larger the market share of a product in the disloyal segment where the competitor is a market leader, the more attractive (from the viewpoint of variance reduction) is trimming that product to the firm.
- *Reduction*(i) increases in p_0. If the firm has low market share in the disloyal segment, then the firm can reduce its variance by trimming the product line.
- *Reduction*(i) decreases in q_{i0} if $y_i = q_{i0} + \alpha_{i0}x_i < 0.5$. So, if the trimmed product is similar to any of competitive products, the firm can experience an increase in the variance. This observation and the fact that *Loss*(i) increases in q_{i0} imply that trimming a product similar to any competitive product reduces the expected value and simultaneously increases the variance of the demand for the entire product line.
- *Reduction*(i) decreases in l_i. If each product has a sufficiently large loyal segment, then *Reduction*$(i) < 0$ for all i; hence it does not pay to trim.

Based on observations 1 and 2, each point T_i would either located at the south-west quadrant or the south-east quadrant from the point S (see Figure 3). For any point T_i located at the south-east quadrant from the point S, there is a decrease in the expected demand and an increase in the variance of the demand if the firm trims product i from its product line. Clearly, it is undesirable to trim a product i when the corresponding point T_i is located at the south-east quadrant from the point S. This observation enables us to specify a *necessary condition* under which the same line strategy is optimal as follows:

10. Note that $x(1 - x)$ increases in x for $x < 0.5$ and decreases in x for $x > 0.5$.

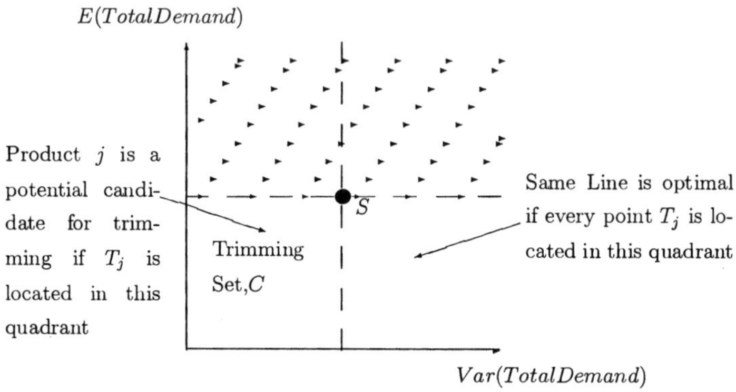

Figure 4. Optimality condition for same line strategy.

OBSERVATION 3: *If Reduction(i)* < 0 *for each i, where* $i = 1, \ldots, N$, *then the Same Line strategy is optimal.*[11]

If all T_i, $(i = 1, \ldots, N)$, are located at the south-east quadrant, the Same Line strategy is optimal (see Figure 4). When the above condition does not hold, then there exists a set of products, denoted by C, where

$$C = \{i : Reduction(i) > 0, \ i = 1, \ldots, N\} \tag{3.4}$$

such that there will be a reduction in the variance of the demand as a result from trimming product $i \in C$ from the line. In this case, for each product i in the set C, the corresponding point T_i will be located at the south-west quadrant as depicted in Figure 4. Even though there is a reduction in variance when we trim a product $i \in C$, an expected loss, *Loss*(i), would result from trimming product i. Hence, it is necessary to evaluate the trade-off between expected loss in demand and the reduction in the variance when selecting a product to trim. The actual trade-off would depend on the price/cost structure, which is beyond the scope of this paper.

In light of Observation 3, we now investigate the impact of substitutability, α_{ij}, on the Same Line and Trim line strategies. Since $\Sigma_{j=0, j \neq i}^{N} \alpha_{ij} = 1$, we can express α_{i0}, the substitutability of the competitive product for the firm's product, as $\alpha_{i0} = 1 - \Sigma_{j=1, j \neq i}^{N} \alpha_{ij}$. In this case, the substitutability with the competitive product, α_{i0}, decreases linearly as the substitutability with the firm's remaining product line, $\Sigma_{j=1, j \neq i}^{N} \alpha_{ij}$ increases. Hence, studying the impact of the substitutability with the competitive product, α_{i0}, is equivalent to studying the impact of substitutability with the firm's remaining product line, $\Sigma_{j=1, j \neq i}^{N} \alpha_{ij}$. For this reason, it suffices to focus

11. We assume that all benefits derived from trimming the product line are captured by the variance term. This is a reasonable assumption if the manufacturing technology is flexible enough that the unit cost is not affected by the number of products in the product line.

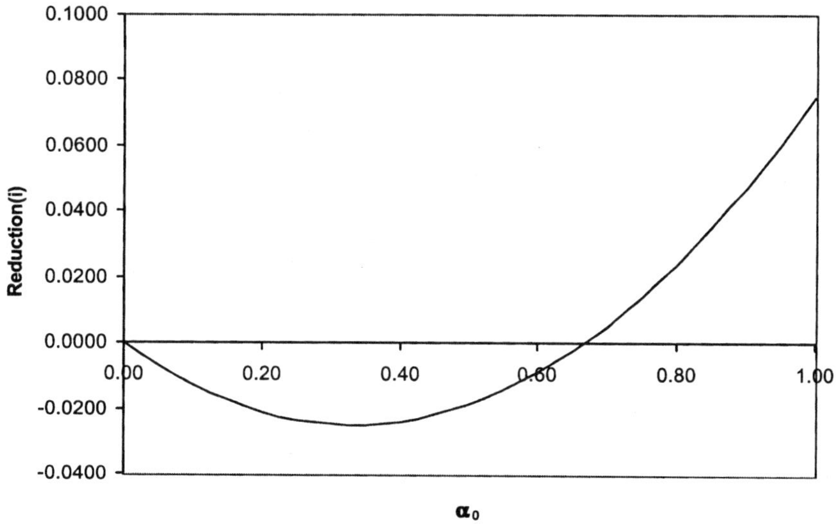

Figure 5. Reduction (i) as a quadratic function in α_0.

on the impact α_{i0} on the Same Line and Trim Line strategies. In order to obtain some insights, we assume that there are no loyal segments (i.e., $l_i = 0, \forall i$).

Using Observation 3, we establish the condition for α_{i0} under which the Same Line strategy is optimal; i.e., the condition for α_{i0} such that *Reduction(i)* < 0 for all $i, i = 1, \ldots, N$. It follows from equation (3.3) and the fact that $l_i = 0, \forall i$, we have:

$$Reduction(i) = d\alpha_{i0}p_i(2p_0 + \alpha_{i0}p_i - 1) \tag{3.5}$$

Recall from the second remark following observation 2 that *Reduction(i)* $> 0, \forall i$ when $p_0 \geq 0.5$, regardless of the value of α_{i0} . Therefore, it suffices to consider the case when $p_0 < 0.5$. Note from (3.5) that *Reduction(i)* is a quadratic function in α_{i0} that has one root for Reduction(*i*) $= 0$ at $\alpha_{i0} = 0$. This implies that there exists another root $\theta_i > 0$ such that *Reduction(i)* $= 0$ at θ_i where $\theta_i > 0$ satisfies the equation $d\theta_i p_i (2p_0 + \theta_i p_i - 1) = 0$ or,

$$\theta_i = \frac{1 - 2p_0}{p_i} \tag{3.6}$$

This implies that *Reduction(i)* < 0 whenever α_{i0} is in the interval $0 < \alpha_{i0} < \theta_i$ (see Figure 5). It follows from Observation 3 that the Same Line strategy is optimal when $0 < \alpha_{i0} < \theta_i$. This leads to:

OBSERVATION 4: *Consider a market consists entirely of disloyal segment (i.e., $l_i = 0, \forall i$). If the firm is a market leader in the disloyal segment (i.e., $p_0 < 0.5$) and the substitutabilities of competitive product for the firm's products are sufficiently low (i.e., when $\alpha_{i0} < \theta_i$, $\forall i$), then the Same Line strategy is optimal.*

We now relate Observation 4 to the case in which product substitutability conforms to the classical attraction model of Bell, Keeney and Little (1975).[12] First, it can be shown (see Appendix) that the substitutability α_{ij} relates to the market share of product j in the disloyal segment p_j and the market share of the trimmed product i in the disloyal segment p_i as follows:

$$\alpha_{ij} = \frac{p_j}{1 - p_i} \tag{3.7}$$

and correspondingly for $\alpha_i 0$,

$$\alpha_{i0} = \frac{p_0}{1 - p_i} \tag{3.8}$$

It follows from (3.6) and (3.8) that the condition $\alpha_{i0} < \theta_i$ can be rearranged into:

$$\frac{1 - 2p_0}{p_0} > \frac{p_i}{1 - p_i} \tag{3.9}$$

In this case, as stated in observation 4, Same Line strategy is optimal if (3.9) is true for all i. In other words, Same Line strategy is optimal if

$$\frac{1 - 2p_0}{p_0} > \max_i\{\frac{p_i}{1 - p_i}\} \tag{3.10}$$

Notice that condition (3.10) is more likely to hold when the term $\max_i \{p_i/(1 - p_i)\}$ is minimized. Suppose that p_0 is fixed. Hence, $\Sigma_i p_i = 1 - p_0$ is also fixed. In this case, it is easy to see that $\max_i \{p_i/(1 - p_i)\}$ is minimized when the p_i's are all equal. This implies that when the competitive market share p_0 is fixed, the Same Line strategy is more likely to be optimal if the market shares of the firm's products do not differ very much among themselves.

12. Assumption A4 of the classical attraction model by Bell, Keeney and Little (1975) assumes that the market share of a seller depends on the magnitude of the change in the attraction of other seller(s) but does not depend on which seller(s) is making the change. In other words, the change in the attraction of a seller does not impact *differentially* on other sellers (i.e. no asymmetry). By construction, our model allows for such an asymmetry because no specific form of substitutability, α_{ij}, is assumed

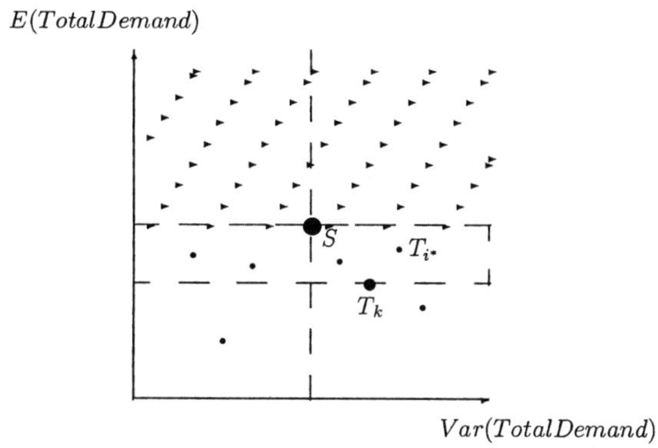

Figure 6. Lame duck strategy does not results in minimum loss.

4. THE LAME DUCK HEURISTIC

In the last section, we have presented a framework to analyze the conditions under which the Same Line strategy is optimal. In the event the same line strategy may not be optimal, the framework has also suggested a potential set of candidates to be trimmed. Now, we apply our framework to analyze the effectiveness of a common product trimming heuristic, i.e., the lame duck heuristic.[13] Specifically, the lame duck heuristic prescribes the elimination of product *k*, where

$$k = argmin \{j : E(S_j), j = 1, \ldots, N\} = argmin\{j : l_j + dp_j, j = 1, \ldots, N\}. \quad (4.1)$$

By (4.1), it is obvious that the lame duck heuristic ignores the demand interactions and the substitutability among products.[14] Without taking into account substitutability, the lame duck heuristic can be flawed for three reasons, as stated in the following three observations:

OBSERVATION 5: *In general, trimming the lame duck will not result in minimum loss in expected sales (i.e.,* $k \neq i^\star$ *where* $i^\star = argmin \{j : loss(j), j = 1, \ldots, N\}$*). However, if* $l_i = 0, \forall i$ *and the classical attraction model for market share holds for the disloyal segment, then* $k = i^\star$.

In Figure 6, the point T_k corresponds the lame duck whereas the point $T_{i^\star}$ corresponds to the product where trimming it results in minimum loss (or maximum expected demand after trimming). This can happen when the lame duck is a closer substitute to the competitive product than some products in the firm's product line.

13. Various surveys in the literature cited in the introduction suggest that this is a common practice.
14. Assumption A4 of the attraction model by Bell, Keeney and Little (1975) basically implies that the impact of trimming is symmetric and linear. That is the impact of trimming product *i* on product *j* is proportionate to the market share of product *j*, which implicitly ignores the differences in substitutability. By trimming the product with smallest market share, the attraction model predicts a minimum loss of market share; hence, the attraction model essentially prescribes a lame duck selection.

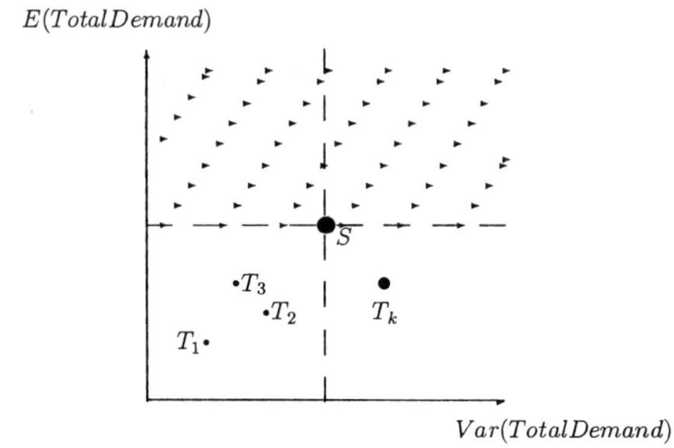

Figure 7. Lame duck strategy results in increased variance.

This can also happen when the lame duck is sufficiently unique that its removal results in a big portion of the loyal customers leaving the market.

Trimming lame duck will result in minimum loss when the attraction model is valid. To elaborate, let us consider the case in which the loyal segments are empty. From (3.1), we have $Loss(i) = d\alpha_{i_0}p_i$. If the substitutability of competitive product for product i, α_{i_0}, conforms to equation (3.8)(i.e., α_{i_0} agrees with the attraction model), then we have:

$$Loss(i) = d\frac{p_0}{1 - p_i}p_i = dp_0\frac{p_i}{1 - p_i},$$

which can be minimized by choosing the product with the smallest p_i. Notice from (4.1) that the lame duck heuristic will select the product with the smallest p_i. In this case, we can conclude that the lame duck heuristic will produce minimum loss when the attraction model is valid.

OBSERVATION 6: *Under certain scenarios (discussed below), trimming the lame duck can result in a loss in sales and a simultaneous increase in demand variability, i.e., Reduction (k)* < 0 *and loss(k)* > 0. *Thus, it is sub optimal.*

Let us consider two cases. First, suppose that $Reduction(i) < 0$ for all i. Recall from the first remark following Observation 2 that this happens when the firm is a market leader (i.e. $p_0 < 0.5$) and each product does not have a significantly larger market share than others in the product line. In this case, Observation 3 implies that the Same Line strategy is optimal. Hence, the firm should not trim any product and the lame duck heuristic is definitely not appropriate. In the second case, suppose there exists some products j such that $Reduction\ (j) > 0$ for each j . Then the set C, as defined in (3.4), is not empty. Hence, it is quite reasonable to trim a product that belongs to the set C such that there is a reduction in the variance of the demand as a result of product trimming. Figure 7 depicts the case when $C = \{1,2,3\}$ and $k \notin C$. Since k

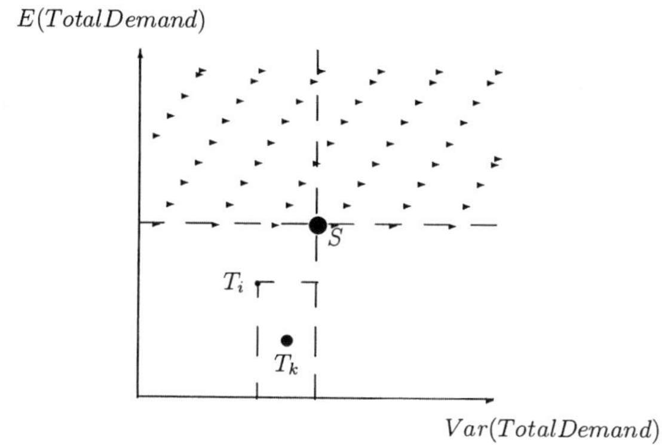

Figure 8. Lame duck strategy is dominated by trim i strategy.

$\notin C$, trimming product k would increase the variance of the demand (in addition to some loss in the expected sales, *loss*(k)). A possible scenario that could result in above situation is when the lame duck is sufficiently unique in the firm's product line, trimming the product results in the loyal customers either leaving the market completely (i.e. $y_k \gg 0$) or joining the disloyal segment (i.e. $x_k \gg 0$).

OBSERVATION 7: *Under certain scenario (discussed below), trimming the lame duck can be dominated even if it belongs to the trim set (i.e., $k \in C$).*

In other words, there exists at least one other product when trimmed is no worse off than trimming the lame duck and is strictly preferred on at least one of the two dimensions, i.e. expected demand and variance of demand. Figure 8 depicts such a case. In Figure 8, the point T_k corresponds the lame duck whereas the point T_i, located north-west from T_k, corresponds to a product when trimmed results in smaller loss and larger variance reduction than trimming the lame duck. One scenario where the lame duck is dominated is when the lame duck is highly substitutable by the competitive product in a market where the competitor has major market share in the disloyal segment.

5. AN ILLUSTRATIVE EXAMPLE

This section presents a numerical example that intends to serve two purposes: (1) to illustrate our findings in previous sections and (2) to show that lame duck heuristic can be flawed.

Consider a firm that offers two products, 1 and 2. Each product has a moderate size of loyal segment ($l_1 = 10, l_2 = 20$). The disloyal segment is large compared to the loyal segments ($d = 400$). Within the disloyal segment, the competitor is the market leader and product 1 has a larger market share than product 2. In our example, we have $p_0 = 0.7$, $p_1 = 0.2$ and $p_2 = 0.1$.

Based on managerial subjective estimates, the parameters associated with trimming product i are the same for both products 1 and 2. Specifically, upon trimming product

Table 1. Impact of same line and trim line strategies

		Scenario 1: $\alpha_{10} = 0.05$ $\alpha_{20} = 0.10$		Scenario 2: $\alpha_{10} = 0.10$ $\alpha_{20} = 0.20$		Scenario 3: $\alpha_{10} = 0.20$ $\alpha_{20} = 0.40$	
Strategy	Same Line	Trim 1	Trim 2	Trim 1	Trim 2	Trim 1	Trim 2
E(Total Demand)	150	143.8	141.2	139.6	136.4	131.2	126.8
Var(Total Demand)	84	84.08	86.01	82.46	84.67	78.98	81.57
Loss(i)	0	6.2	8.8	10.4	13.6	18.8	23.2
Reduction(i)	0	−0.08	−2.01	1.54	−0.67	5.02	2.43

i, $i = 1,2$, each consumer in the loyal segment l_i would either join the disloyal segment with probability $x_i = 0.4$, or become loyal to the competitive product with probability $q_{i0} = 0.1$ (and $q_{12} = q_{21} = 0.4$), or leave the market completely with probability $y_i = 0.1$. Let us consider the following three scenarios with varying degree of substitutability with the competitive product:

Scenario 1: $\alpha_{10} = 0.05$ and $\alpha_{20} = 0.10$
Scenario 2: $\alpha_{10} = 0.10$ and $\alpha_{20} = 0.20$
Scenario 3: $\alpha_{10} = 0.20$ and $\alpha_{20} = 0.40$

Notice that products 1 and 2 become more likely to be substituted by the competitive product as we progress from Scenario 1 to Scenario 3. In addition, the ratio α_{20}/α_{10} is equal to 2 in each scenario, i.e. $\alpha_{20} > \alpha_{10}$. This implies product 2 is more likely than product 1 to be substituted by the competitive product. The impact of the Same Line and Trim Line strategies on the expected demand and the variance of demand for these three scenarios are summarized in Table 1.

Observe from table 1 that as product substitutability α_{i0} increases (as constructed for Scenario 1 through 3), both *Loss*(i) and *Reduction*(i) increase. This phenomenon is consistent with the fact that *Loss*(i) is increasing in α_{i0}, as discussed in the fourth remark following Observation 1. Next, observe that *Reduction*(i) increases as the product substitutability increases (as constructed for Scenario 1 through 3). This observation is consistent with the case depicted in (3.5) in which *Reduction*(i) increases in α_{i0} when $p_0 > 0.5$. As the substitutability of competitive product for product i increases, trimming product i will result in increasing transfer of demand to the competitive product.

Let us plot the expected demand and the variance of demand for the three scenarios (as depicted in Table 1) in Figure 9. Each point $T_j^{(s)}$ in Figure 9 corresponds to the case where we trim product j in Scenario s. First, observe that all points $T_j^{(s)}$, $j = 1,2$, $s = 1,2,3$ are located below the point S. It means that expected sales for Trim Line strategies are lower than that of Same Line strategy. This illustrates Observation 1. Second, each point $T_j^{(s)}$ either locates in the south-east quadrant or the south-west quadrant. This implies that the variance of demand could either increase or decrease as a result of trimming the product line. This illustrates Observation 2. Third, observe

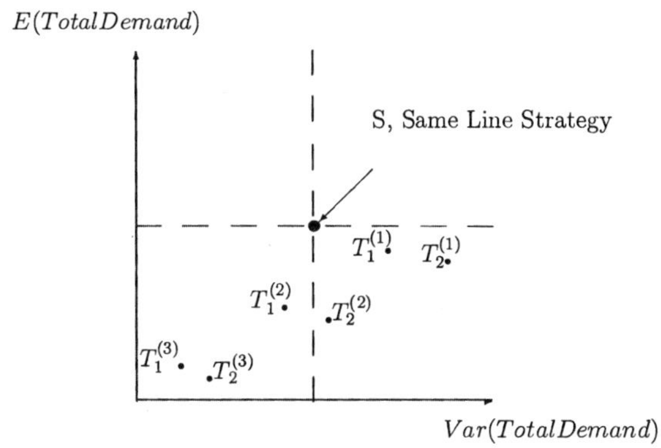

Figure 9. Comparing same line and trim line strategies.

that in Scenario 1, both $T_1^{(1)}$ and $T_2^{(1)}$ are located in the south-east quadrant. This implies that trimming either product 1 or product 2 would reduce expected demand and increase variance of demand; hence, Same Line strategy is optimal. This illustrates Observation 3.

We now illustrate when the lame duck heuristic could be flawed. In our example, product 2 is the lame duck, since $l_1 + d_{p1} = 90 > l_2 + d_{p2} = 60$; hence $k = 2$. Observe from Figure 9 that in each Scenario s, $T_1^{(s)}$ is located above $T_1^{(s)}$. This implies that the expected demand associated with trimming product 1 is consistently higher than that of product 2. This illustrates Observation 5 on why the lame duck heuristic could result in a lower sale. Next, consider Scenario 2. Observe that $T_1^{(2)}$ is located in the south-west quadrant from S while $T_2^{(2)}$ is located in the south-east quadrant from S. Hence, trimming product 1 would reduce the variance of total demand while trimming product 2 would increase the variance of total demand. Again, the lame duck heuristic leads to the wrong selection. This illustrates Observation 6. Finally, consider Scenario 3. Observe that $T_1^{(3)}$ is located north-west from $T_2^{(3)}$. This implies that trimming product 1 would results in a higher expected demand and lower expected variance than that of trimming product 2. In other words, product 2 is dominated by product 1. This illustrates Observation 7.

6. CONCLUDING REMARKS

Due to the increasing rate of new product introduction over the last two decades, many product lines have become over-crowded. Firms may improve their profits for its entire product line by trimming some of the products. In this paper, we developed a model to examine the impact of product trimming on the expected demand and demand variability for the entire product line. Our model is based on the consumer choice model developed by Grover and Srinivasan (1987). It captures the product substitution phenomenon. In addition, our model includes demand variability as an important determinant in making product trimming decision.

Our model exhibits the common observation that product line trimming leads to increased individual demands but to a reduced demand for the entire product line. Contrary to conventional wisdom, our model suggests that trimming the product with the minimal expected demand (i.e. the lame duck heuristic) can be flawed. We showed that this heuristic is deficient in three respects. First, it may not lead to minimal revenue loss. Second, even if it leads to minimal revenue loss, it can be sub-optimal if it leads to an increase in variance. Third, the lame duck heuristic may be dominated. Trimming another product may yield a lower loss and a higher variance reduction in expected demand. The heuristic appears to work well only when there is no demand substitution among products.

Carrying a smaller product line may not be strategically desirable for some firms. Smaller product line may translate to reduced presence at the retail front. Consider the grocery industry. Due to limited shelf space, supermarkets often allocate a fixed shelf space for carrying and displaying products from a particular manufacturer. Thus, the manufacturer may have to keep the breath of product line fixed to justify the shelf space. Consequently, the firm must determine which product to eliminate and which one to introduce. This product replacement decision is like a 'football-team' problem in which the 'coach' has to configure the best team to compete in the field.

Chong, Ho and Tang (2001) proposed a framework to capture the effect of product replacement on sales. Since their model conforms to the classical attraction model of Bell, Keeney and Little (1975), product substitutability implied by their model is similar to equation (3.7). Furthermore, they did not use sales variance in their modeling framework. By hybridizing our model with that of Ho and Tang (1995)[15], we can have a model setup more general than Chong, Ho and Tang (2001). This general hybridized model seems like a logical next step for this research.

APPENDIX A: PROOF OF EQUATION (3.8)

By the definition of market share from Bell, Keeney and Little (1975), we have,

$$p_j = \frac{A_j}{\sum_{l=0}^{N} A_l}, \quad \forall_j,$$

where A_l is the level of attraction of product $_l$, as defined in Bell, Keeney and Little (1975). If product i is trimmed, we have the corresponding revised market share:

$$p_j + \alpha_{ij} p_i = \frac{A_j}{\sum_{l=0, l \neq i}^{N} A_l c,} \quad \forall_j,$$

From the two equations above, we have,

15. The authors examined the issue of product line extension. The paper suggested conditions under which line extension is beneficial and related the benefits of line extension to market leadership and manufacturing capability.

$$
\begin{aligned}
\alpha_{ij} &= \left\{\frac{A_j}{\sum_{l=0,l\neq i}^{N} A_l} - \frac{A_j}{\sum_{l=0}^{N} A_l}\right\} \Big/ \frac{A_i}{\sum_{l=0}^{N} A_l}, \\
&= \frac{A_j}{A_i}\left\{\frac{\sum_{l=0}^{N} A_l}{\sum_{l=0,l\neq i}^{N} A_l} - 1\right\}, \\
&= \frac{A_j}{A_i}\left\{1 + \frac{A_i}{\sum_{l=0,l\neq i}^{N} A_l} - 1\right\}, \\
&= \frac{A_j}{\sum_{l=0,l\neq i}^{N} A_l}, \\
&= p_j + \alpha_{ij} p_i.
\end{aligned}
$$

This implies that,

$$
\alpha_{ij} = \frac{p_j}{1 - p_i}
$$

and also,

$$
\begin{aligned}
\alpha_{i0} &= 1 - \sum_{j=1, j\neq i}^{N} \alpha_{ij}, \\
&= 1 - \sum_{j=1, j\neq i}^{N} \frac{p_j}{1 - p_i}, \\
&= \frac{p_0}{1 - p_i}.
\end{aligned}
$$

REFERENCES

1. Alexander, Ralph S., 'The Death and Burial of Sick Products,' *Journal of Marketing*, Vol 28, (April 1964), pp. 1–7.
2. Avlonitis, G.J., 'Production Elimination Decision Making: Does Formality Matter?' *Journal of Marketing*, Vol. 49, (Winter 1985), pp. 41–52.
3. Avlonitis, G.J. and James, B.G.S., 'Some Dangerous Axioms of Product Elimination Decision-Making,' *European Journal of Marketing*, Vol. 16, No. 1 (1982), pp. 36–48.
4. Bell, D., Keeney, R. and Little, J., 'A market share theorem,' *Journal of Marketing Research*, Vol. 12 (August 1975), pp. 136–141.
5. Berenson, Conrad, 'Pruning the Product-Line,' *Business Horizons*, Vol 6, (Summer 1963), pp. 63–70.
6. Browne, W.G. and Kemp, P.S., 'A Three Stage Product Review Process,' *Industrial Marketing Management*, 5, (Dec 1976), pp. 333–342.
7. Bayus, B. and Putsis, W., 'Product Proliferation: An Empirical Analysis of Product Line Determinants and Market Outcomes,' *Marketing Science*, Vol. 18, No. 2, (1999), pp. 137–153.
8. Chong, J.K., Ho, T.H. and Tang, C.S., 'A Modeling Framework for Category Assortment Planning,' *Manufacturing & Service Operations Management,* Vol. 3, No. 3, (2001), pp. 191–210.
9. Corstjens, M. and Weinstein, D., 'Optimal Strategic Business Units Portfolio Analysis,' *TIMS Studies in the Management Sciences*, 18, (1982), pp. 141–160.
10. Green, Paul E. and Krieger, Abba M., 'Conjoint Analysis with Product Positioning Applications,' in

J. Eliashberg and G.L. Lilien, Eds., *Handbooks in Operations Research and Management Science, Vol 5: Marketing,* North-Holland, Amsterdam, (1993), pp. 467–515.

11. Greenley, G., and Bayus, B., 'A Comparative Study of Product Launch and Elimination Decision in UK and US companies,' *European Journal of Marketing,* vol. 28, (1994), pp. 5–29.
12. Grover, R. and Srinivasan, V., 'A Simultaneous Approach to Market Segmentation and Market Structuring,' *Journal Of Marketing Research,* Vol. 24 (may 1987), pp. 139–153.
13. Hamelman P.W. and Mazze, E.M., 'Improving Product Abandonment Decisions,' *Journal Of Marketing,* Vol. 36, (April 1972), pp. 20–26.
14. Hart, Susan J., 'The Causes of Product Deletion in British Manufacturing Companies,' *Journal of Marketing Management,* Vol. 3 No. 3, (1988), pp. 328–343.
15. Hays, L., 'Too Many Computer Names Confuse Too Many Buyers,' *Wall Street Journal,* (June 29, 1994), B1, B6.
16. Hise, R.T. and McGinnis, M.A., 'Product Elimination: Practices, Policies, and Ethics,' *Business Horizons,* 18, (June 1975), pp. 25–32.
17. Hise, R.T., Parasuraman, A. and Viswanathan, R., 'Product Elimination: The Neglected Management Responsibility,' *The Journal of Business Strategy,* Vol. 4, No. 4, (Spring 1984), pp. 56–63.
18. Ho, T and Tang, C.S., 'When is Product Line Extension Beneficial?' UCLA Working Paper, (1995)
19. Khermouch, G., 'Amid Rampant Product Proliferation, Beverage Marketers Desperately Seeking Simplicity,' *Brandweek,* (May 29, 1995), 1, 9.
20. Kotler, P., 'Phasing-out Weak Products,' *Harvard Business Review,* Vol. 43, (March-April, 1965), pp. 108–118.
21. Larréché, J.C. and Srinivasan, V., 'STRATPORT: A Model for the Evaluation and Formulation of Business Portfolio Strategies,' *Management Science,* Vol. 28, (1982), pp. 979–1001.
22. Mahajan, V. and Wind, Y., 'Integrating Financial Portfolio Analyses with Product Portfolio Models,' in H. Thomas and D. Garner, Eds., *Strategic Marketing and Management,* John Wiley & Sons, (1985), pp. 193–212.
23. Mahajan, V., Wind, Y. and Bradford, J.W., 'Stochastic Dominance Rules for Product Portfolio Analysis,' *TIMS Studies in the Management Sciences,* 18, (1982), pp. 161–183.
24. Narisetti, R., 'P&G, Seeing Shoppers Were Being Confused, Overhauls Marketing,' *Wall Street Journal,* (January 15, 1997), A1, A8.
25. Putsis, W. and Bayus, B., 'An Empirical Analysis of Firms' Product Line Decisions,' *Journal of Marketing Research,* Vol. 38 (February 2001), pp. 110–118.
26. Quelch, J. and Kenny, D., 'Extend Profits, Not the Product Lines,' *Harvard Business Review,* Vol. 72, (Sep–Oct, 1994), pp. 153–160.
27. Silver, E.A. and Peterson, R., *Decision Systems for Inventory Management and Production Planning.* John Wiley & Sons, (1985).
28. Teinowitz, Ira and Lawrence, Jennifer, 'Brand Proliferation Attacked,' *Advertising Age,*1, (May 10 1993), pp. 49.
29. Urban, G.L., 'A Mathematical Modeling Approach to Product Line Decisions,' *Journal of Marketing Research,*Vol. 6, (Feb. 1969), pp. 40–47.

II. SUPPLY CHAIN COORDINATION

3. COORDINATED PRICING AND PRODUCTION/ PROCUREMENT DECISIONS: A REVIEW

CANDACE A. YANO and STEPHEN M. GILBERT

ABSTRACT

It has been nearly 50 years since researchers began to develop analytical models to aid in simultaneous decisions regarding pricing strategy, which influences demands, and production/procurement decisions, which determine the cost of satisfying those demands. In this chapter, we provide a comprehensive review of analytical models on this topic, focusing on models in which external demand is price-sensitive. We review models in both continuous and discrete-time frameworks, considering both constant and time-varying demand functions, with and without demand uncertainty. Although our emphasis is on decision problems facing a single firm, we also provide a brief overview of models involving multiple firms or multiple decision-makers. We also offer suggestions for future research.

1. INTRODUCTION

In recent years, the general perception of the role played by operations management has been elevated from one of cost control to one of fundamental value creation. In practice, there is widespread recognition that *business models*, also known as operational process designs, can determine the success or failure of a firm. Paralleling this increasing recognition of the critical role of operations, we have witnessed an expansion of the academic literature on integrating traditional operational decisions (e.g., production scheduling, inventory planning, etc.) with other types of decisions made by the firm. To the extent that a firm's operations are where value creation occurs, it seems quite natural to want to understand how operational decisions interact with pricing decisions.

The purpose of this paper is to review the literature on integrated pricing and production/procurement decisions involving a single firm. Although some of the articles we survey consider other types of operational decisions, production or procurement decisions and the consequent inventory stocking decisions, are central in all of them. For this reason, we do not include models involving pure service systems without inventory. The centrality of production and procurement decisions leads us to models in which the operational costs of setups, inventory, production smoothing, etc., play an important role. Similarly, although a few of the articles incorporate other marketing-related issues, our focus is on models in which pricing is the key marketing decision. We focus exclusively on situations in which price has a direct effect upon the firm's instantaneous or aggregate rate of demand. Therefore, we do not include articles in which transfer pricing between two entities affects their costs and profits but not the external demand. Our focus on decisions involving a single firm necessarily excludes the literature on supply chain coordination and models with competition. However, later in the paper, we provide a brief overview of research on coordination issues in models with price-sensitive external demand, where coordination is sought through simple pricing decisions alone, not via more complex contract arrangements. We also briefly mention several papers, including some very recent ones, in which price competition and inventory decisions are considered simultaneously. For surveys of research on supply chain coordination with an emphasis on contractual agreements, see Cachon (2001), Tsay et al. (1999) and Lariviere (1999). A broader review of integrative research in marketing and operations management can be found in Karmarkar (1996) and a thorough review of some of the early work on integrated pricing and production models appears in Eliashberg and Steinberg (1993).

The models that we study are at the boundary between economics/marketing and operations. Viewed from the perspective of economics, these models introduce more accurate representations of the firm's cost function and thereby lead to better pricing decisions. On the other hand, viewed from the perspective of operations management, these models emphasize how pricing influences operational policies. Clearly, these two different frames of reference add to the richness of the literature.

The papers discussed in this chapter differ much more with respect to the modeling of operational costs than they do in their modeling of the relationships between price and demand. Consequently, we decided to organize this review according to the key operational cost drivers in the models, which, together with assumptions regarding whether demand is deterministic or stochastic, also have strong implications for the methods used to analyze the various models.

Traditionally, the operations management literature has been concerned primarily with understanding and managing a firm's costs. Depending on the environment in which the firm operates, several fundamental types of cost trade-offs have been identified. These usually balance the cost of holding inventory with one or more of the following:

- The cost of setting up/ordering to initiate an inventory cycle.
- The cost of varying the rate of production.
- The cost of producing/ordering too little when demand is uncertain.

Because each of these has different implications for the firm's cost function, it is natural to organize our review according to the type of operational trade-offs that are involved. We first consider models in which setup costs play a significant role in the firm's cost function. The simplest models with setup costs assume time-invariant demand and are referred to as Economic Order Quantity models, which are typically modeled in continuous time. However, when demand is allowed to vary over time, most of the models that emphasize set-up cost issues are based on a discrete time framework.

We then consider models of situations in which the need for demand smoothing to facilitate production smoothing, e.g., due to convexity of the instantaneous production cost function, plays a dominant role in cost management. Most of these models are based on a continuous time framework.

Following this, we discuss models of situations in which demand uncertainty is the dominant operational concern. These models typically utilize a discrete-time framework and assume that in each of one or more periods, production must occur prior to the realization of demand.

Finally, we provide a brief review of representative articles that address coordination issues within the scope of our survey, and provide a discussion of fruitful avenues for future research. Although we have attempted to be thorough, we may have inadvertently omitted relevant articles and regret any such oversights.

2. SETUP COST EMPHASIS WITH TIME INVARIANT DEMAND

The seminal paper on the economic order quantity (EOQ) model with pricing is by Whitin (1955). He studies the problem faced by a single decision-maker who orders a product from a supplier who charges a constant unit cost. The decision-maker's annual demand is assumed to be a decreasing linear function of price. The problem is to:

$$\text{maximize } pD(p) - cD - \sqrt{2D(p)Sic} \tag{1}$$

$$\text{s.t. } 0 \leq p \leq a/b \tag{2}$$

where p is the price (decision variable), S is the order setup cost, i is the annual inventory holding cost rate (\$/\$/year), c is the unit purchase cost, and the annual demand rate is $D(p) = a - bp$ for some a, $b > 0$. The square root expression is the annual setup and inventory cost for the selected price if the optimal order quantity is selected for that price. Whitin derives a cubic expression, one of whose roots defines the optimal price, and the optimal order quantity is derived from the consequent demand rate.

Evidently unaware of the work of Whitin described above, Kunreuther and Richard (1971) analyze the same model, focusing on the difference between the centralized solution and the solution from a sequential solution procedure in which marketing sets price and the purchasing department determines the order quantity. In the sequential solution, the marketing department assumes the unit cost is the sum of the variable manufacturing cost and the average cost per unit due to EOQ-related costs, and the purchasing department uses the resultant demand rate to determine the order quantity. In a numerical example, the centralized solution produces a higher profit with a higher price and smaller lot size. Kunreuther and Richard also extend Whitin's model to allow a finite production rate (i.e., an economic production quantity framework) and show that the optimal lot size is larger and the optimal price is smaller than in the EOQ framework.

A linear demand function is assumed in several other articles, also. Porteus (1985) extends the basic centralized model to also consider investments in reducing setup costs. He shows that setup cost reductions are worthwhile if the demand exceeds a threshold, and price reductions contribute to demand increases. Chen and Min (1994a) derive optimal solutions for two different objectives: one for maximizing return on (inventory) investment, and one for maximizing profits. They show that the optimal markup is higher for the return on investment objective than for the profit objective. Their results also demonstrate that the solutions for the two models become more similar as the holding, setup, or variable cost increases.

In an attempt to incorporate demand uncertainty into an EOQ model, Cheng (1984) considers a situation in which the price-sensitive distribution of demand is stationary. The seller incurs setup costs, linear procurement costs, inventory holding costs charged on average inventory and a fixed cost for offering the product. Cheng seeks to find the economic order quantity that optimizes the seller's expected utility, where the utility is expressed as the mean profit less a constant multiplied by the standard deviation of profit. He shows how to obtain optimal solutions using a Lagrangian approach. Demand uncertainty affects only the annual revenue in this model, as (short-term) shortage costs are not included.

Abad (1988a) shows that if the optimal price for a given lot size is unique, then for a broad class of demand curves that are decreasing in price, the profit, expressed as a function of the lot size (with the corresponding optimal price implicit in the formula), is a convex-concave function. Consequently, the optimal lot size is either the larger of the two stationary points if the objective function is somewhere positive, or zero if the objective function is non-positive everywhere. Although many researchers have provided detailed analysis for specific demand functions, Abad's result suggests that the qualitative results and solution procedures may extend to other demand curves as well.

Another line of research focuses on a demand curve with constant elasticity for the objectives of maximizing profit and return on inventory investment, among others. The constant elasticity demand function has the form:

$$D(p) = kp^{-\alpha} \tag{3}$$

where $k > 0$ and $\alpha > 1$ are parameters. Kim and Lee (1998b) consider a variant of the profit maximization problem in which capacity investment and disinvestment decisions must be made in addition to price and lot size decisions. Arcelus and Srinivasan (1987) derive the optimal markup for the profit and return on (average) inventory investment objectives, as well as for the so-called residual income objective, which is the difference between profit and the opportunity cost of the inventory investment.

They show that the optimal multiplicative markup factor for the profit objective is greater than or equal to the markup factor for the residual income objective, which in turn is greater than or equal to the markup factor for the return on inventory investment objective. Not surprisingly, the optimal lot sizes have the reverse relationships. These results imply that the return on inventory investment objective leads to solutions with high prices and small lot sizes because such solutions increase the return on investment, but they also lead to foregone profit (mostly in the form of lost revenue) if the true opportunity cost of capital is less than the maximal return on inventory investment.

Still another demand curve—one that is logarithmically concave in price, is analyzed by Rosenberg (1991), who derives and compares solutions for the profit and ROI objectives. Many of the insights are similar to those obtained for other convex demand functions. He observes that for the ROI model, the sequential solution in which price is set first is optimal because the pricing decision does not depend on inventory-related costs.

2.1 Quantity discounts

Several researchers (Ladany and Sternlieb 1974, Brahmbhatt and Jaiswal 1980, Subramanyam and Kumaraswamy 1981, Arcelus and Srinivasan 1985, Lee and Rosenblatt 1986, and Abad 1988b) have generalized various models mentioned above to incorporate purchase cost functions that are concave in the purchase quantity. In effect, these models assume that incremental quantity discounts are available from the supplier. A variety of different cost structures are modeled, and solutions are obtained from first-order optimality conditions. Abad (1997) considers the case of a temporary discount offered by the supplier to encourage forward-buying. The buyer faces price-sensitive demand and seeks to optimize both his selling price and the order quantity. A similar scenario is studied by Ardalan (1994, 1995) in a Stackelberg framework with the supplier setting the price in his first-mover role. (See Section 5 for further details.)

Incorporating all-units-discounts from the supplier is more complicated. Abad (1988a) develops a heuristic for optimizing a single price for a seller facing a linear or constant-elasticity demand function. The proposed procedure takes advantage of what is known about the structure of the buyer's objective function, iterating between optimizing the price and optimizing the buyer's lot size and associated (marginal) unit cost. In an extensive computational study, Burwell et al. (1990) found that Abad's procedure performed near-optimally for the case of linear demand, and on average, within about 6% of optimal for constant elasticity demand functions. Lee (1993) subsequently showed that Abad's procedure may produce suboptimal solutions and proposed an optimal procedure based on geometric programming. Burwell et al. extend Abad's model to allow for shortages in their 1991 paper and to allow for mixed types of purchase and freight discounts (all units or incremental units discounts) in their 1997 paper. Shinn et al. (1996) incorporate freight costs with a step function structure, as well as trade credit offered by the supplier, which reduces the effective cost to the customer.

In addition to these models of how a buyer should respond to a quantity discount, there is a closely related line of research that considers quantity discounts from the perspective of the supplier. Quantity discounts are, in themselves, pricing mechanisms that a supplier can use to optimize its profits. Dolan (1987) provides an excellent summary of research that characterizes the impact of different discount pricing structures on the decisions and profits of the buyer and supplier for EOQ-like frameworks. Of particular interest are two-part (fixed charge plus linear) and two- or n-block (all-units-discounted) tariffs. Dolan's discussion provides insight into the tradeoffs faced by a supplier when selecting the discount schedule to offer. Although there is a long-standing and extensive literature on the use of wholesale discounts when the external demand is insensitive to price, the research on structuring such discounts in the presence of price-sensitive external demand is fairly recent. Weng and Wong (1993) derive the optimal all-units-discount price schedule for a supplier whose customers face price-sensitive demand. Unlike the models described above, there are multiple types of setup costs in their model: order setup for the buyer, outbound order handling for the supplier and (inbound) order processing for the supplier. This leads to another decision for the supplier—the (integer) ratio of its purchase quantity to the buyer's lot size. Parlar and Wang (1994) examine a special case of this problem in which the buyer faces a linear demand curve and uses a constant-multiple markup policy, so the supplier influences the demand directly. The authors show that the supplier cannot benefit from the introduction of (all-units) discounts unless the buyer is provided sufficient incentive to increase his order quantity from the status quo. Later, Weng (1995b) investigates both all-units and incremental units discount schedules, each with one price break, and shows that the optimal solutions under the two discount structures produce the same results. He also shows that using the same lot size for the supplier and buyer may be suboptimal.

2.2 Decaying and perishable inventory

Interestingly, the pricing issue has received a good deal of attention in the context of EOQ models with decaying or perishable inventories. In most of these models, the product is assumed to be saleable and to have the same perceived value until it perishes, at which point it is disposed at no cost. The earliest paper in this area is by Cohen (1977) who assumes exponential decay and a constant price (to be selected). For a model with an arbitrary downward-sloping demand curve and without shortages, he obtains a nearly-closed form solution for the optimal order cycle (time between orders), expressed in terms of the price, based on a Taylor series approximation of the exponential function. Utilizing this result, he then derives an expression that enables computation of the optimal price. In a numerical example, although the optimal price is not monotonic in the decay rate, it is surprisingly stable, whereas the order cycle decreases rapidly—although less than proportionally—with the decay rate. Cohen extends the analysis to allow backlogging, and the analysis shows that the fraction of the cycle with backlogging declines as the decay rate declines. Cohen's model is extended by Kang and Kim (1983) to allow production at a finite rate (rather than procurement). In their model, they assume that shortages will not occur and enforce this constraint approximately by requiring expected end-of cycle inventory levels to be non-negative. Aggarwal and Jaggi (1989) suggest a more accurate representation of inventory for Cohen's model and demonstrate that it leads to slightly shorter ordering cycles.

Rajan et al. (1992) generalize Cohen's model to allow for more general inventory decay, where the decay rate may change over time. Among other things, they show that the optimal dynamic price is independent of the cycle length (time between orders), and the optimal price maximizes the instantaneous margin, trading off change in revenue with the change in inventory holding and wastage costs, at each point in time. They also provide an upper bound on the cycle length and show that in the case of linear and exponential demand functions, the optimal cycle length is decreasing in the market potential (maximum demand at a price of zero upon receipt of fresh product). A Weibull distribution is used to model deterioration in Wee (1999), who assumes a linear demand curve. His model allows incremental units discounts from the supplier, backlogging of a known fraction of unsatisfied demand, and lost sales for the remaining shortages. He approaches the problem by fixing the order cycle and optimizing the selling price and the portion of the cycle without shortages. In a later paper (Wee and Law 2001), the incremental units discount aspect of the model is omitted and the authors concentrate on representing the net present value of revenues and costs.

Goyal and Gunasekaran (1995) extend Cohen's model to consider raw materials inventory and decay, assuming a single raw material is purchased in a batch that is an integer multiple of the production batch size. In addition to pricing and lot sizing decisions, the authors consider advertising expenditures as a decision variable. A constant elasticity demand function is assumed, and the impact of advertising on demand also has a constant elasticity. The authors use an ascent search method and recommend multiple starting points because of the non-concavity of the objective. Also incorporating a constant elasticity demand function, Shinn (1997) and Hwang and Shinn (1997) seek to optimize the price and lot size when the supplier allows the buyer to delay payment for a fixed duration. Like advertising, such trade credit also stimulates demand, but unlike advertising, it does so by reducing the buyer's costs. In two papers, Abad (1996, 2001), examines models with a constant-elasticity demand function and partial backordering, where the fraction of excess demand that is backordered depends upon the time until the next shipment arrives. In the 1996 paper, which generalizes the model of Rajan (1992) to include partial backordering, the price is allowed to change continuously, and this permits an optimal sequential solution in which the optimal price trajectory is determined first, and the optimal order cycle is determined using the optimal price trajectory. In the 2001 paper, a constant price (to be decided) is maintained, so the price and order frequency decisions are intertwined. This problem may have multiple local optima, but for a fixed price, the objective function is concave in the in-stock and out-of-stock durations within the order cycle. Similarly, for fixed in-stock and out-of-stock durations, the objective function is pseudo-concave in the price. With these two results, the author develops an iterative improvement heuristic.

Several researchers (McTavish and Goyal 1989, Urban 1992 and Lee and Kim 1993) jointly optimize prices, advertising or marketing expenditures that increase demand,

and lot sizes within an EOQ-type framework. In these models, both marketing expenditures and price reductions serve to stimulate demand with decreasing marginal returns. Thus, the optimal prices typically are higher than in models without other sources of demand stimulation. Kim et al. (1995) introduce a trade credit interval as a decision variable, in addition to price. Trade credit stimulates demand by reducing the buyer's cost of holding inventory. They derive results for the cases of linear and constant-elasticity demand functions.

2.3 Multiple products / multiple customers

All of the papers discussed above pertain to a single product and either a single customer or an aggregate market with price-sensitive demand. In the remainder of this section, we discuss papers that treat multiple products or multiple customers. Cheng (1990) considers a multi-product version of the problem, assuming downward sloping demand curves. He seeks a solution in which all products are ordered with the same frequency and the lot sizes satisfy storage capacity constraints. Later, Chen and Min (1994b) point out that the solutions proposed by Cheng does not guarantee the same order frequency for all products. They characterize the Karush-Kuhn-Tucker conditions for the problem. They also derive closed-form expressions for the solution in the case of linear demand and associated conditions on the problem parameters that guarantee optimality. Lee (1994) generalizes this line of research by allowing constraints on storage capacity and inventory investment. He solves the problem via Lagrangian methods combined with geometric programming.

Morgan, Daniels and Kouvelis (2001) model a multi-product situation in which the products are produced with the same frequency on the same production facility. Each product is produced to stock and has a linear demand curve, and the authors present a method to select prices to maximize profit for a fixed production interval. Dobson and Yano (2002) generalize the framework to permit decisions of which (of a set) of products to offer and whether each should be offered on a make-to-stock (MTS) or make-to-order (MTO) basis. While the MTS demand declines with the price, the MTO demand curve declines with both price and the production interval; the latter factor reflects customer's impatience and their need to hold more inventory. Dobson and Yano derive conditions in which MTS (or MTO) is preferred, and develop an algorithm that simultaneously optimizes which products to offer, their prices, and the production interval.

Two patterns emerge from the research on EOQ-based models with pricing decisions. The first is that, because the basic framework is simple, researchers have been able to explore a variety of different demand functions and objectives. Second, including price as a decision variable causes the objective function to lose its unimodality. Consequently, both structural results and closed-form solutions are elusive, so researchers have concentrated on the development of algorithms to find optimal solutions. It is also clear that EOQ-based models with pricing decisions have been very well-studied, with several dozen articles published over the past five decades.

In relaxing the time-invariant assumption regarding demand, two streams of research have emerged. The first, and perhaps most natural, is to simply extend the inquiry into the interactions among setup costs, inventory costs and pricing to allow for time-varying demand. A significant portion of this stream of research is based on a discrete time framework. The other stream involves introducing the costs associated with varying the rate of production. Most of these models are set in a continuous time framework.

2.4 Setup cost emphasis with time varying demand

A number of authors have introduced pricing decisions into traditional discrete-time deterministic production planning models. The problem is generally formulated over a finite horizon of T periods, in which the firm must determine a vectors of prices (p_t, $t = 1, \ldots, T$) and production quantities (x_t, $t = 1, \ldots, T$), which jointly determine end-of-period inventory levels (I_t, $t = 1, \ldots, T$). The most general version of the single-product problem involves holding costs (h_t, $t = 1, \ldots, T$), non-decreasing production cost functions ($c_t(\cdot)$, $t = 1, \ldots, T$), setup costs (S_t, $t = 1, \ldots, T$) and capacity (C_t, $t = 1, \ldots, T$) limits. The problem can be formulated as follows:

$$\operatorname*{Max}_{\mathbf{x},\mathbf{p},\delta} \left\{ \sum_{t=1}^{T} (p_t\, D_t\, (p_t) - \delta_t S_t - c_t(x_t) - h_t I_t) \right\} \tag{4}$$

subject to:

$$I_{t-1} + x_t - D_t(p_t) - I_t = 0 \qquad t = 1, \ldots, T \tag{5}$$

$$x_t - \delta_t C_t \leq 0 \qquad t = 1, \ldots, T \tag{6}$$

$$x_t,\ p_t,\ I_t \geq 0 \qquad t = 1, \ldots, T \tag{7}$$

$$\delta_t \in \{0,1\} \qquad t = 1, \ldots, T \tag{8}$$

where $D_t(\cdot)$ is a non-increasing demand function and δ_t is a binary variable that is equal to 1 if production occurs in period t. In all of the articles surveyed here, each demand function is assumed to be independent of demands and prices in other periods. Constraint (5) is the inventory balance equation, which, together with the non-negativity constraint on inventory, ensures that demand is satisfied on time. Constraint (6) ensures that the volume of production is less than the capacity available in each period. Although we have included this constraint for the sake of generality, it is worth noting that most of the research in this area treats capacity as infinite.

Wagner and Whitin (1958a) were the first to introduce price as a decision variable in a discrete-time model. They present a thoughtful discussion of the modeling issues that arise when demand and cost functions vary across time periods. Under the assumption of constant marginal costs and infinite capacity, they prove that their two well-known planning horizon results (1958b) generalize to cases with price-dependent demands: First, there exists an optimal solution in which no inventory is carried into a period in which production occurs. Second, if an optimal production pattern over periods 1, . . . ,τ includes two production setups, then all periods prior to the last set-up can be ignored in determining the optimal policy for periods $t > \tau$. Based on these planning horizon results, they describe a dynamic programming approach for solving the problem when there are setup costs. They also discuss how calculus can be used to solve the problem in the absence of setup costs. This latter discussion provides nice insight into the fundamental economic trade-offs.

These foundational results are generalized in Wagner (1960) to marginal costs that are either non-increasing or non-decreasing in the production quantity. For non-increasing marginal costs, he shows that, as in the case of constant marginal costs, we need not carry inventory into a period in which there is positive production. Hence, the original dynamic programming algorithm can be applied to situations with non-increasing costs. However, he provides an example that shows that if setup costs are present and cost functions are not identical in all periods, then even though it may be optimal to produce in period $t^\star$ when periods 1, . . . ,$t^\star$ are considered by themselves, it is not necessarily optimal to produce in period $t^\star$ when considering the problem involving periods 1, . . . ,T. For non-decreasing marginal costs, he describes a recursive algorithm for identifying the optimal solution.

Thomas (1970) builds upon the planning horizon results for the uncapacitated case with linear variable costs. In his model, setup, variable costs, and inventory holding costs may differ from period to period. Perhaps the most significant result is that the prices in periods between setups can be optimized independently, and this provides computational advantages in solving the dynamic program.

Bhattacharjee and Ramesh (2000) study an uncapacitated model in which demand is assumed to have constant elasticity. They show that for a given production plan, the profit is concave in the price vector, and when considering alternative production plans, the profit is piecewise concave in the price vector. They then exploit this result to develop heuristic search procedures to overcome the computational burden of the 2^{T-1} possible production plans associated with a given pricing vector.

Kunreuther and Schrage (1973) consider a variation of the problem in which a single price must be chosen over the entire planning horizon. This constraint is motivated by the problem faced by a catalogue retailer that lacks the flexibility to change its price frequently. They develop an iterative approach for finding the optimal production schedule for a given price and then optimizing the price based on the marginal costs that are associated with that production schedule. Although their approach provides bounds on the optimal price, and performs well in computational experiments, it does not necessarily identify the optimal solution.

The variant of the problem in which a single price must be chosen for the entire horizon was also studied by Gilbert (1999). Using a slightly less general model of demand and imposing the restriction that setup and holding costs be time-invariant, he shows that the total cost of production and inventory is a piecewise linear function of price. He exploits this property to develop a simple algorithm that identifies an optimal solution in $O(T^3)$ time.

In another paper, Gilbert (2000) considers a situation in which pricing and production decisions must be made for multiple products that share the same production equipment. A unique feature of this paper is that it explicitly considers multiple products as well as limited production capacity. To solve the problem, he develops a procedure that identifies an optimal basis for the network-flow problem that minimizes the production costs for a given demand vector. He then uses the dual variables associated with this basis to formulate the pricing problem as a non-linear program (NLP) that has a concave objective function and linear constraints. By iterating between identifying an optimal basis for the network flow problem and solving the simple NLP, his algorithm identifies an optimal solution. He then performs numerical experiments to explore how the relative seasonalities of various products affect the optimal pricing decisions.

It is worth noting that, with the exception of Gilbert (2000), the literature on discrete time, deterministic pricing and production models has not addressed the issue of finite capacity. There are two working papers that have attempted to fill this void.

Chan et al. (2000) study a capacitated model of a single product in which setup costs are assumed to be zero. The primary objective of this paper is to characterize a general class of problems for which the optimal solution can be found by a greedy algorithm. Specifically, they apply and extend the theory of polymatroids to the pricing and production problem. They then present a series of numerical experiments that demonstrate, among other things, the value of being able to vary the price of a product over time. They conclude with a discussion of how the theory can be extended to deal with multiple products, lead times, and set-up costs.

Deng and Yano (2000) are unique in considering both setup costs and capacity constraints for a single product. The model that they propose is general in the sense that all parameters associated with either demand or production / inventory costs need not be time-invariant. They develop a solution procedure that exploits the fact that the inclusion of price as a decision variable in each period does not alter the well-known properties of the capacitated lot-sizing problem. After discussing how the complexity of this procedure is affected by the stability of costs and capacity over time, they present some numerical experiments. Among other things, their numerical results demonstrate how it is possible for a decrease in capacity to lead to a decrease in the optimal prices. This counter-intuitive effect arises when the smaller capacity leads to a different production pattern.

One clear pattern from the discrete-time models with deterministic demand functions is that, even with the introduction of prices as decisions, the lot sizing decisions still contribute heavily to problem complexity. Indeed, if the lot *timing* is known, the pricing decisions are relatively simple. As a consequence, much of the research has focused on the development of algorithms to integrate pricing decisions into the classical discrete time lot-sizing models, and this is a relatively well studied area.

3. DEMAND SMOOTHING

Another stream of research that is based on time-varying demand emphasizes demand smoothing rather than setup costs. This stream is concerned with demand trajectories that change continuously as a function of price, so the instantaneous demand rate is a deterministic function of both time and price (and possibly other factors). This line of research is motivated by products with seasonal demand or short product life cycles. Pekelman (1974) was the first to present such a model, as formulated below:

$$\text{Max} \int_0^T (p(t)[a(t) - b(t)p(t)] - hI(t) - f[q(t)])dt \tag{9}$$

$$\text{s.t.} \ \frac{dI(t)}{dt} = q(t) - [a(t) - b(t)p(t)] \tag{10}$$

$$I(t) \geq 0 \tag{11}$$

$$p(t) \leq a(t)/b(t) \tag{12}$$

$$q(t),\ p(t) \geq 0 \tag{13}$$

$$I(0) = 0 \tag{14}$$

where p(t) is the price at time t, demand at time, $d(t)$ is expressed as $a(t) - b(t)p(t)$ for some non-negative continuous functions $a(\cdot)$ and $b(\cdot)$, h is the inventory holding cost rate per unit per unit time, $I(t)$ is the inventory at time t, $q(t)$ is the production rate at time t and $f(\cdot)$ is a convex increasing production cost function. Note that, in contrast to the EOQ models, this form of the production cost function encourages the supplier to smooth production. He approaches the problem by dualizing constraints (10) and (11) with multipliers $\lambda(t)$ and $\rho(t)$, respectively. Careful analysis of the Karush-Kuhn-Tucker conditions leads to a characterization of the optimal solution:

$$p = \begin{cases} 0 & \text{if } \lambda + \rho < -a/b \\ 0.5(\lambda + \rho + a/b) & \text{if } -a/b \leq \lambda + \rho \leq a/b \\ a/b & \text{if } a/b < \lambda + \rho \end{cases} \tag{15}$$

and

$$q = \begin{cases} 0 & \text{if } \lambda + \rho \leq f'(0) \\ f'^{-1}(\lambda + \rho) & \text{otherwise} \end{cases} \tag{16}$$

From this, it is clear that the solution depends on the function $\lambda(t) + \rho(t)$. The above results imply that during a time interval with zero initial and ending inventory and with strictly positive inventory in between, the optimal price depends only on $\lambda(t)$ because $\rho(t) = 0$. Then because inventory holding costs do not depend on the price of the product in this model, $\partial\lambda(t)/\partial t = h$, so one only needs to determine the optimal value of λ at the beginning of such an interval.

Pekelman also discusses conditions for the existence of planning horizons for this problem—points in time when demand information beyond that time does not influence (optimal) production decision prior to that time. Thompson et al. (1984) impose an upper bound on the inventory level and characterize planning horizons in relation to times at which the inventory level reaches the upper and lower (zero) inventory limits.

Several other researchers have extended Pekelman's model to include other factors. Gaimon (1988) also uses a time-dependent linear demand function and generalizes the model to consider capacity acquisition, where new capacity with advanced technology leads to lower marginal costs. This, in turn allows the firm to reduce prices and thereby increase demand. Increased capacity also allows the firm to be more responsive to demand and to carry less inventory. Feichtinger and Hartl (1985) generalize Pekelman's model to allow shortages (incurring time-weighted shortage costs). They provide interesting insights by applying their model to a dynamic situation in which there is excess initial inventory. While inventory remains excessive, there is no production and the product is sold at a price of zero in order to deplete inventory as quickly as possible. When inventory has fallen to a (reasonable) threshold level, the price rises but production remains at zero, then at a smaller inventory threshold the production rate increases from zero and the price continues to increase. In the final phase of a finite horizon problem, both the production rate and price continue to increase, and shortages may start to occur near the end of the horizon. In an infinite horizon problem, the final phase does not occur and the production rate and price converge to the long-run optima. The authors also explore characteristics of the solution when demand is seasonal. Jorgensen et al. (1999) further generalize Pekelman's model to allow for increased product diffusion as cumulative sales increase as well as unit cost reductions from accumulated production experience. They show that the pattern of phases characterized by Feichtinger and Hartl also hold in their more general setting, although not all phases have non-zero duration depending upon the initial inventory. Jorgensen and Kort (2002) analyze a finite-horizon model in which the demand rate is the product of the level of on-display inventory at the retailer and a base demand curve that declines linearly with price. The inventory, thus, has an effect that is similar to advertising. They investigate a system in which the retailer incurs linear purchase and holding costs (no shortages are allowed) and the warehouse incurs convex procurement costs. Although the focus of the paper is on the comparison of decentralized and centralized policies, their analysis of the retailer's problem is pertinent to our survey. They show that if the retailer starts the horizon with a high inventory, the initial price is low (perhaps zero), and if the initial inventory is low, the price is high. The optimal rate of replenishment depends on the tradeoff between the cost of holding inventory and the revenue benefits from the demand-stimulating effect of on-display inventory, and the concomitant ability to raise prices.

Eliashberg and Steinberg (1987) study a system with a distributor who faces a seasonal demand pattern (increasing then decreasing). The distributor optimizes his price and ordering trajectories to maximize profit (revenue less linear procurement and inventory costs and convex holding costs). The manufacturer chooses a constant price as well as a production rate trajectory, facing an analogous cost function. Although the authors pose the problem in a Stackelberg game context, the distributor's problem (with a given unit procurement cost) is itself an interesting problem. The authors show that the optimal policy has a two-part processing strategy, a two-part (two-phase) pricing strategy and a three-part (three-phase) inventory policy. The first two phases of the inventory policy coincide with the first phases, respectively, of the processing and pricing strategy. During the first phase of the inventory policy (and the early portion of the first phase of strategies for the other decisions), price is increasing at a decreasing rate, processing is increasing at an increasing rate, and inventory is increasing at a decreasing rate. At this point, the inventory level reaches its peak and the inventory strategy changes while the other strategies have the same structure. In particular, inventory declines initially at an increasing rate, then at a decreasing rate, reaching zero. The third phase of the inventory policy coincides with the second phase of the other two strategies. Here, price declines at an increasing rate, production exactly satisfies demand, and inventory remains constant at zero.

Kumar et al. (2000) study coordination issues in a system with a manufacturer and distributor, and as part of their analysis, examine the impact of decentralized decision-making at the distributor, with the marketing department setting prices and the processing department making processing (production) decisions. The other modeling assumptions are similar to those in Eliashberg and Steinberg. The authors show that the distributor follows a two-part pricing policy, first increasing his prices at a decreasing rate, then decreasing his price at an increasing rate. The optimal processing policy has two parts, first processing at an increasing rate, then at exactly the demand rate. In the first processing phase, inventory accumulates at an increasing rate, then depletes initially at an increasing rate then a decreasing rate. During the second processing phase, there is no inventory.

Another paper focuses on a scenario in which the supplier chooses to terminate production prior to the end of the horizon. Chen and Chu (2001) present a finite-horizon, continuous-time model with a time-invariant linear demand curve in which the price depends on the production rate, which may differ from the sales rate. Their model is based on the assumption that production occurs at constant rate (to be chosen) for the initial portion of the horizon, and during this period, the price is set so that demand is equal to the production rate. At this point, production continues at the same rate, but the price is allowed to change so that demand is less than the production rate. Then, at some point production terminates and the some or all of the remaining inventory is sold while the price is allowed to vary. The authors develop a control policy that seeks to maximize revenue less the production costs, which is assumed to be a convex function of the instantaneous rate, and inventory holding costs.

One paper in this category (Wee 1995) focuses on deteriorating inventory. He treats a finite-horizon, continuous-time model in which the product deteriorates at a constant rate and demand declines linearly with price and exponentially with time. A known fraction of shortages is backordered and the remainder constitute lost sales. A constraint is imposed on the fraction of demand satisfied from stock. The author seeks to divide the horizon into ordering intervals and to find the order quantity for each interval, as well as a selling price that is held constant throughout the horizon. The goal is to maximize expected revenue less the cost of the product itself, inventory costs, and shortage (backlogging and lost sales) costs. The solution procedure involves searching over the number of order cycles and using first-order necessary conditions for the other decisions.

4. DEMAND UNCERTAINTY

Models of pricing and production under uncertain demand can be divided into three broad categories. In the next subsection , we focus on single-period models in which there is truly a single point in time when decisions are made. Subsequently, we discuss models in which there is demand updating and/or the opportunity for recourse (in pricing or production/procurement) in the context of a problem with one or a few periods of demand. Later, we turn to models with multiple time periods and repeated decision-making.

4.1 Single-period models with uncertain demand

One of the most significant issues modeling uncertain demand when price is endogenous is that of whether the uncertainty is additive or multiplicative. In additive models, demand is represented as the sum of a deterministic, price-dependent demand function and a random noise term that is independent of price. For example, if $X(p)$ is a price-dependent demand curve, and μ is a random noise term with $E[\mu] = 0$, then the demand can be represented as $X(p) + \mu$. One obvious but important implication of this additive model is that the uncertain component of demand is insensitive to price. In such models, the variance of demand is constant, while the coefficient of variation tends to increase in price.

In multiplicative models, demand is represented as the product of a deterministic, price-dependent demand function and a random noise term that is independent of price. For example, if α is a non-negative random variable with $E(\alpha) = 1$, then demand can be represented as $X(p)\alpha$. In these models, the variance of demand is decreasing in price, while the coefficient of variation is constant. A consequence of this important difference between additive and multiplicative models is that the specific form of the model of demand uncertainty matters a great deal.

Most of the early work on single period models is based on the newsvendor framework in which both the price (p) and the quantity (q) must be specified prior to the realization of demand. Under the assumption that the per-unit cost of procurement is c and that the per-unit salvage value for excess stock is s, the optimization problem can be specified as follows:

$$\operatorname*{Max}_{p,q}\left\{ -cq + \int_0^q (px + s(q-x))f(x,p)dx + pq(1 - F(q,p))\right\} \tag{17}$$

or

$$\underset{p,q}{\text{Max}}\left\{ -cq + \int_0^q (p-s)xf(x,p)dx + pq - qF(q,p)(p-s) \right\} \tag{18}$$

where $f(x,\text{p})$ is the conditional density of demand given price p, and $F(x,\text{p})$ is the conditional cumulative distribution of demand. Additionally, some formulations include a cost of goodwill for unsatisfied demand.

Whitin (1955) was the first to consider both pricing and stocking decisions in a single-period model with uncertain demand. He provides a qualitative discussion of how a change in price shifts both the marginal revenue and marginal liquidation (salvage) cost curves. To illustrate, he derives the optimal price for a particular representation of demand in the class of multiplicative demand uncertainty: demand is modeled as the product of a linear deterministic demand function and a U(0,2) random variable.

Several researchers have explored how the introduction of uncertainty affects traditional, single-period, monopolistic pricing models. Mills (1959) uses an additive model of demand uncertainty, and assumes only that the various costs (e.g., inventory and shortage costs) are increasing in their arguments. In his model, the firm must make both its pricing and quantity decisions prior to observing demand. Hence, it is a classical newsvendor problem with the addition of a pricing decision. The main result that Mills obtains is that, when demand uncertainty is additive, the firm will optimally set its price below the optimal price in the absence of demand uncertainty.

The first paper to characterize the differential effects of additive and multiplicative models of demand uncertainty was Karlin and Carr (1962). Focusing on a model with constant marginal costs, they show that Mills' result is reversed when uncertainty is multiplicative instead of additive. That is, the presence of demand uncertainty increases the optimal price for the firm facing multiplicative uncertainty, while it lowers the optimal price of the firm facing additive uncertainty.

Zabel (1970) extends the results that Karlin and Carr obtained for the multiplicative model to a more general, convex holding and shortage cost function. In addition, Zabel obtains optimal prices and production quantities as functions of the initial starting inventory. He demonstrates that, when the level of inventory is sufficiently low, the price that maximizes the expected revenue from uncertain demand is less than the price that would maximize the revenue from deterministic demand. This result facilitates Zabel's later (1972) extension of the model to multiple periods.

Young (1978) develops the following generalized demand model that subsumes both the additive and the multiplicative models:

$$X = \alpha(p)\varepsilon + \beta(p)$$

where ε is a random variable and $\alpha(p)$ and $\beta(p)$ are general functions of the price p. She shows that the existence of an optimal price and inventory policy and the exclusion of boundary solutions can generally be inferred from analysis of the corresponding deterministic demand function. However, establishing the uniqueness of an optimal solution depends on the form of the demand density. Young demonstrates that a sufficient condition for uniqueness of the optimal solution is that either the logarithm of the demand density is concave or the density is lognormal. She also generalizes the results of Mills, Karlin and Karr, and Zabel by showing that if the coefficient of variation of demand is non-increasing with price then the optimal price will be higher than the corresponding deterministic price. Alternatively, if the variance of demand is non-increasing with price, then the optimal price will be below the corresponding deterministic price.

Petruzzi and Dada (1999) make a significant step toward resolving the long-standing conflict and inconsistency between models with additive demand uncertainty and those with multiplicative demand uncertainty. Specifically, they transform their model by using safety stock (z) as a surrogate for the stocking quantity (q), and show that this transformation provides analytical tractability. They then define $p_B(z)$ as a base price that maximizes the function $J(z, p) = (p - c)\ \mathrm{E}[Sales(z, p)]$, where c is the unit production cost. For additive demand uncertainty, they show that $p^0 \geq p^\star = p_B(z)$, where p^0 is the optimal price for the deterministic problem. They then show that for multiplicative demand uncertainty, $p^\star \geq p^0 = p_B(z)$. Thus, for both multiplicative and additive demand uncertainty, the optimal price associated with a given quantity (as represented by the surrogate z) is larger than the base price.

A few papers seek to generalize the basic additive and multiplicative models of demand uncertainty to include other factors. Young (1979) extends her earlier model to incorporate a competitiveness parameter into the term $\beta(p)$ so that an increase in the competitiveness parameter corresponds to a decrease in $\beta(p)$. In this way, a change in competitiveness affects not only the demand elasticity, but also the coefficient of variation. With this model, she is able to demonstrate that increased competitiveness can have the opposite effect on the optimal price and quantity relative to what is predicted by deterministic models. In particular, she shows that if shortage and overage costs are significant, increased *competitiveness* can lead to higher prices.

In a similar attempt at generalization, Polatoglu (1991) proposes a relationship between price and demand uncertainty that is not necessarily either additive or multiplicative. His primary concern is determining conditions in which the expected profit function is unimodal in the production quantity, assuming that the optimal price is selected for each candidate production quantity. He utilizes his results to analyze the simpler cases of additive and multiplicative uncertainty, and confirms the results of Mills, Karlin and Carr, and Zabel.

In an attempt to extend this line of analysis beyond the assumption of risk-neutrality, Agrawal and Seshadri (2000) investigate how risk aversion affects the pricing and ordering decisions of a retailer. They show that the effect depends upon whether the uncertainty of demand is additive or multiplicative. For multiplicative demand uncertainty, they show that risk aversion tends to increase the price and decrease the quantity relative to the decisions of a risk-neutral newsvendor. For additive demand uncertainty, they show that risk aversion tends to result in lower prices, but the effect on the quantity ordered depends on the elasticity of demand.

A unifying characteristic of the models above is that they begin with a deterministic model of price-dependent demand, then introduce uncertainty. In contrast, Lau and

Lau (1988) start with a single-period model with uncertain demand, i.e., the classic newsvendor problem, then introduce price dependency of the demand. Specifically, they use an additive model of uncertainty and develop solution procedures for the objective of profit maximization as well as for the objective of maximizing the probability of attaining a target profit. In numerical experiments for the profit maximization objective, they confirm the Mills result that increasing the variance of the random error term tends to decrease the optimal price. For the objective of maximizing the probability of attaining a given profit level, they show analytically that when the shortage cost is zero, the variance of demand affects only the value of the objective function, and does not affect the optimal price and order quantity.

Three recent papers have studied the single-period pricing and production / inventory problem for multiple products with partial substitution or demand cannibalization. Birge et al. (1998) consider a situation in which a firm offers two partially substitutable products in a single-period context. They assume that the products are substitutable only in the sense that the mean demand for each product depends on the prices of both of them. Specifically, they do not allow for unsatisfied demand for one product to spill-over to the other one. They characterize optimal prices and capacity (production) levels for the two products for various scenarios in which different combinations of the prices and capacity levels are controllable.

Aydin and Porteus (2003) study a single-period, multi-product model in which demand for each product is influenced by the prices for all products (indexed by j). They consider two models of demand (for product i):

(1) $D_i = \alpha_i - \Sigma_j \beta_{ji} p_j + \varepsilon_i$ where $\beta_{ii} > 0$, $\beta_{ji} < 0$ for $j \neq i$, and ε_i has an IFR c.d.f.;

and

(2) $D_i = \mu q_i\ (p) \varepsilon_i$ where the values $q_i(p)$ are multinomial logit probabilities.

The goal is to select inventory quantities and prices to maximize expected profit. The authors show that although the objective function is not jointly quasi-concave, for these two representations of demand, the optimal prices satisfy the first order conditions of the induced profit function (i.e., at the corresponding optimal inventory levels). Consequently, the jointly optimal prices and inventory quantities can be computed easily. Using comparative statics, they provide the managerially useful insight that, as the cost for product i increases, p_i will increase and p_j will decrease for $j \neq i$.

Dana and Petruzzi (2001) describe a situation in which utility-maximizing consumers choose between attempting to purchase the newsvendor's product and an exogenous alternative. Because their expected utility is a function of the firm's fill-rate, this leads to interaction between the stocking decision and the distribution of demand. It is shown that this interaction leads to higher fill-rates, but the effect of price is ambiguous.

Other recent work in this area includes a paper by Raz and Porteus (2002), who develop a model of a profit-maximizing newsvendor making joint pricing and quantity decisions, with the added restriction that the firm must choose among a finite set of service levels. Among other things, this restriction allows them to show the counter-intuitive result that an increase in marginal costs can result in a decrease in the optimal price if either the demand distribution becomes stochastically worse or if the optimal service level changes.

Another recent paper by Cachon and Kok (2002) recognizes that the salvage value parameter in a newsvendor model is, in reality a function of the decision variable. That is, the average price at which a firm can liquidate excess inventory depends critically on the initial stocking decision. The main purpose of the paper is to introduce a concept that the authors call a heuristic equilibrium for settings in which a decision variable interacts with a parameter value. In a heuristic equilibrium, the relation between the decision variable and parameter is expressed only approximately and an equilibrium is obtained accordingly. The authors use the newsvendor model as their primary motivating example. Among their results that are specific to the newsvendor model is that even a heuristic equilibrium solution tends to overstock relative to a stocking decision that is optimal with respect to the true non-linear cost of liquidation.

A very strong theme in the research on newsvendor models with pricing is the exploration of different relationships between the distribution of demand and the pricing decision. Of course, this not surprising in the analysis of a single-period model in which the demand distribution ultimately determines the optimal policy. Although some research has been done on models for risk-averse decision makers in this context, there is more opportunity for research along these lines, as risk aversion weighs more heavily in one-time decisions than in a repeated decision-making environment.

4.2 Models with learning and/or recourse for short-life-cycle products

In this subsection, we first discuss models of learning where early observations of demand provide information about future demand in a Baysian sense. We then turn to models in which the distribution of demand is known, but there are opportunities for recourse with respect to production/procurement and/or pricing as additional information about the realization of demand becomes available.

The speed with which a firm can obtain information, i.e., learn, about demand is an important aspect of managing short life cycle products that has recently been addressed by a few researchers. Because this research is germane to short-life-cycle products, it is closely related to single-period models. However, in order to capture the effects of learning, these models necessarily require the consideration of more than one period.

Alpern and Snower (1988) were among the first to recognize the role that pricing and production decisions play in influencing the rate of learning about demand. In their model, the only source of uncertainty pertains to the relationship between price and demand. They show that it can benefit a firm to increase production in the first period in order to increase the chance of observing actual, instead of truncated, sales. This information helps it to *fit* the unknown function that maps price to demand. Subrahmanyan and Shoemaker (1996) also address the interaction between price and production decisions and the rate of learning about demand. In a discrete time framework, they use brute-force dynamic programming with Bayesian updating to compute optimal solutions. Petruzzi and Dada (2001) investigate the same issue in a two-period model in which demand is assumed to be a linear function of price in which the intercept is a random variable, i.e., an additive model of demand uncertainty. They show how this problem can be reduced to a search for a single decision variable and provide the insight that a firm can benefit from over-pricing in the first period in order to avoid censored demand information and learn more about second period demand. In all three of these papers, one of the significant results is that a firm can benefit from making decisions that sacrifice short-term profits in order to obtain better information about demand.

A slightly different situation involving learning is examined in a paper by Burnetas and Smith (2000). They consider a situation in which inventory cannot be held from one period to the next and all excess demand is lost, so that the firm faces a series of identical newsvendor problems. They adopt a very general model of demand uncertainty, assuming only that the firm has access to past ordering and sales information. Within this framework, they develop a simple ordering policy and show that it converges to the optimal newsvendor quantity for a given price. They then extend their policy to include a pricing decision, where they assume a discrete set of possible prices. This pricing and ordering policy is also shown to converge to the optimal one.

A number of researchers have investigated how a firm's opportunities for recourse, with respect to capacity (production), sales, or price affect the pricing and capacity (production) decisions in a single-period model. Leland (1972) examines four "modes of behavior" for the firm: *Certainty*: where price (p), production quantity (X), and sales quantity (q_s) decisions are all postponed until demand uncertainty is resolved; *Quantity Setting*: where X is determined ex ante, but decisions regarding p and q_s are postponed; *Price Setting*: where p is determined ex ante, but X and q_s are postponed; and *Price/Quantity Setting*: where p and X are determined ex ante, but q_s is postponed. Leland's analysis is based on a very general demand model. He assumes only that there is an implicit function, $f(p,q,u) = 0$, among price (p) and quantity (q) and the random parameter (u). He assumes that for any realization of the random parameter u, the relationship between p and q is downward sloping and that larger values of u correspond to larger demand. Leland shows that, when uncertainty is considered, the firm is not indifferent between quantity setting and price setting. He also considers how the extent of the firm's risk aversion influences the decisions in each of his four modes of behavior.

Van Mieghem and Dada (1999) undertake a similar study, but assume a linear demand function and additive uncertainty. These more restrictive assumptions allow them to obtain more specific results about the relative rankings of the modes of firm behavior than did Leland. In addition, they show that, for a broad class of distributions for the additive demand shock, the optimal levels of capacity investment and production tend to increase when the firm has more flexibility to postpone pricing and quantity decisions. They also examine the effect of competition upon the modes of behavior.

Three recent articles examine clearance pricing decisions in settings where, unlike in traditional revenue management models, there is latitude in setting the initial inventory, along with the opportunity to adjust prices. Khouja (2000) considers a model in which, after ordering and observing demand, the retailer can offer a series of discounts to liquidate any excess stock. A fixed cost is incurred for each discount level (for advertising and marking down the product). Khouja derives the optimal discounting scheme, then analyzes the problem of optimizing the initial order quantity for a given number of discount levels, and finally studies the problem of optimizing the initial price. In a numerical example, he illustrates how ex-post pricing flexibility (number of discount levels) affects the initial pricing and ordering decisions. Chun (2001) considers a model of inventory liquidation with a negative binomial distribution of demand, and derives a simple characterization of the optimal price and quantity for a single period problem. He then considers the possibility of dividing the sales period into sub-periods and demonstrates the intuitive result that the profits are higher when there are opportunities for recourse. Netessine (2002) considers inventory liquidation in an environment in which the number of price changes is limited, perhaps due to costs associated with changing prices. He characterizes the dependence among prices, the number of price changes allowed, and the capacity.

Recent research is beginning to explore resource decisions that are complementary to pricing decisions. Chod and Rudi (2002) examine a scenario with two products that are produced using the same resource. Flexible resource capacity must be purchased before demand is realized, but price setting and the allocation of capacity to products take place after the demand curves are known. It is assumed that the demands for the products have a deterministic component (i.e., mean values) that depend linearly upon the prices for the two products, the same coefficient of sensitivity of prices to demand, and an additive uncertain component expressed via a variance-covariance matrix. Most of the results are obtained for the special case of a bivariate Normal distribution with equal demand variances. Under this assumption, among other things, the authors show that both the optimal resource level and the expected profit are increasing in the demand variance. The latter effect arises because the flexibility to set prices in response to the realization of demand is similar to the exercise of an option. They also show that while the optimal resource level is increasing in the demand correlation, the expected profit is decreasing, as a result of risk pooling.

4.3 Multi-period models with stochastic demand and without setup costs

Eilon and Mallya (1966) consider a setting in which products have a two-period lifetime. For a scenario with ordering opportunities in two consecutive periods, they explore differences in the expected profit as a function of how prices are set for the fresh and second-class (one-period-old) goods, the issuing policy (e.g., first-in-first-out or last-in-first-out) and the order quantities. Although this paper is exploratory in nature and there is no attempt to optimize the decisions, to the best of our knowledge, it is the only paper to treat the issues of pricing and ordering for deteriorating goods in a discrete-time context.

Mills (1962) provides an insightful discussion of the issues and tradeoffs in the multi-period problem with stochastic demand. He also contrasts traditional longer-term economic models that tend to be static and deterministic and therefore do not include inventory, with shorter-term models in which uncertainty and price changes give rise to inventory considerations. In addition to the usual tradeoffs, he mentions (but does not address) the effect of one period's price on customer perceptions that then influence demands in subsequent periods. Under the assumption of additive demand uncertainty, Mills proposes adaptations of the single-period problem as approximations for the multi-period problem. One adaptation involves a terminal value function based on the level of inventory and shortages at the end of the period. He also presents a model that, in addition, includes the costs of changing the production level and the price from one period to the next.

Karlin and Carr (1962) extend their single-period results to the infinite-horizon problem. They examine the so-called economic version of the model in which neither inventory nor shortage costs are incurred. They assume demands are i.i.d., unsatisfied demands are lost sales, and that only a single price can be selected which is maintained throughout the horizon. For the case of multiplicative demand uncertainty, they show that the optimal order-up-to point satisfies:

$$y^\star = g(p^\star)\, \Phi^{-1}[(p^\star - c)/(p^\star - \theta c)] \tag{19}$$

where g is the expected demand function, $p^\star$ is the optimal price, c is the unit cost and θ is the one-period discount factor. The optimal price is obtained by substituting the "newsboy" quantities into the objective function and finding the stationary point of the resulting expression. They show that the optimal price is larger than that for the infinite-horizon model with deterministic demand, but that no general relationship exists between the optimal prices for the single-period and infinite-horizon stochastic models.

For the model with additive demand uncertainty, under certain regularity conditions, Karlin and Carr obtain an analogous result:

$$y^\star = g(p^\star) + \Phi^{-1}[(p^\star - c)/(p^\star - \theta c)] \tag{20}$$

They show that the optimal price for the infinite-horizon stochastic problem is bounded below by the optimal prices for the single-period stochastic problem and bounded above by the optimal price for the infinite horizon deterministic problem.

Observe that the expressions in (19) and (20) represent the optimal order-up-to points as the expected demand at the optimal price with a multiplicative or additive adjustment for demand uncertainty. In both cases, the adjustment factor has a newsvendor-like form.

In an extension of the single-period models of Mill, Karlin and Carr, and others, to a multi-period setting, Nevins (1966) introduces the concept of an "equilibrium stock of inventory," which he defines as the level toward which the inventory is expected to move under an optimal pricing and stocking policy. Using this concept, he uses numerical analysis to show that the gap between the optimal deterministic and stochastic prices depends upon the interaction among demand elasticity, shortage costs, and the discount rate.

Zabel (1972) considers a finite horizon model under the same demand representations as those in the Karlin and Carr models. He allows linear or convex production costs and linear holding costs, but retains the assumption of lost sales. He shows that in the case of multiplicative demand uncertainty, extension of the single-period results (obtained by Karlin and Carr) to two periods requires stronger assumptions, namely concavity of the demand function with respect to price and certain conditions on the distribution of demand uncertainty. He states that the question of uniqueness of the optimal solution is unresolved. However, if the largest optimal production quantity and corresponding price are chosen, the results carry through to multiple periods under the same regularity conditions as in the two-period problem.

For the case of additive demand uncertainty, concavity of the demand function and conditions on the distribution of demand uncertainty are required for uniqueness of the optimal solution. With the imposition of certain regularity conditions on problem parameters, Zabel obtains results for the case of uniform and exponential demand uncertainty distributions (which satisfy one of the conditions for uniqueness of the optimal solution). The main results include: (1) for "low" base stock levels, the optimal price increases with more periods remaining, given the same base stock level; and (2) the optimal base stock level similarly increases with more periods remaining, given the same initial inventory. Zabel also observes that changes in the holding cost and discount rate mostly affect inventory decisions, with a lesser effect on prices and output.

Zabel also provides some interesting comparisons between the results for the two models of demand uncertainty, noting that the multiplicative model tends to induce high prices and low output, and the reverse is true for the additive model. This is the consequence of the demand variance, which is constant for the additive model, but varies inversely with price for the multiplicative model. Recall that these phenomena have also been observed in single-period models.

Thowsen (1975) examines a model with additive demand uncertainty with finite support, linear procurement costs and convex shortage and holding costs. He also allows a deterministic fraction of ending inventory to deteriorate (become unusable) each period, as well as a fixed delay in payment after the arrival of each order. His focus is on obtaining conditions to ensure that single-period results carry over to multiple periods. He shows that if all excess demand is backlogged, mean demand is a concave function of price, and revenue is collected a fixed time after each order, then no additional assumptions are needed to satisfy these conditions. Although these conditions are sufficient, many other realistic situations satisfy the required conditions, such as the case with linear expected demand, linear stockout costs, convex holding costs and a log-concave distribution of demand uncertainty. Thowsen first analyzes the problem with an unconstrained order-up-to point. He shows that under certain regularity conditions, most of which serve to eliminate speculative motive for holding inventory, the optimal safety stock quantity (order-up-to point less expected demand) lies between the lower and upper limits of the support of the demand uncertainty random variable, and that the upper bound on price (price at which demand is no longer always positive) is not a binding constraint. With this result and certain not-easy-to-interpret conditions that ensure unimodality of the dynamic programming value function, he shows that the optimal solution to the original problem is to order up to a period-dependent base stock level and set the price accordingly, if the on-hand inventory is less than the base stock level. On the other hand, if the on-hand inventory exceeds the period-dependent optimal base stock level, no order is placed and the optimal price depends on the inventory level. He further shows that the optimal price is decreasing in the inventory level. Under the assumption of additive demand uncertainty and complete backlogging, the optimal price is independent of the moments of the demand distribution. This result is a consequence of the assumption that all demand is eventually satisfied, and that customers pay the price in effect at the time of their demand, not the time at which the demand is satisfied. The optimal price is that which equates marginal discounted revenue (with respect to quantity) to the marginal ordering cost in a deterministic one-period problem. On the other hand, if there are full or partial lost sales, the optimal price is smaller.

Federgruen and Heching (1999) consider both the discounted and time-average profit criteria. Their models allow either bi-directional price changes or markdowns only. They assume full backlogging, linear procurement costs and convex inventory and shortage costs. They further assume that the one-period expected inventory cost function is jointly concave in the order-up-to-level and price—an assumption that is stronger than those imposed by Thowsen—which allows them to obtain results that are either elusive or not always true under more general assumptions. For the finite horizon problem with bi-directional price changes, they obtain results that are analogous to those of Thowsen, i.e., that a so-called "base stock list price" policy is optimal. That is, one orders up to $y_t^\star$ and prices at $p_t^\star$ if the on-hand inventory, x, is less than $y_t^\star$, where t is the number of periods remaining in the horizon. On the other hand, if $x > y_t^\star$, no order is placed, and the optimal price is a monotonically non-increasing function of x. For the finite horizon model with markdowns only, the price in each period is constrained by the price in the previous period, and this causes the order-up-to levels to be price-dependent.

Federgruen and Heching go on to show that for the infinite-horizon discounted problem, a base stock list price policy is optimal for both bi-directional price changes and markdowns only. For the average cost criterion, they show that the base stock list price policy remains optimal in the case of bi-directional price changes, but for the case of markdowns only, the optimal policy reduces to one with a constant price and an order-up-to point. The various results obtained by Federgruen and Heching extend to more general situations including those in which backorders are satisfied by a secondary source and those with production capacity constraints.

Federgruen and Heching (2002) extend their earlier model to allow multiple retailers and coordination of procurement through a warehouse and to allow random demands. The authors assume that all retailers offer the same price (to be determined) in a given period. Demands (for product j in period t) are modeled as:

$$d_{jt}(p_t) = [\gamma_{jt} + \delta_{jt} p_t]\, \varepsilon_{jt} + [\eta_{jt} + \theta_{jt} p_t]$$

where p_t is the price in period t, the Greek letters are coefficients of the demand function, and the vector of ε's has a general multivariate distribution. This model of demand admits both additive and multiplicative demand uncertainty as special cases. All costs (procurement, shipping, inventory and shortage) may be time-dependent. They present a model that is based on a deterministic approximation of the problem using expected demands as well as expected values of the ending inventory one lead time in the future (i.e., the inventory level affected by the current decision), computed assuming that the same price is maintained throughout the lead time. They show that for this approximate model, there exists an optimal policy with the base stock-list price form. They also provide some comparative statics results for changes in the optimal order quantities and prices as the (mean) demand curve shifts or as the demand uncertainty of a single retailer in a single period changes. In addition, they present and test several heuristic procedures based on variants of the approximate model and demonstrate numerically how various problem parameters (e.g., lead time) affect the optimal profits.

Chan et al. (2001) consider a finite horizon setting with limited capacity, lost sales, and the possibility of rationing. Demand in each period is random and the random error may depend upon the selected price. They study control policies in which either a pricing or a production decision is made at the beginning of the horizon; this decision is held fixed throughout the horizon and decisions in the other category are allowed to change in each period. For the so-called "delayed production" scenario in which prices are held fixed but production quantities may vary from period to period, they show that the optimal policy has an order-up-to and save-up-to level (for rationing purposes) in each period. For this scenario, they also propose pricing heuristics based on the assumption of deterministic demand. For the "delayed pricing" scenario in which production decisions are made at the beginning of the horizon, but the price can be adjusted in each period, they show that the optimal prices are not necessarily monotonic in the inventory levels. They also present heuristics for setting prices. They obtain a variety of insights from a computational study, some intuitive (e.g., the benefit of dynamic pricing increases with demand seasonality and with tightness of capacity) and others less intuitive (e.g., delayed production alone achieves a significant portion of an upper bound on profit).

All of the aforementioned papers in this section are based on the assumption that the customer will not choose a substitute product from the same vendor. Moinzadeh and Ingene (1994) examine a situation in which each shortage of the preferred product ("good 1") in stock may result in a lost sale, a backorder for good 1, or an order for a substitute ("good 2") with delayed delivery. The retailer chooses the stocking level for good 1 and the markup for good 2, assuming the price of good 1 is fixed. The demands for both products (prior to substitution) are assumed to be Poisson with rates that depend on the markup for good 2. The probability of a lost sale and the probability of switching given that the customer places an order for one of the two products, are also functions of the markup for good 2. It is assumed that there is an alternate distributor, and customers' utilities for the products may be a function of the distributor or its location, but the competitive aspects of the model do not change the fundamental nature of the results. The authors characterize the optimal base stock level of good 1 and discuss how to find the optimal solution (because the objective function may not be jointly concave). For the special case in which the demand rate for good 1 and the lost sales probability do not depend on the markup for good 2, they show that the optimal order-up-to level of good 1 is non-increasing in the contribution margin of good 2 up to a threshold, then non-decreasing above that threshold. When the margin for good 2 is low, the distributor has incentive to stock a high level of good 1 to avoid reduced revenue from switching. At intermediate values, the distributor is less concerned about switching because of the improved markup on good 2. At high markups, the per-unit margin from good 2 is high, but fewer customers are inclined to switch, so it becomes more important to stock good 1.

All of the papers mentioned above seek to characterize the structure of the optimal policy. Other research has applied both optimization-based and heuristic approaches to problems of this type in real-world settings. A paper by Lodish (1980) presents a dynamic programming formulation of the multi-period problem with pricing and production decisions in each period, then goes on to apply the method to three real broadcast spot advertising scenarios. (Here, only pricing decisions are pertinent.) Barbarosoglu (2000) proposes a heuristic approach for a multi-product, multi-customer scenario. The approach consists of three parts: (1) an optimization-based production planning model for the supplier in which demand uncertainty is handled using chance constraints; (2) a heuristic step to adjust the production quantities to account for buyer commitments; and (3) an optimization-based pricing model by which the supplier maximizes profit subject to cost expectations of the buyers. The method was implemented at a manufacturing company in Turkey.

Although most of the research that has introduced pricing decisions into stochastic demand models has been done in a discrete time framework, a noteworthy exception to this is Li (1988) who considers a continuous-time model in which the underlying production process is represented as a queue. He studies the problem of making a static capacity decision at the beginning of the horizon, along with dynamic production rate and price decisions to optimize discounted profit. He assumes that excess demands are lost sales and that the production and demand processes are independent Poisson processes with time-varying intensities. The firm chooses a price to control the average demand rate, and may also reject customers. The revenue function is assumed to be concave in the demand rate. The production costs and the per-unit, per-unit-time holding costs are assumed constant, and the capacity investment cost is increasing in the capacity. Li shows, both for the case in which a single price is chosen and in the case where price may be changed dynamically, that the optimal policy has a single barrier (produce-up-to) quantity. In the constant-price case, the optimal price is equal to the full marginal cost, which includes not only the marginal cost of production, but also marginal capacity and inventory costs. In the dynamic price case, the optimal price decreases with the on-hand inventory level and is always higher than the optimal price under certainty. Li extends his model to allow for production learning effects: the capacity is concave and increasing in the cumulative production, and has a finite upper bound (to be decided). Li shows that a barrier policy is still optimal in this case, with the barrier declining as cumulative production increases. He also shows that the price decreases in the inventory level but is not always higher than the optimal price under certainty, because of the moderating effects of learning. Although there is a substantial literature on pricing issues in queueing models, with very few exceptions, the papers focus on pure service operations (without inventory).

A large majority of papers reviewed in this section are concerned with obtaining characterizations of the optimal policy. Many of these papers derive results under infinite horizon assumptions, which simplifies the analysis and leads to results that are qualitatively similar to those derived for analogous single-period settings. The simplifications from the infinite horizon assumptions have permitted researchers to obtain strong and definitive characterizations of the optimal solutions under a variety of assumptions.

4.4 Multi-period models with stochastic demand and setup costs

Thomas (1974) was the first to introduce setup costs into multi-period models with stochastic demand. In addition to setup costs, his model considers linear inventory and backorder costs; variable production costs are not included. Revenue in each period is a linear function of unit sales. He assumes a finite number of feasible prices and conjectures that a (S,s,p) policy will perform well, where S is the order-up-to point, s is the order threshold, and the price, p, depends on the current inventory level. He provides an example to show that the policy is not always optimal: multiple local maxima may occur when the feasible prices are too far apart. In a computational study, the proposed policy is shown to perform well.

Over a quarter-century later, Polatoglu and Sahin (2000) consider the discounted profit version of the problem, assuming linear production costs, linear inventory costs and lost sales. In the first portion of their analysis they assume that demand is i.i.d., and that demand in each period is a continuous random variable distributed over $[L(p), U(p)]$, where $L(p)$ and $U(p)$ are differentiable functions such that $F(x; p_1) \leq F(x; p_2)$ for $p_1 < p_2$. The latter condition ensures a stochastic ordering of demands with respect to p. Like many other researchers, they also impose the condition that the single-period cost function is unimodal with respect to the procurement (or correspondingly, the order-up-to) quantity. Even under these assumptions, their analysis shows that there may be multiple order-up-to points and for each order-up-to point, there may be multiple reorder intervals (i.e., initial inventory levels at which ordering should take place). Not surprisingly, the value functions tend to have an increasing number of local maxima (with respect to the order quantity) as the number of periods in the horizon increases. Optimal prices also exhibit a non-monotonic behavior with respect to the order quantities. Because of the irregular behavior of the various functions, the authors focus on obtaining bounds on the various functions that permit more efficient computation of the optimal policy, but also note that the optimal policy is not easy to compute. They also derive conditions on the value functions that ensure optimality of a (S_n, s_n) policy for the case of positive setup costs, or a single-critical-number policy for the case of zero setup costs.

Feng and Chen (2002) study an infinite horizon setting in which demands are Poisson, but with a parameter that depends upon the price. In addition to variable costs, there is a setup cost for each order. They focus on situations in which there is a finite number of price alternatives, and introduce the concept of a *maximum increasing concave envelope* via which dominated prices can be eliminated from consideration. They show that the optimal policy has the form $(s = d_n, d_{n-1}, \ldots .d_1, D_1, \ldots .D_{n-1}, D_n = S)$, where s is the reorder level, S is the order-up-to-point, and $d_{n-1}, \ldots, d_1, D_1, \ldots D_{n-1}$ define the beginning-of-period inventory levels at which each price is optimal. In particular, for initial inventory levels between d_i and d_{i-1} and between D_{i-1} and D_i, it is optimal to charge price p_i. Although the second part of this result is intuitive (to charge a higher price when there is less inventory), the first part of the result may appear counterintuitive. It arises because when inventory levels are relatively low (and in particular, if they are negative and some of the d_i values are negative), it may be preferable to charge a lower price to stimulate demand and deplete enough inventory to trigger an order, which then reduces shortage costs.

Chen and Simchi-Levi (2002) examine a slightly more general model than that considered by Feng and Chen, allowing continuous pricing decisions and with a demand function:

$$d_t = \alpha_t D(p_t) + \beta_t,$$

where α_t and β_t are random variables satisfying $\alpha_t \geq 0, \geq E\{\alpha_t\} = 1$ and $E\{\beta_t\} = 0$, and are independent across time. (Observe that this functional form to represent demand admits both the additive and multiplicative models of demand uncertainty discussed earlier in this paper.) They show for the finite-horizon, undiscounted version of the problem that the (S,s,p) policy proposed by Thomas is optimal in the case of additive demand uncertainty, and that it is not necessarily optimal in the case of more general demand uncertainty. They introduce the concept of symmetric-k-convexity, which is a weaker condition than k-convexity, and use it to show that the optimal policy is characterized by two parameters in each period (an order-up-to point and an ordering threshold), but there is an intermediate range of inventory levels at which an order may or may not be placed. They also extend these results to more general settings including the infinite horizon problem and Markovian demand.

The papers discussed in this section show very clearly that the introduction of prices into discrete-time models with setup costs and uncertain demands leads to optimal policies that are significantly more complicated than their counterparts without pricing decisions. Much progress has been made within the past few years on characterizing and computing optimal policies for this class of problems. We expect that as researchers continue to relax the restrictive assumptions that have been made in some of these papers, our understanding of how the various modeling assumptions affect the structure of the policies will become more complete.

5. COORDINATION UNDER PRICE-DEPENDENT EXTERNAL DEMAND

A great deal of the current research effort on supply chains is concerned with coordination, particularly the design of contracts that achieve coordination. For recent surveys, see Cachon (2001), Tsay et al. (1999) and Lariviere (1999). This line of research concentrates on the issues of wholesale or transfer prices, buy-back provisions, franchise fees, and other financial agreements between the buyer and seller. Very few of these papers consider situations with price-dependent external demand, which is the focus of this section.

It is not surprising that pricing decisions play a prominent role in coordinating supply chains. Naturally, in settings requiring coordination, one or more upstream suppliers also must make decisions that usually include pricing and may include production/procurement decisions. Here, we provide a brief overview of representative research on coordination in settings that share the essential elements of the other articles in this survey: price-sensitive external demand, as well as pricing and production or procurement decision-making by the party that satisfies external demand. We discuss models in which only simple pricing decisions are implemented, not more complex contracts. Our intent is not to be comprehensive, but to provide the reader a sense of the types of issues that have been addressed in this emerging research area. We discuss papers in this section using the same sequence of modeling frameworks as we used earlier in the paper.

Coordination for EOQ-like supply chains was the earliest to be addressed in the research literature. Kunreuther and Richard (1971) compare a centralized policy with sequential policy in which the price is determined before the lot size. They develop analytical models that show when it is preferable for the manufacturer to offer a prespecified discounted price to induce the retailer to change his order frequency to the benefit of the manufacturer. Kim and Lee (1998a) present a method for identifying equilibria for systems with either the marketing organization (which determines the price) or the manufacturing organization (which determines the marginal cost, including inventory-related cost) as the leader. In Ertek and Griffin (2002), the buyer sets the retail price equal to a (selected) multiple of the wholesale price plus a (selected) constant. The supplier sets the wholesale price and determines production quantities. The authors characterize the equilibrium solutions for the buyer-driven and supplier-driven versions of the game and show that in the former, it is optimal to use only a multiplier to determine the markup.

Ardalan (1994, 1995) examines situations in which the supplier seeks to find the optimal price for a one-time discount, taking into account that the (single) retailer may choose to purchase more than the usual batch size and to reduce his selling price for all units in the corresponding batch and thereby temporarily increase the demand rate. The one-time offer causes a change in what is otherwise a standard EOQ framework. Ardalan develops models and solution procedures for both the average profit and net present value of profit objectives. In the latter case, the effect of the decisions during the special cycle on the present value of future profit is considered. From numerical results, he concludes that the present value approach leads to smaller prices and larger order quantities.

Abad (1994) studies a cooperative game in an EOQ context in which the buyer chooses the retail selling price and his own order quantity and the vendor chooses the wholesale price and the production quantity as a multiple of the buyer's order quantity. He characterizes the Pareto efficient and Nash bargaining solutions for this problem. He uses these results to propose pricing schedules for the multiple-buyer case, modeling the problem as a Stackelberg game with the vendor as the leader. He shows that coordination can be achieved either by a profit-sharing arrangement or an all-units quantity discount schedule. Weng (1995a) shows that, in the case of homogeneous buyers, there exist equivalent all-units and incremental discount schedules that induce the buyer(s) to purchase the system-optimal quantity, but they are not sufficient to induce the buyer(s) to set his selling price to optimize the system profit when demand is a decreasing function of the selling price.

Two papers analyze two-stage supply chains in an EOQ-like framework. Chakravarty and Martin (1991) study a system in which the retailer places m (to be determined) orders during each production cycle (which is determined by the manufacturer). They develop analytical models that show when it is preferable for the manufacturer to offer a pre-specified discounted price to induce the retailer to change his order frequency to the benefit of the manufacturer. Reyniers (2001) examines a situation in which both parties incur setup and inventory holding costs, and use a common production/ordering interval. She demonstrates that if the optimal delivery cycle differs between the centralized and decentralized systems, it is possible for the optimal price to be larger in the centralized system than in the decentralized system.

The continuous-time, dynamic, deterministic version of the problem with a single buyer and a single supplier is modeled as an open-loop Nash game by Jorgensen (1986), who characterizes the possible patterns of optimal policies that may be undertaken by both parties over a finite horizon. By contrast, Desai (1996) derives the Stackelberg and Pareto efficient solutions. He considers not only the general case in which prices and production/procurement decisions may change continuously, but also the case in which the supplier chooses a constant production rate and the case in which the buyer chooses a constant "processing" (procurement) rate.

As noted earlier, the continuous-time, dynamic models of Eliashberg and Steinberg (1987) and Kumar et al. (2000) were motivated by coordination issues. Eliashberg and Steinberg are interested in coordination at an operational level, i.e., the degree of coupling between the distributor's production and pricing policies and those of the supplier that manufactures the product. Kumar et al. focus on a different aspect of coordination: the decentralization of decision-making between the marketing (price-setting) and procurement departments of the distributor. They obtain the counterintuitive result that decentralization of decision-making at the distributor may improve the performance of the supply chain if it results in reducing the distributor's sensitivity to increases in the manufacturer's wholesale price.

Freeland (1981) analyzes coordination between the marketing and production departments of a firm in a deterministic, single-period setting. In his model, there is a single product for which the price, level of promotion, and production quantity must be decided under the assumption of convex production costs. He emphasizes the importance of correctly estimating the marginal production cost in achieving a coordinated solution, and discusses the extension of the model to multiple periods.

Zhao and Wang (2002) examine a discrete-time setting in which the distributor faces time-varying, deterministic demand curves. The supplier selects a price for each period, the distributor chooses prices and purchase quantities, and finally, the supplier chooses production quantities to satisfy the buyer's orders. The authors derive the equilibrium solution and devise an algorithm to determine the optimal prices and purchase quantities for the distributor given any vector of wholesale prices. They prove that there exists a wholesale price schedule that will induce the distributor to choose the optimal coordinated solution, and devise an incentive scheme that leads to channel coordination.

One of the only articles to consider multiple customers in a coordination context is by Chen et al. (2001), who study a single-distributor, multi-retailer system in which the retailers serve independent markets. Each retailer faces a downward sloping demand curve (which defines the annual demand rate) and incurs order setup and inventory holding costs, in addition to variable costs based on a pre-specified wholesale price. The distributor incurs setup, inventory and unit purchase costs. Each retailer chooses a price and an order quantity while the distributor chooses its own order quantity. The authors assume that the system operates using a power-of-two lot ratio policy. They characterize and/or develop solution procedures for (i) the optimal centralized solution, (ii) the solution in which retailers first set prices to maximize gross margin then the procurement department sets all lot sizes simultaneously, and finally (iii) the Stackelberg solution with the distributor as the leader. In the Stackelberg game, the distributor optimizes the wholesale prices and his own lot size fully anticipating the retailers' reactions. The retailers then select their prices and lot sizes. The authors compare solutions for the three scenarios and find that the non-centralized solutions, either from the sequential procedure or from the Stackelberg game, produce profits that are far from the system-optimal. Although the focus of this paper is on supply chain coordination issues, the distributor's problem in the Stackelberg game is also of significance because it represents a situation in which the supplier's pricing decision only indirectly affects his final demand, a situation commonly faced by wholesalers and manufacturers.

Weng (1997) studies coordination in the newsvendor framework. The supplier chooses a price, and the distributor (who fulfills external demand) chooses a price and an order quantity. Weng derives the optimal decentralized and coordinated solutions for a specific representation of demand and shows that the supplier's pricing policy alone is not sufficient to achieve coordination; a fixed payment from the distributor to the supplier is required to ensure participation of the supplier.

6. FUTURE RESEARCH DIRECTIONS

The literature on joint production/procurement and pricing decisions spans nearly 50 years, and well over 100 articles have been published on the topic. Notably, a very significant portion of the papers are based on either an EOQ or a newsvendor framework. The fact that these two, most basic, models of production and inventory have spawned so much research is an indication of the richness of joint pricing and production decisions. Subtle changes in modeling assumptions can often lead to dramatically different mathematical structures and give rise to entirely new streams of research. As a result, the most significant research in this area has, more often than not, come from the adoption of new and broader perspectives, to reveal previously unrecognized managerial issues. We believe that this pattern will continue to hold in the future, where the most significant contributions will come not from continued examination of recognized trade-offs, but from expansive inquiry that breaks down traditional boundaries and identifies new issues. Moreover, we believe that there is abundant opportunity for such contributions to be made, and we will attempt to discuss these opportunities below.

One of the most obvious patterns to emerge from this survey is the scant attention that has been paid to the effects of pricing decisions and price changes on intertemporal substitution by customers. Virtually all of the papers are based on the assumption of independence of demand across time, in both a deterministic and a stochastic sense. There is a longstanding literature on intertemporal substitution at a very aggregate level (e.g., current consumption versus saving for future consumption) in the economics literature, and the marketing literature contains a wealth of studies and models on the effects of promotions, including price promotions, on forward-buying by consumers. The former effects are difficult to incorporate in short- to medium-term models of the type discussed here, but the latter effects are relevant and of practical significance, yet they are virtually unaddressed in the literature. The effects of customers anticipating future price changes also appears to be unexplored in the joint production-pricing literature. The fact that prices do change over time also raises the question of how best to adjust prices over time to learn about the demand curve without unduly sacrificing profits. Only a few articles (discussed at the beginning of Section 4.2) have addressed this issue.

Another important practical consideration that has not received much attention is the effect of pricing on customer substitution within a set of similar products. Customer choice among products is a key factor in models of strategic decisions regarding product offering and pricing, but only one article in our survey (Moinzadeh and Ingene 1993) explicitly models product substitution and its effect on procurement decisions. With the growth of business-to-consumer e-commerce, it is now possible for firms to observe and collect data on both substitution behavior when a customer's preferred product is out of stock, and changes in the relative demand for products as prices change. As such data become available, appropriate models of consumer behavior may emerge, and these models could be incorporated into production/price decision-making. Similar observations hold for complementary products, as well. Whether or not a customer purchases a set of complementary products depends not only upon their aggregate price, but also upon the availability of the set of products he/she wishes to purchase. We are not aware of any papers in the literature that address joint inventory and pricing decisions for complementary products.

With the exception of several papers that propose quantity discounts (usually in an EOQ framework) as a means for achieving price discrimination among heterogeneous customers, virtually none of the literature models heterogeneity of customers' preferences in a general and realistic way. Various representations of demand uncertainty can be viewed as methods to reflect this heterogeneity, but such representations can be approximate, at best. Models of customer choice from the marketing literature that account for customer heterogeneity are complex, however, and careful modeling will be required to integrate them into decision models involving pricing and production.

A fourth factor that has been considered in only a small portion of the literature is capacity constraints. Although virtually all of the continuous-time models in Section 3 incorporate capacity constraints either implicitly or explicitly, only a few of the discrete-time deterministic and stochastic models do so. (See also Gelders and de Roose (1981) for a case study of a single-period problem with multiple products and multiple types of capacity constraints.) Capacity limitations may lead to short-supply conditions, which, in turn, affect how prices should be chosen. In industries with oligopolies or regulated monopolies, suppliers may limit production, thereby creating the appearance of tight capacity limits that allow them to raise prices (see, for example, Lippman and Winston 1981). Although the effect of market power on prices is beyond the scope of our survey, it is clear that supply availability is influenced by an array of factors that should not be ignored if one wishes to make realistic pricing decisions.

Another consideration that is becoming increasingly important as product life cycles become shorter is the integration of product life cycle planning with pricing decisions. As old product lines are phased out and new ones are phased in, pricing and inventory decisions can be critical in achieving high profits and managing demand vis-à-vis production capacity. Several of the articles discussed in Section 3 can be adapted to model the dynamics of the life cycle for a single product, but significant extensions would be required to handle multiple, related products.

A sixth factor that may play an even more important role than all of the others mentioned above is the effect of competition, although we acknowledge that it is the most difficult to integrate with models of short-term operational decisions such as production or procurement. There is an enormous literature on price setting in the presence of competition, but to the best of our knowledge, relatively few papers consider detailed operational costs and detailed operational decisions, such as production/purchase quantities or short-term production rates. (See Eliashberg and Steinberg 1991, Lederer and Li 1997, Tsay and Agrawal 2000, Lovell 2003, Atkins and Zhao 2002, Chen, Yan and Yao 2002 and references therein.) This line of research is rapidly expanding at this writing, however, so we expect significant developments within the next several years.

A very closely related issue to that above is the impact of industry structure on production and pricing decisions. The economics literature has much to say about how industrial market structure affects pricing and production decisions in an aggregate sense, but it offers little regarding these same decisions in an operational sense. These linkages appear to be relatively unexplored, and may offer opportunities to relate the operationally-focused models described in this review to broader economic frameworks.

The various issues discussed above highlight the very dichotomy that motivated us to survey the literature on joint pricing and production/procurement decisions: the economics and marketing literatures treat the impact of prices on consumer behavior in some detail but either ignore cost or capacity considerations or assume very simplistic cost structures, while the operations management literature treats the operational costs and decisions in detail, but takes a simplistic view of price and its effect on demand. Some of the generalizations and extensions mentioned above will present major challenges with respect to data gathering, modeling, and/or analysis. Nevertheless, our hope is that this survey motivates more research that treats both consumer behavior and operational costs realistically when modeling joint pricing and production decisions.

Acknowledgement

The authors are grateful to Richard Steinberg and to two anonymous referees for their thoughtful comments on an earlier version of this paper.

REFERENCES

Abad, P.L. (1988a). Determining Optimal Selling Price and Lot Size when the Supplier Offers All-Unit Quantity Discounts, *Decision Sciences* 19 (3), 622–634.

Abad, P.L. (1988b). Joint Price and Lot-Size Determination when Supplier Offers Incremental Quantity Discounts, *Journal of the Operational Research Society* 39 (6), 603–607.

Abad, P.L. (1994). Supplier Pricing and Lot Sizing when Demand is Price Sensitive, *European Journal of Operational Research* 78 (3), 334–354.

Abad, P.L. (1996). Optimal Pricing and Lot-Sizing under Conditions of Perishability and Partial Backordering, *Management Science* 42 (8), 1093–104.

Abad, P. L. (1997). Optimal Policy for a Reseller when the Supplier Offers a Temporary Reduction in Price, *Decision Sciences* 28 (3), 637–653.

Abad, P.L. (2001). Optimal Price and Order Size for a Reseller under Partial Backordering, *Computers and Operations Research* 28 (1), 53–65.

Aggarwal, S.P. and C.K. Jaggi (1989). Ordering Policy for Decaying Inventory, *International Journal of Systems Science* 20 (1), 151–155.

Agrawal, V. and S. Seshadri (2000). Impact of Uncertainty and Risk Aversion on Price and Order Quantity in the Newsvendor Problem, *Manufacturing and Service Operations Management* 2 (4), 410–423.

Alpern, S. and D.J. Snower (1989). A Search Model of Optimal Pricing and Production, *Engineering Costs and Production Economics* 15, 279–284.

Arcelus, F.J. and G. Srinivasan (1985). A ROI-Maximizing EOQ Model under Variable Demand and Markup Rates, *Engineering Costs and Production Economics* 9 (1–3), 113–117.

Arcelus, F.J. and G. Strinivasan (1987). Inventory Policies under Various Optimizing Criteria and Variable Markup Rates, *Management Science* 33 (6), 756–762.

Ardalan, A. (1994). Optimal Prices and Order Quantities when Temporary Price Discounts Result in Increase in Demand, *European Journal of Operational Research* 72 (1), 52–61.

Ardalan, A. (1995). A Comparative Analysis of Approaches for Determining Optimal Price and Order Quantity when a Sale Increases Demand, *European Journal of Operational Research* 84 (2), 416–430.

Atkins, D. and X. Zhao (2002). Supply Chain Structure under Price and Service Competition. Working Paper, Faculty of Commerce and Business Administration, University of British Columbia, Vancouver, British Columbia, Canada.

Aydin, G. and E.L. Porteus (2003). Inventory and Pricing for Multiple Products with Demand Cannibalization. Working paper, Graduate School of Business, Stanford University, Stanford, CA.

Barbarosoglu, G. (2000). An Integrated Supplier-Buyer Model for Improving Supply Chain Coordination. *Production Planning and Control* 11 (8), 732–741.

Bhattacharjee, S. and R. Ramesh (2000). A Multi-Period Profit Maximizing Model for Retail Supply Chain Management: An Integration of Demand and Supply-Side Mechanisms, *European Journal of Operational Research* 122, 584–601.

Birge, J., J. Drogosz, I. Duenyas (1998). Setting Single-Period Optimal Capacity Levels and Prices for Substitutable Products, *International Journal of Flexible Manufacturing Systems* 10, 407–430.

Brahmbhatt, A.C. and M.C. Jaiswal (1980). An Order-Level-Lot-Size Model with Uniform Replenishing Rate and Variable Mark-up of Prices, *International Journal of Production Research* 18 (5), 655–664.

Burnetas, A. and C. Smith (2000). Adaptive Ordering and Pricing for Perishable Products, *Operations Research* 48 (3), 436–443.

Burwell, T.H., D.S. Dave, K.E. Fitzpatrick and M.R. Roy (1990). A Note on Determining Optimal Selling Price and Lot Size under All-Unit Quantity Discounts, *Decision Sciences* 21 (2), 471–474.

Burwell, T.H., D.S. Dave, K.E. Fitzpatrick and M.R. Roy (1991). An Inventory Model with Planned Shortages and Price Dependent Demand, *Decision Sciences* 22 (5), 1187–1191.

Burwell, T.H., D.S. Dave, K.E. Fitzpatrick and M.R. Roy (1997). Economic Lot Size Model for Price-Dependent Demand under Quantity and Freight Discounts, *International Journal of Production Economics* 48 (2), 141–155.1

Cachon, G. (2001). Supply Chain Coordination with Contracts, Working Paper, Wharton School, University of Pennsylvania, Philadelphia, PA.

Cachon, G. and A.G. Kok (2002). Heuristic Equilibrium in the Newvendor Model with Clearance Pricing, Working Paper, The Wharton School, University of Pennsylvania, Philadelphia, PA.

Chakravarty, A.K. and G. E. Martin (1991). Operational Economies of a Process Positioning Determinant. *Computers and Operations Research* 18 (6), 515–530.

Chan, L.M.A., D. Simchi-Levi and J. Swann (2000). Flexible Pricing Strategies to Improve Supply Chain Performance, Working Paper, Department of Industrial Engineering and Management Science, Northwestern University, Evanston, IL.

Chan, L.M.A., D. Simchi-Levi and J. Swann (2001). Dynamic Pricing Strategies for Manufacturing with Stochastic Demand and Discretionary Sales, Working Paper, School of Systems and Industrial Engineering, Georgia Institute of Technology, Atlanta, GA.

Chen, Y., H. Yan and L. Yao (2002). Newsvendor Pricing Game, Working Paper, Department of Systems Engineering and Engineering Management, Chinese University of Hong Kong, Shatin, New Territories, Hong Kong.

Chen, C.-K. and K.J. Min (1994a). An Analysis of Optimal Inventory and Pricing Policies under Linear Demand, *Asia Pacific Journal of Operational Research* 11 (2), 117–129.

Chen, C.-K. and K.J. Min (1994b). A Multi-Product EOQ Model with Pricing Consideration—T.C.E. Cheng's Model Revisited, *Computers and Industrial Engineering* 26 (4), 787–794.

Chen, F., A. Federgruen and Y. Zheng (2001). Near-Optimal Pricing and Replenishment Strategies for a Retail/Distribution System, *Operations Research* 49(6), 839–853.

Chen, M.-S. and M.-C. Chu (2001). The Analysis of Optimal Price Control Model in Matching Problem between Production and Sales, *Asia Pacific Journal of Operational Research* 18 (2), 131–148.

Chen, X. and D. Simchi-Levi (2002). Coordinating Inventory Control and Pricing Strategies with Random Demand and Fixed Ordering Cost: The Finite Horizon Case, Working paper, Operations Research Center, Massachusetts Institute of Technology, Cambridge, MA.

Cheng, T.C.E. (1990). An EOQ Model with Pricing Consideration, *Computers and Industrial Engineering* 18 (4), 529–534.

Cheng, T.C.E. (1984). An Integrated Product Pricing and Batch Sizing Model with Stochastic Demand, *Engineering Costs and Production Economics* 8, 27–31.

Chod, J. and N. Rudi (2002). Resource Flexibility with Responsive Pricing, Working Paper, Simon Graduate School of Business, University of Rochester, Rochester, NY.

Chun, Y.H. (2003). Optimal Pricing and Ordering Policies for Perishable Commodities. *European Journal of Operational Research* 144 (1), 68–82.

Cohen, M.A. (1977). Joint Pricing and Ordering Policy for Exponentially Decaying Inventory with Known Demand, *Naval Research Logistics Quarterly* 24 (2), 257–268.

Dana, J. and N. Petruzzi (2001). Note: The Newsvendor Model with Endogenous Demand, *Management Science* 47 (11), 1488–1497.

Deng, S. and C.A. Yano (2000). Joint Production and Pricing Decisions with Setup Costs and Capacity Constraints, Working Paper, Department of Industrial Engineering and Operations Research, University of California at Berkeley.

Desai, V.S. (1996). Interactions between Members of a Marketing-Production Channel under Seasonal Demand, *European Journal of Operational Research* 90 (1), 115–141.

Dobson, G. and C.A. Yano (2002). Product Offering, Pricing and Make-To-Stock/Make-To-Order Decisions with Shared Capacity, *Production and Operations Management* 11(4), 293–312.

Dolan, R.J. (1987). Quantity Discounts: Managerial Issues and Research Opportunities, *Marketing Science* 6, 1–22.

Eilon, S. and R.V. Mallya (1966). Issuing and Pricing Policies for Semi-perishables, in *Proceedings of the Fourth International Conference on Operational Research* (Actes de la Quatrième Conference Internationale de Recherche Operationnelle), D.B. Hertz and J. Melese (eds.). New York : Wiley-Interscience.

Eliashberg, J. and R. Steinberg (1987). Marketing-Production Decisions in an Industrial Channel of Distribution, *Management Science* 33 (8), 981–1000.

Eliashberg , J. and R. Steinberg (1991). Competitive Strategies for Two Firms with Asymmetric Production Cost Structures, *Management Science* 37 (11), 1452–1473.

Eliashberg, J. and R. Steinberg (1993). Marketing-Production Joint Decisions, Chapter 18 in *Handbooks in Operations Research and Management Science: Marketing,* North-Holland, 827–880.

Ertek, G. and P.M. Griffin (2002). Supplier and Buyer Driven Channels in a Two-Stage Supply Chain, *IIE Transactions* 34 (8), 691–700.

Federgruen, A. and A. Heching (1999). Combined Pricing and Inventory Control under Uncertainty, *Operations Research* 47 (3), 454–475.

Federgruen, A. and A. Heching (2002). Multilocation Combined Pricing and Inventory Control. *Manufacturing and Service Operations Management* 4 (4), 275–296.

Feichtinger, G. and R. Hartl (1985). Optimal Pricing and Production in an Inventory Model, *European Journal of Operational Research* 19 (1), 45–56.

Feng, Y. and F.Y. Chen (2002). Joint Pricing and Inventory Control with Setup Costs and Demand Uncertainty, Working Paper, Department of Systems Engineering and Engineering Management, Chinese University of Hong Kong, Shatin, New Territories, Hong Kong.

Freeland, J.R. (1980). Coordination Strategies for Production and Marketing in a Functionally Decentralized Firm, *AIIE Transactions* 12 (2), 126–132.

Gaimon, C. (1988). Simultaneous and Dynamic Price, Production, Inventory and Capacity Decisions, *European Journal of Operational Research* 35 (3), 426–441.

Gelders, L.F. and F.H. deRoose (1981). Determining Price Levels and Product Mix in a Rolling Mill: A Case Study, *AIIE Transactions* 13 (4), 315–324.

Gilbert, S.M. (1999). Coordination of Pricing and Multi-Period Production for Constant Priced Goods, *European Journal of Operational Research* 114, 330–337.

Gilbert, S.M. (2000). Coordination of Pricing and Multiple-Period Production Across Multiple Constant Priced Goods, *Management Science* 47 (12), 1602–1616.

Goyal, S.K. and A. Gunasekaran (1995). An Integrated Production-Inventory-Marketing Model for Deteriorating Items, *Computers and Industrial Engineering* 28 (4), 755–762.

Hwang, H. and S.W. Shinn (1997). Retailer's Pricing and Lot Sizing Policy for Exponentially Deteriorating Products under the Condition of Permissible Delay in Payments, *Computers and Operations Research* 24 (6), 539–547.

Jorgensen, S. (1986). Optimal Production, Purchasing and Pricing: A Differential Game Approach, *European Journal of Operational Research* 24 (1), 64–76.

Jorgensen, S., P.M. Kort, G. Zaccour (1999). Production, Inventory, and Pricing under Cost and Demand Learning Effects, *European Journal of Operational Research* 117 (2), 382–395.

Jorgensen, S. and P.M. Kort (2002). Optimal Pricing and Inventory Policies: Centralized and Decentralized Decision Making, *European Journal of Operational Research* 138 (3), 578–600.

Kang, S. and I.-T. Kim. (1983). A Study on the Price and Production Level of the Deteriorating Inventory System, *International Journal of Production Research* 21 (6), 899–908.

Karlin, S. and C.R. Carr. (1962). Prices and Optimal Inventory Policy, Chapter 10 in *Studies in Applied Probability and Management Science,* K. Arrow, S. Karlin and H. Scarf (eds.), Stanford University Press, Stanford, CA.

Karmarkar, U. S. (1996). Integrative Research in Marketing and Operations Management, *Journal of Marketing Research* 33(2), 125–133.

Khouja, M.J. (2000). Optimal Ordering, Discounting, and Pricing in the Single-Period Problem, *International Journal of Production Economics* 65, 201–216.

Kim D.-S. and W.J. Lee. (1998a). Optimal Coordination Strategies for Production and Marketing Decisions, *Operations Research Letters* 22 (1), 41–47.

Kim, D.-S. and W.J. Lee. (1998b). Optimal Joint Pricing and Lot Sizing with Fixed and Variable Capacity, *European Journal of Operational Research* 109 (1), 212–227.
Kim, J., H. Hwang and S. Shinn. (1995). An Optimal Credit Policy to Increase Supplier's Profits with Price-Dependent Demand Functions, *Production Planning and Control* 6 (1), 45–50.
Kumar, R.K., A.P.S. Loomba and G.C. Hadjinicola. (2000). Marketing-Production Coordination in Channels of Distribution, *European Journal of Operational Research* 126 (1), 189–217.
Kunreuther, H. and J.F. Richard. (1971). Optimal Pricing and Inventory Decisions for Non-Seasonal Items, *Econometrica* 39 (1), 173–175.
Kunreuther, H. and L. Schrage. (1973). Joint Pricing and Inventory Decisions for Constant Priced Items, *Management Science* 19 (7), 732–738.
Ladany, S. and A. Sternlieb. (1974). The Interaction of Economic Ordering Quantities and Marketing Policies, *AIIE Transactions* 6 (1), 35–40.
Lariviere, M. (1999). Supply Chain Contracting and Coordination with Stochastic Demand, in *Quantitative Models of Supply Chain Management,* S. Tayur, M. Magazine, and R. Ganeshan (eds.), Kluwer Academic Publishers, Norwell, MA.
Lau, H.L. and H. Lau (1988). The Newsboy Problem with Price Dependent Demand Distribution, *IIE Transactions* 20 (2), 168–175.
Lederer, P.J. and L. Li (1997). Pricing, Production, Scheduling, and Delivery-Time Competition, *Operations Research* 45(3), 407–420.
Lee, H.L. and M.J. Rosenblatt (1986). The Effects of Varying Marketing Policies and Conditions on the Economic Ordering Quantity, *International Journal of Production Research* 24 (3), 593–598.
Lee, W.J. (1993). Determining Order Quantity and Selling Price by Geometric Programming: Optimal Solution, Bounds, and Sensitivity, *Decision Sciences* 24 (1), 76–87.
Lee, W.J. (1994). Optimal Order Quantities and Prices with Storage Space and Inventory Investment Limitations, *Computers and Industrial Engineering* 26 (3), 481–488.
Lee, W.J. and D.-S. Kim (1993). Optimal and Heuristic Decision Strategies for Integrated Production and Marketing Planning, *Decision Sciences* 24 (6), 1203–1213.
Leland, H.E. (1972). Theory of the Firm Facing Uncertain Demand, *American Economic Review* 62, 278–291.
Li, L. (1988). A Stochastic Theory of the Firm, *Mathematics of Operations Research* 13, 447–466.
Lippman, S.A. and W.L. Winston (1981). Optimal Control of Price through Restricted Production. *Naval Research Logistics Quarterly* 28 (3), 463–74.
Lodish, L. (1980). Applied Dynamic Pricing and Production Models with Specific Application to Broadcast Spot Pricing, *Journal of Marketing Research* 17, 203–211.
Lovell, M. (2003). Optimal Lot Size, Inventory, Prices and JIT under Monopolistic Competition. *International Journal of Production Research* 81–82 (Issue C).
McTavish, R. and S.K. Goyal (1989). Integrating Marketing and Production Decisions, *Engineering Costs and Production Economics* 15, 387–90.
Mills, E.S. (1959). Uncertainty and Price Theory, *Quarterly Journal of Economics* 73, 117–130.
Mills, E.S. (1962). *Price, Output, and Inventory Policy, A Study in the Economics of the Firm and Industry,* New York, Wiley.
Moinzadeh, K. and C. Ingene (1993). An Inventory Model of Immediate and Delayed Delivery, *Management Science* 39 (5), 536–548.
Morgan, L.O., R.L. Daniels and P. Kouvelis (2001). Marketing/Manufacturing Trade-Offs in Product Line Management, *IIE Transactions* 33 (11), 949– 962.
Netessine, S. (2002). Dynamic Pricing of Inventory/Capacity when Price Changes are Costly, Working Paper, The Wharton School, University of Pennsylvania, Philadelphia, PA.
Nevins, A.J. (1966). Some Effects of Uncertainty: Simulation of a Model of Price. *Quarterly Journal of Economics* 80, 73–87.
Parlar, M. and Q. Wang (1994). Discounting Decisions in a Supplier-Buyer Relationship with a Linear Buyer's Demand, *IIE Transactions* 26 (2), 34–41.
Pekelman, D. (1974). Simultaneous Price-Production Decisions, *Operations Research* 22 (4), 788–794.
Petruzzi, N. and M. Dada (1999). Pricing and the Newsvendor Problem: A Review with Extensions, *Operations Research* 47 (2), 183–194.
Petruzzi, N. and M. Dada (2001). Information and Inventory Recourse for a Two-market, Price Setting Retailer, *Manufacturing and Service Operations Management* 3 (3), 242–263.
Polatoglu, L.H. (1991). Optimal Order Quantity and Pricing Decisions in Single-Period Inventory Systems, *International Journal of Production Economics* 23, 175–185.
Polatoglu, H. and I. Sahin (2000). Optimal Procurement Policies under Price-Dependent Demand, *International Journal of Production Economics* 65 (2), 141–171.

Porteus, E.L. (1985). Investing in Reduced Setups in the EOQ Model, *Management Science* 31 (8), 998–1010.

Rajan, A., Rakesh and R. Steinberg (1992). Dynamic Pricing and Ordering Decisions by a Monopolist, *Management Science* 38 (2), 240–262.

Raz, G. and E. Porteus (2002). A Discrete Service Levels Perspective to the Newsvendor Model with Simultaneous Pricing, Working Paper, Graduate School of Business, Stanford University, Stanford, CA.

Reyniers, D.J. (2001). The Effect of Vertical Integration on Consumer Prices in the Presence of Inventory Costs. *European Journal of Operational Research* 130 (1), 83–89.

Rosenberg, D. (1991). Optimal Price-Inventory Decisions-Profit vs. ROII, *IIE Transactions* 23 (1), 17–22.

Shinn, S.W., H. Hwang , S.S. Park (1996). Joint Price and Lot Size Determination under Conditions of Permissible Delay in Payments and Quantity Discounts for Freight Cost, *European Journal of Operational Research* 91 (3), 528–542.

Shinn, S.W. (1997). Determining Optimal Retail Price and Lot Size under Day-Terms Supplier Credit, *Computers and Industrial Engineering* 33 (3–4), 717–720.

Subramanyam, E.S. and S. Kumaraswamy (1981). EOQ Formula under Varying Marketing Policies and Conditions, *AIIE Transactions* 13 (4), 312–314.

Subrahmanyan, S. and R. Shoemaker (1996). Developing Optimal Pricing and Inventory Policies for Retailers Who Face Uncertain Demand, *Journal of Retailing* 72 (1), 7–30.

Thomas, J. (1970). Price-Production Decisions with Deterministic Demand, *Management Science* 16 (11), 747–749.

Thomas, L.J. (1974). Price and Production Decisions with Random Demand, *Operations Research* 22 (3), 513–518.

Thompson, G.L., S.P. Sethi and J. Teng (1984). Strong Planning and Forecast Horizons for a Model with Simultaneous Price and Production Decisions, *European Journal of Operational Research* 16, 378–388.

Thowsen, G.T. (1975). A Dynamic, Non-Stationary Inventory Problem for a Price/Quantity Setting Firm, *Naval Research Logistics Quarterly* 22 (3), 461–476.

Tsay, A.A., S. Nahmias and N. Agrawal (1999). Modeling Supply Chain Contracts: A Review, Chapter 10 in *Quantitative Models for Supply Chain Management,* S. Tayur, R. Ganeshan, and M. Magazine (eds.), Kluwer Academic Publishers, Norwell, MA.

Tsay, A.A. and N. Agrawal (2000). Channel Dynamics under Price and Service Competition. *Manufacturing and Service Operations Management* 2 (4), 372–391.

Urban, T.L. (1992). Deterministic Inventory Models Incorporating Marketing Decisions, *Computers and Industrial Engineering* 22 (1), 85–93.

Urban, T.L. and R. C. Baker (1997). Optimal Ordering and Pricing Policies in a Single-Period Environment with Multivariate Demand and Markdowns, *European Journal of Operational Research* 103, 573–583.

Van Mieghem, J.A. and M. Dada (1999). Price Versus Production Postponement: Capacity and Competition, *Management Science* 45 (12), 1631–1649.

Wagner, H.M. (1960). A Postscript to Dynamic Problems in the Theory of the Firm, *Naval Research Logistics Quarterly* 7 (1), 7–12.

Wagner, H.M. and T.M. Whitin (1958a). Dynamic Problems in the Theory of the Firm, *Naval Research Logistics Quarterly* 5 (1), 53–74.

Wagner, H.M. and T.M. Whitin (1958b). Dynamic Version of the Economic– Lot-Size Formula, *Management Science* 5(1), 89–96.

Wee, H.-M. (1995). Joint Pricing and Replenishment Policy for Deteriorating Inventory with Declining Market, *International Journal of Production Economics* 40 (2–3), 163–171.

Wee, H.-M. (1999). Deteriorating Inventory Model with Quantity Discount, Pricing and Partial Backordering, *International Journal of Production Economics* 59 (1–3), 511–518.

Wee, H.-M. and S.-T. Law (2001). Replenishment and Pricing Policy for Deteriorating Items Taking into Account the Time-Value of Money, *International Journal of Production Economics* 71 (1–3), 213–220.

Weng, Z.K. (1995a). Channel Coordination and Quantity Discounts, *Management Science* 41 (9), 1509–1522.

Weng, Z.K. (1995b). Modeling Quantity Discounts under General Price-Sensitive Demand Functions: Optimal Policies and Relationships, *European Journal of Operational Research* 86 (2), 300–314.

Weng Z.K. (1997). Pricing and Ordering Strategies in Manufacturing and Distribution Alliances, *IIE Transactions* 29 (8), 681–692.

Weng, Z.K. and R.T. Wong (1993). General Models for the Supplier's All-Unit Quantity Discount Policy, *Naval Research Logistics Quarterly* 40 (7), 971–991.

Whitin, T.M. (1955). Inventory Control and Price Theory, *Management Science* 2, 61–68.
Young, L. (1978). Price, Inventory and the Structure of Uncertain Demand. *New Zealand Operations Research* 6 (2), 157–177.
Young, L. (1979). Uncertainty, Market Structure and Resource Allocation. *Oxford Economic Papers* 31, 47–59.
Zabel, E. (1970). Monopoly and Uncertainty, *Review of Economic Studies* 37, 205–219.
Zabel, E. (1972). Multi-Period Monopoly under Uncertainty, *Journal of Economic Theory* 5, 524–536.
Zhao, W. and Y. Wang (2002). Coordination of Joint Pricing-Production Decisions in a Supply Chain, *IIE Transactions* 34 (8), 701–715.

4. CONTRACTUAL RELATIONSHIPS AND COORDINATION IN DISTRIBUTION CHANNELS

GANESH IYER and V. PADMANABHAN

ABSTRACT

Marketing Channels refer to the set of interdependent organizations involved in taking a product or service from its point of production to its point of consumption. Channels are increasingly being viewed as a key strategic asset because of their ability to create a sustainable competitive advantage relative to competition. This chapter reviews the growing literature in marketing, operations and industrial organization that deals with the role of contracts and other institutional arrangements in channel organization and management. The focus is on understanding the implications of strategic interactions among institutions in a distribution channel and the efficiency of different contractual mechanisms. The chapter closes with suggestions for future research in this domain.

1. INTRODUCTION

Marketing channels refer to the set of interdependent organizations that are involved in making a product or service available to the consumers. These are the set of institutions that are involved in taking a product or service from its point of production to the point of consumption (Coughlan, Anderson, Stern and El-Ansary 2001). Channels are increasingly being viewed as a strategic asset because of their ability in creating a sustainable differential advantage relative to competition (Stern and Weitz 1997). The durability of the competitive advantage stems from the fact that a properly designed and managed channel can confer advantages running the gamut from lower costs to value added through product and service differentiation which are difficult to imitate.

The early research in marketing channels (Alderson 1965, Bucklin 1969) was largely descriptive and developed the concepts of the functions and flows of a distribution channel. The ideal channel provided value to the consumer beyond the utility derived from the consumption of the product and/or service (i.e., beyond the consumption utility from the product, consumers also value the benefits from assortment, convenience, after-sales service, etc.). The literature has since grown substantially and we now have a large body of analytical research in marketing, industrial organization as well as in operations management. The focus of a large part of this literature has been to understand the role of contracts and other institutional arrangements on channel organization and management. Much of the analytical research in this domain has been based on agency theoretic models of distribution channels which recognizes that a channel consisting of independent institutions will reflect divergent interests and focuses on identifying contracts that can generate channel performance equivalent to the integrated channel. The intent of this article is to review the analytical literature on strategic interactions in channels of distribution, highlight its implications for contractual relationships between independent parties at successive stages in a vertical channel and their consequences for performance of various channel institutions.

This review looks at situations where the firm under study does not supply it's goods to the final consumers directly. Rather, the focus is on situations where the "upstream" firm supplies a "downstream" intermediary which in turn supplies the final consumer. Thus, the "upstream" firm operates in an intermediate goods market.[2] An example of a "downstream" firm could be a retailer who buys the intermediate good from the upstream firm (either a manufacturer or a wholesaler). Retailers sell directly to final consumers and thereby also operate in the final goods market. In many cases, retailers sell a final good which is often close or identical to the intermediate good which they brought from their upstream firm. In such cases a retailer merely makes the manufacturer's good easily available to the final consumer. The retailer here only produces "product distribution" for which he charges a retail margin. However in many other cases, the retailer also provides point of sale services which are valuable to the consumer. For example, authorized dealers (e.g., consumer electronics, durables, etc.) have counter salesmen who provide consumers with valuable product information. Retailers could also provide repackaging, product display, repair and maintenance and a host of other services.

The distinction between the market for intermediate goods and final goods is important for the purpose of this review because they may have very different implications for contractual relationships between agents. By and large, it is the case that the contractual issues between firms in an intermediate goods market (e.g., a manufacturer and retailer) are often more complex than those between a firm and its consumers in the final goods market. The complexities are due to important differences between these two markets and we highlight some of these distinctions below:

2. The usage of the terms "upstream" and "downstream" indicate the position of a firm in a successive chain ending with the final consumer of the good.

- Consumers often just consume goods but downstream firms (retailers) could transform an intermediate good by adding value (e.g., retailer service). Thus the downstream firm could make further decisions on retail price and promotional services on the intermediate good brought from the upstream firm. Typically, these decisions made by the downstream firm could affect the profitability of the upstream firm. However the downstream firms may disregard the effect their decisions might have on the profitability of the upstream firm and might thus exert an externality on the upstream firm.
- Typically, the demands of retail buyers of intermediate goods are related to each other. For example, in a market with multiple competing retailers, the demand for a product at any one retailer would depend on its actions as well as the actions of all other retailers.
- Contracts between self interested parties like retailers and manufacturers involve problems that arise because of actions that are taken by channel members after the contract is in place. For example, say a retailer contracts with a manufacturer to provide a certain level of service s which increases the demand for the manufacturer's good. The manufacturer offers the retailer a wholesale price w. Now, suppose that retail demand is uncertain and can be high or low depending upon a random parameter. If the actions of the retailer in terms of providing the contracted level of service are not fully observable by the manufacturer, then the manufacturer may not be able to clearly tell whether a high sales realization was because of the correct level of service or because of a high draw of the random variable. In such circumstances the retailer may have an incentive to provide less than the contracted level of service. Contracts in the intermediate goods market would typically have to account for these types of moral hazard problems.

These differences between the intermediate and final goods markets obviously have a bearing on the type of contracts which will be optimal in the intermediate goods market. We elaborate on these implications in the sections that follow. Clearly, there are instances where more complex pricing schemes may also be seen in the final goods market (e.g., retailers offer consumers discounts for buying larger quantities or rebates on future purchases). But, it is important to note that these pricing contracts arise more out of retailers wanting to provide consumers incentives to buy larger volumes and not because of any moral hazard or externality problems. Moreover, transactions between consumers and retailers (especially in the consumer goods industry) are rarely mediated by non-price contractual provisions. Any single transaction between a consumer and retailer usually involves the consumer simply paying the retail price for the title of the goods. In short, uniform price contracts are commonly the norm for final goods markets. As a consequence, this review will focus solely on contracting between manufacturers and retailers in the intermediate goods market.

Channel relationships involve problems of externalities and moral hazard. Since these problems affect manufacturer profits, the manufacturer has an incentive to use contractual mechanisms to make the retailer behave in a fashion which is optimal (for the manufacturer). Vertical contracts in the literature typically refer to the private incentives which a manufacturer might have to use mechanisms which would help control retailers who sell her product. Thus, vertical channel contracts are seen as instruments which help the manufacturer to (a) maximize channel profits, and, (b) to appropriate these profits. In this sense, most of the work done in this area implicitly assumes that all bargaining power (i.e., power to appropriate excess channel profits) is vested in the manufacturer. We recognize this but also discuss situations where upstream and downstream parties bargain to reach an agreement to trade the good.

1.1 Channel Contracts and Vertical Integration as the Benchmark

One can view the use of channel contracts as an intermediate case falling between the two extremes of simple uniform pricing and vertical integration. In comparing vertical restraints with vertical integration the existing literature usually views integration as giving one party (usually the upstream party) complete control of all channel decision variables. The vertically integrated profit then is the maximum aggregate (manufacturer plus retailer) profit that can be obtained from the vertical structure. This is why the integrated solution is a useful benchmark because in a decentralized situation it indicates which decisions a manufacturer would like the downstream retailer to make.

Given that channel contracts can be responses to problems of externalities, moral hazard, adverse selection or the need for risk-sharing, the natural question to ask is whether vertical integration solves these problems. Our view is that as long as there are self interested agents some degree of "incentive incompatibility" between them could persist in spite of their being labeled as vertically integrated. Grossman and Hart (1986) in fact take the more radical view that vertical integration does nothing to alter either the set of feasible contracts or even the objectives of the self-interested decision makers at the manufacturing and retailing level. Tirole (1988) also cautions on the misleading use of the term vertical integration. On the one extreme the upstream manufacturer who vertically integrates may still not be able to exert total control due to the existence of some decisions which must be delegated even under integration. In contrast, full control of the vertical structure can sometimes be achieved even in a non-integrated set-up through contracts which aligns the retailer's interest with that of the channel.

The rest of this paper is organized as follows. Section 2 presents the contractual problem in the basic setting. Section 3 elaborates on this to consider issues raised by competition in the retail market. Section 4 reviews the literature that considers competition in both the upstream and downstream market and highlights their implications. Section 6 concludes with a brief summary and a discussion of directions for future research.

2. THE BASIC SETTING: BILATERAL MONOPOLY

We start with the simplest possible model to develop the study of contracting between parties in a distribution channel. There is a single upstream provider of some product which may be either a service or a good. We will refer to the upstream entity as a "manufacturer." The manufacturer must use the services of a downstream retailer to reach the final consumers.

Assume that the retailer's problem is limited to the choice of retail price, p. The manufacturer's problem is choice of a wholesale price, w. Let c denote the marginal cost of production of the good. If $D(p)$ denotes the demand function, the retailer's problem can be represented as

$$\max_{p} \Pi^{R} = (p - w)D(p). \tag{1}$$

The retailer's optimal retail price, $p^\star$, obtains from the first-order condition for this problem, i.e.,

$$(p^\star - w) = \frac{-D(p^\star)}{D'(p^\star)}. \tag{2}$$

If we assume that retail demand is specified as a linear demand function which depends only on primary demand and retail-level factors that influence consumers' retail price elasticity, then $D(p) = \alpha - \beta p$ where $\alpha > 0$ represents primary demand and $\beta > 0$ captures retail-level factors. Then, it follows that the optimal retail price is

$$p^\star = \frac{\alpha + \beta w}{2\beta}. \tag{3}$$

Given this, the manufacturer's problem can now be represented as the choice of a wholesale price, w that maximizes

$$\max_{w} \Pi^{M} = (w - c)D(p^\star(w)). \tag{4}$$

The manufacturer's optimal wholesale price, $w^\star$ obtains from the first-order condition for this problem, i.e.,

$$w^\star = \frac{\alpha + \beta c}{2\beta}. \tag{5}$$

Plugging these back into the profit functions of the two agents implies that $\Pi^{R} = \frac{(\alpha - \beta c)^2}{16\beta}$ and $\Pi^{M} = \frac{(\alpha - \beta c)^2}{8\beta}$. Let Π^{NI} denote the total channel profits for this bilateral monopoly. Therefore,

$$\Pi^{NI} = \Pi^{R} + \Pi^{M} = \frac{3(\alpha - \beta c)^2}{16\beta}. \tag{6}$$

Now, consider the case where the channel is integrated. In other words, a monopolist who faces a marginal cost, c and maximizes $(p - c)(\alpha - \beta p)$ through choice of a retail price. It follows that the optimal price for the integrated channel is

$$p^{\star} = \frac{\alpha + \beta c}{2\beta}. \tag{7}$$

and the profits for the integrated channel, Π^{I}, is

$$\Pi^{I} = \frac{(\alpha - \beta c)^2}{4\beta} > \Pi^{NI}. \tag{8}$$

Profits are higher in the integrated channel. If the manufacturer uses a simple uniform wholesale price, then to earn a positive profit the manufacturer must set $w > c$. This forces the retailer to face a marginal cost of w. The retailer when faced with a wholesale price $w > c$ will maximize his profits by choosing a retail price $p(w)$ which will obviously be higher than the monopoly price which he would have set had the retailer faced a wholesale price of c. This higher price will result in a smaller level of demand. The reduction in demand affects manufacturer profits through the $(w - c)$ margin. It is also clear that due to this distortion in the retailer's incentive, the aggregate channel profit is also less than the vertically integrated one. The vertically integrated case is the one where there is just one party controlling the manufacturing and the retailing functions. Such a party will face no incentive distortions since it sets the monopoly price based on the marginal cost c. Thus, the correct monopoly price which is based upon the marginal cost c will be set resulting in the correct demand which maximizes channel profit.

The above vertical price externality problem was first studied by Spengler (1950). Spengler called this problem "double marginalization" in the context of a chain of monopolies (in our case the manufacturer-retailer chain). In the presence of such a chain, the retail price is higher in a decentralized structure than in an integrated one because of the two successive mark-ups or marginalizations (one by the manufacturer and the other by the retailer). The retailer while choosing her price does not take into account the effect this would have on manufacturer profits thereby generating an externality.

How can this vertical price externality or double marginalization problem be solved? To understand the solutions proposed in the literature to the above-mentioned problem and also the other types of channel contractual problems which will be encountered in this review it is necessary to define a few relevant concepts (see also Tirole 1988). Usually observable and verifiable channel variables are called instruments. Next the sum of the profits at each level of the vertical structure is defined as the aggregate profit. The subset of channel decision variables which directly affect aggregate profit are called targets (e.g., retail price). The channel control problem then consists of knowing how to use the instruments to achieve the desired values of the targets which then helps one to maximize the vertical structure's aggregate channel profit. The maximized profit is nothing else but the vertically integrated profit. Mathewson and Winter (1984) define a set of instruments to be sufficient if they enable the attainment of the vertically integrated profit.

It is useful to define some of the basic contractual arrangements that are relevant for the bilateral channel prior to understanding the efficacy of these arrangements in different settings and circumstances. Note that the following typology applies to settings wherein prices and/or quantities are the instruments for contracting between the manufacturer and retailer. We will broaden the space of contracting instruments as we progress through the paper.

- The simplest contract is a *linear price* "spot" contract that specifies that the retailer pays a wholesale price w for each unit of the good ordered from the manufacturer. So, if the retailer orders q, then he repatriates a payment pq back to the manufacturer. This will be termed as equivalent to the decentralized channel outcome in this paper.
- A *two-part tariff* is the simplest example of a non-linear price contract. Here, the retailer pays a fixed fee F for buying any amount of the good plus a variable charge w per unit of the good. So, if the retailer orders q, then he pays $F + wq$ back to the manufacturer. Note that there is no a-priori restriction on F. An instance where $F > 0$ is when the retailer pays a franchise fee to the franchisor. An example of $F < 0$ is where the manufacturer pays the retailer a slotting allowance for carrying her good.
- *Resale price maintenance (RPM)* refers to provisions in contracts wherein the choice of the retail price, p is specified to the retailer. Variants of RPM include impositions of a price ceiling (i.e., $p \leq \bar{p}$), or a price floor (i.e., $p \geq \underline{p}$).
- *Quantity fixing* specifies the amount of stock that can be obtained by the retailer from the manufacturer. Variants of this include *quantity forcing* contracts where the retailer is required to lift stocks above a minimum quantity (i.e., $\underline{q}$), and *quantity rationing* where $q \leq \bar{q}$.

2.1 Coordinating the Benchmark Setting

The double marginalization problem in the benchmark setting can be mitigated in a variety of ways. A two-part tariff is sufficient. The manufacturer by setting the wholesale price at marginal cost (i.e., $w = c$) removes the downstream distortion and thus provides full incentive for the retailer to choose the correct price (i.e., the monopoly price $p = \frac{\alpha + \beta c}{2\beta}$). In addition, she appropriates the retailer's profit by setting the franchise fee F equal to the retailer's profit (i.e., $F = \frac{(\alpha - \beta c)^2}{4\beta}$). The intuition is that setting the wholesale price at the marginal cost makes the retailer the residual claimant of the aggregate profit. Since there is no distortion to the marginal cost (i.e., retailer faces the true marginal cost), he takes the correct decision with regard to the retail price and channel profits are maximized. The franchise fee enables full extraction of retailer profits. Thus, the franchise fee and the wholesale price are sufficient instruments to coordinate the channel in the benchmark setting.

An alternative sufficient restraint in the face of a vertical price externality is RPM (resale price maintenance). The manufacturer can impose a retail price ceiling,

$p = \frac{\alpha + \beta c}{2\beta}$, the correct downstream monopoly price. The manufacturer can then set the wholesale price so as to extract all the excess profits generated at the retail level. The price restraint removes the possibility of distortion in the retail price when the manufacturer sets the wholesale price $w > c$. It is important to note the correspondence between the number of instruments used in the above-mentioned channel control problem and the number of functions which needs to performed. The two functions relevant to the above-mentioned problem are obvious (i.e., the provision of incentives to the retailer to set the correct retail price and the transfer of all excess profits to the manufacturer). With the two-part tariff the variable fee w can be set equal to c to provide the correct incentives and the fixed franchise fee F can be used to effect the transfer. However as will be seen later on in this paper with more complicated control problems, this correspondence between the number of instruments and the number of functions need not always hold.

Jeuland and Shugan (1983) suggest another possible means for obtaining the coordinated solution. Their approach relies on the manufacturer and retailer agreeing to share the channel profits. If we restrict transfers to simple linear functions then the manufacturer in their scheme received $k_1(p - c)D(p) + k_2$ and the retailer received $(1 - k_1)(p - c)D(p) - k_2$ where k_1 was the fraction of channel profits received by the manufacturer and k_2 was the fixed fee paid by the retailer to the manufacturer. They argue this structure is analogous to a quantity discount schedule. A linear pricing scheme implies that the retailer needs to decrease the retail price (and hence, retail margin) to sell additional units of the product. A quantity-discount scheme in effect drops the wholesale price for additional unit of the product and can thereby subsidize the drop in retail margin needed to sell the additional unit.

2.2 Moral Hazard: The Case of Retailer Promotional Effort

We now broaden the benchmark setting to consider situations wherein retailer investments in promotional effort (e.g., point-of-sale services, merchandising, etc.) influence the retail demand in the final goods market in addition to the retail price. The reason why retailers invariably provide these services in most markets is because the manufacturer finds decentralization more efficient for such services. Since these services increase demand, a positive externality is bestowed by retail services on the manufacturer and so the manufacturer has a private incentive to ensure that the retailer provides the correct level of services. However, the manufacturer cannot ensure this by simply writing the optimal level of services into the contract since the provision of services is generally not verifiable. This is a typical example of a moral hazard problem since the agent (retailer) can take an action which is not verifiable by the principal (manufacturer) but which affects the principal's profit. Let $D(p, s)$ denote the downstream demand when the retailer invests in effort, s. The demand is concave in effort and let the cost to the retailer of providing this effort be $\phi(s)$ which is convex in s. The vertically integrated channel would set p^m and s^m such that they maximize $\Pi^I = ((p - c - \phi(s))D(p, s)$. If the manufacturer charges a wholesale price w, then $\Pi^R = (p - w - \phi(s))D(p, s)$ and $\Pi^M = (w - c)D(p, s)$.

Consider the situation where the manufacturer uses a linear price contract. This implies that the optimal wholesale price $w^\star > c$. Since this implies that the retailer's margin, $p - w$ is strictly less than that of the integrated channel (i.e., $p - c$), the negative externality affects the retailer's choice of both the retail price and retail service. The retailer will charge a retail price higher than p^m and provide retail service lower than s^m. To ensure that the retailer makes the correct choice of p and s, the manufacturer can use a two-part tariff and set $w = c$. The transfer of excess profit can then be achieved using the fixed fee. However unlike section 2.1, RPM is no longer sufficient in this model. Using RPM to set $p = p^\star$ only solves the price externality while the service externality remains. Interestingly, quantity forcing turns out to be a sufficient instrument. The intuition behind this result is that once the quantity is fixed, the retailer is forced to provide the correct level of service (s^m) or else the required quantity will not be attained at the monopoly price.[3]

2.3 Bilateral Promotional Effort

Holmstrom (1982) provides us with a model where both parties involved can take unverifiable/unobservable actions which affects the total profits as also the profits of individual parties. This situation is usually termed as double moral hazard. In the context of this paper, double moral hazard arises when in addition to the retailer providing services which are not verifiable by the manufacturer, the manufacturer provides services, A, which increase demand but which are not easily verifiable by the retailer. Such services, for instance, can be product advertising or product quality. In such cases the manufacturer's choice of A will also impose an externality on the retailer. In a decentralized set up since no party gets the full benefit of their personal investments the correct level of investments will not be made by either party. In this case again a simple two-part tariff is no longer sufficient. If the manufacturer sets the variable fee equal to c the retailer no doubt becomes the residual claimant and gets the incentive set the correct level of s but since the manufacturer makes a zero profit margin he has no incentive to provide the correct level of A.

Holmstrom (1982) shows that this double moral hazard problem can be solved by simultaneously making both the parties residual claimants through the use of a third party. Thus, if the retailer in a decentralized set-up chooses p and s and the manufacturer A, then both of them can jointly contract with a third party in such a way that she (the third party) does not make any excess profit and both the manufacturer and the retailer get the correct incentives. A problem with the above solution needs to be pointed out. Post-signing the contract with the third party, there is always an incentive for the manufacturer and retailer to collude and produce a greater output. This leads to a negative profit for the third party. In channel structures with multiple retailers, the "third party" may be interpreted as an intermediary between the manufacturer and retailer, namely, a wholesaler/distributor. However such an interpretation is possible only if there exist credible mechanisms which guarantee that there will be

3. Of course, this assumes that the demand function is such that there is a unique combination of p and s that is optimal.

no collusion between the manufacturer and retailer once the contract is signed with the wholesaler.

Lal (1990) examines this bilateral effort problem in the context of franchising. Traditionally, franchising contracts have two components, a fixed initial payment paid by the franchisee to the franchisor for acquiring the franchise and a royalty payment specified as a percentage of retail sales. Recall, that a two part-tariff coordinates the channel in section 2.2 (i.e., in the case of franchisee moral hazard). This raises the question as to the role of a royalty payment in the franchising context. Lal expands the setting of section 2.2 to consider a franchisor using two franchisees to market its product to the consumer. Franchisee effort is productive but is costly to the franchisees. He shows that if costs of service provision is sufficiently large, the franchisees prefer to free ride on the other franchisee's service and in equilibrium neither franchisee invests in service. One way to mitigate the free riding problem is through monitoring by the franchisor such that if the desired level of service is not provided, then the franchisee pays a penalty. Since, effort is costly to the franchisee and monitoring is costly to the franchisor, the equilibrium of this game is a mixed strategy. The franchisor monitors each of the franchisee with some probability and franchisees invest in effort with some probability as well. The point to be noted is that even in this setting there is no role for a royalty payment from the franchisee. The rationale for a royalty payment emerges when one allows for the possibility that both the franchisor and the franchisee contribute costly inputs that positively impact the demand for the product. Now the royalty payments serves as an incentive to both agents to contribute non-zero levels of effort. Too small a royalty payment provides little incentive for the franchisor to invest in brand development efforts. Too large a royalty payment provides little incentive for the franchisee to invest in service. However, intermediate levels of royalty payments provide incentives to both agents to invest in effort.[4]

2.4 Uncertain Demand

The preceding sections derived implications for contractual relationships assuming that there was no uncertainty with regard to the parameters of the problem. The most common form of uncertainty in a channel is demand uncertainty. The simplest setting to examine the impact of demand uncertainty is the benchmark setting with a single manufacturer and a single retailer. If the retailer is risk-neutral then the intuition about making the retailer the residual claimant goes through in this setting as well. So, the manufacturer in this instance transfers the product to the retailer at a price equal to marginal cost (i.e., $w = c$) and appropriates the profits through the appropriate fran-

4. This intuition for the incentive effect of contractual terms is observed in a variety of contexts. For instance, Priest (1971), Cooper and Ross (1985) and Padmanabhan and Rao (1993) illustrate how product warranties can serve as a mechanism to induce the manufacturer to invest in higher levels of product quality and the consumer to invest in higher levels of product care and maintenance. For other papers on franchising contracts see Caves and Murphy (1976), Rubin (1978), Mathewson and Winter (1985) and Rao and Srinivasan (1995).

chise fee (i.e., $F = \max_p \Pi^R = E[(p - c)D(p)]$). It is important to note that RPM is not a good solution in this setting because it destroys the retailer's ability to adjust retail prices to demand conditions. However, a quantity-forcing contract can work since the manufacturer can require the retailer to stock the quantity $q^\star$ that equals the integrated channel's order quantity under uncertain demand.

The marketing and operations management literatures have considered a variety of instruments that could be used to coordinate channels in the context of uncertain demand. We review this stream in the following paragraphs.[5] The take-off point for these literatures is the coordination problems created by linear price contracts in the context of uncertain demand. A fixed price contract implies that the retailer will face the possibility of being either over-stocked or under-stocked based on actual demand realization. What type of contracts can help in hedging this inventory problem?

Returns policies are common in the distribution of a variety of products (e.g., books, music, clothing, etc.). Manufacturer's in these industries will accept unsold merchandise from the retailer at the end of the selling season. Pasternack (1985) is one of the earliest attempts at highlighting the role of returns policies in channel coordination. He utilized the classical single period inventory model (*newsboy problem*). The newsboy problem has a long history (e.g., Arrow, Karlin and Scarf 1958) and has been widely used to study inventory management problems.[6] Applications of this model to marketing include Goodman and Moody (1970) in the context of promotions management and Pasternack (1980) in the context of pricing. To incorporate returns policies in the benchmark setting, we now assume that the manufacturer in addition to posting a wholesale price, w, announces that any unsold stock at the retailer will be purchased back at a price, r, per unit (where $r \leq w$).[7] The retailer observing this policy orders a quantity, y. Pasternack (1985) restricts the retailer's decision making to the choice of order quantity alone. The retail price of the good is assumed exogenous and fixed at p independent of the terms of trade offered by the manufacturer.[8]

Let ξ denote the realized demand (where $F(\xi)$ represents the distribution of ξ on the non-negative real line with density $f(\xi)$). If realized demand ξ is less than the order quantity, y, then the retailer receives $r(y - \xi)$. Assume that the manufacturer can salvage the unsold goods at v per unit while the retailer can salvage at an amount less than v. With these assumptions, the retailer's optimization problem is

5. A detailed review of the operations literature is beyond the scope of this article. However, there are several excellent reviews of this literature. See for example, Cachon (2002), Cachon and Lariviere (2002) and Tsay and Agarwal (2002).
6. The interested reader is referred to Silver, Pyke and Peterson (1998) or Nahmias (1993) for detailed treatments of the newsboy problem.
7. Restricting the credit for unsold merchandise to be less than the wholesale price eliminates arbitrage opportunities for the retailer.
8. This is largely typical of the operations literature as well. See Cachon (2002) for a review. Emmons and Gilbert (1998) and Petruzzi and Dada (1999) are instances where the retailer's pricing decision is considered endogenous in the newsboy context.

$$\max_{y} \Pi^R = -wy + \int_0^y [px + r(y - x)] f(\xi)\, d\xi + \int_y^\infty [py] f(\xi)\, d\xi, \tag{9}$$

which simplifies to

$$\max_{y} \Pi^R = (p - w)y - (p - r)\int_0^y F(\xi)\, d\xi. \tag{10}$$

The optimal solution is

$$F(y^\star(w, r)) = \frac{(p - w)}{(p - r)}. \tag{11}$$

In the case of a linear price contract, the manufacturer's profit was deterministic since all the uncertainty was transferred to the retailer. However, with a returns policy the manufacturer is exposed to the demand uncertainty. The manufacturer's optimization problem is given by

$$\max_{w,\, r} \Pi^M = (w - c)y(w, r) - (r - y)\int_0^{y(w,r)} F(\xi)\, d\xi. \tag{12}$$

As shown in Lariviere (1999), the manufacturer's objective function is not well-behaved. In particular, the second order conditions for a local maximum are never satisfied. Pasternack (1985) shows that even though it is not possible to use the standard approach to obtain the optimal contract, channel coordination can be obtained through an appropriate structuring of returns policies. Prior to detailing Pasternack's result, it is useful to note that the integrated channel's profit maximization problem can be represented as

$$\max_{y\, I} \Pi^I = (p - c)y - (p - v)\int_0^y F(\xi)\, d\xi. \tag{13}$$

hence, the optimal stocking quantity for the integrated channel is

$$F(y_I^\star) = \frac{(p - c)}{(p - v)}. \tag{14}$$

Pasternack (1985) shows that the policy of a manufacturer accepting full returns for full credit or accepting no returns whatsoever generates a suboptimal outcome. The intuition is that any wholesale price above marginal costs leads to double marginalization and hence the retailer faces a higher marginal cost relative to the integrated channel. This coupled with demand uncertainty implies the retailer will order a quantity lower than the integrated channel. The only way for the manufacturer to raise the retailer's profit in the low demand states is to subsidize the unsold inventory. This leads to the central result in his paper that a policy that allows for unlimited returns at partial credit will obtain channel coordination. For instance, the manufacturer can obtain the coordinated outcome by offering a contract $w(\epsilon, b\epsilon)$ for $\epsilon \in (0, p - c]$ where $w(\epsilon) = p - \epsilon$ and $r(\epsilon) = p - \frac{\epsilon\,(p - v)}{p - c}$. Note that the coordinating contract is not unique. Since the system profits are fixed, $\Pi^R = \frac{\epsilon}{p - c}\Pi^I$ and $\Pi^M = (1 - \frac{\epsilon}{p - c})\Pi^I$. Clearly, the manufacturer would prefer to set ϵ as close to zero as possible and appropriate all the channel profits. However, setting $\epsilon = 0$ is sub-optimal. One interesting aspect of Pasternack's analysis is that the channel coordinating contract does not depend on the demand distribution. As Lariviere (1999) notes, this does not mean that the manufacturer need not know about the demand distribution. Note that Π^I depends on the demand distribution. Another important observation is that the returns policy obtains channel coordination in a setting where both agents are risk-neutral. In other words, the rationale for returns is not based on any insurance arguments. Kandel (1996) expands the basic setting of Pasternack (1985) to also have the retail price affect market demand. He shows that channel coordination through a returns policy cannot be obtained if the demand distribution depends on the retail price.

2.5 Learning about Demand Realization and Coordinating Contracts

We now consider situations where one or both of the agents obtain some additional information that allows a better estimate about the realized demand. For instance, the retailer being closer to consumer and market may learn about the demand trends for the product earlier than the manufacturer. This allows them to update their demand forecasts and adjust their decision-making accordingly. The question is how can the contractual mechanisms between the agents in the channel leverage this new learning?

Donohue (2000) obtains the coordinating contracts that encourage the coordination of upstream production and downstream demand learning when the manufacturer operates in a dual mode production environment. The first mode of production is cheap but requires a long lead time while the second mode is expensive but offers a quick turnaround.[9] She shows that a contract of the form w_1, w_2, r) where w_i is the wholesale price for production mode i and r is the price for returned merchandise will coordinate the channel.

Iyer and Padmanabhan (2002) examine a setting wherein the uncertainty about demand for a new product unfolds over time. Learning about demand for a product

9. Fashion apparel, book publishing, etc, are industries that exhibit these characteristics. For instance, the first production of a run is cheaper since they are manufactured in long runs due to high fixed costs. However, there is considerable demand uncertainty about the title since the books are produced long before demand realization. A second production run to eliminate inventory underage is possible but involves higher production costs.

can happen at very different instances in product's lifecycle. For instance, a manufacturer can learn about demand potential for a product very early in it's lifecycle or relatively late in it's lifecycle. How should its channel policies respond to what it learns about demand potential as well as when it learns about demand potential?[10] The available evidence suggests that manufacturers use different trading mechanisms to respond to uncertain demand environments. Some manufacturer's contract on prices and quantities with their downstream resellers at a single point in time before product introduction and allows no further modifications on the terms-of-trade. Therefore, this type of rigid contractual arrangement has the flavor of an arms-length spot market relationship. In contrast, others are much more flexible in their contracting policies: they contract on prices and quantities at the start of the selling season and modify these terms over time with changes in the demand conditions.

A key result of the paper is that flexibility is not necessarily optimal from a manufacturer's perspective even if it is costless to acquire. Rigid contracts can, in fact, be beneficial in uncertain demand environments. The basic trade-off in using a flexible as opposed to rigid contract is as follows: The flexibility of a pay-as-you-go contract allows the manufacturer to adjust the terms of trade to the realization of demand. But it also comes with an associated cost for the manufacturer: In the initial period when the demand is uncertain to the parties, the flexible contract makes the retailer less aggressive in that the retailer's stocking and pricing decisions under uncertainty are influenced primarily by the expected demand. In contrast, a rigid contract that is fixed ex-ante makes the retailer more aggressive in terms of pricing and stocking to take advantage of the upside potential for the product. But the disadvantage of this contract is that the retailer might under or over stock the product. Given this, the relative optimality of flexible versus rigid contracts depends on when in a product's lifecycle does the manufacturer learn about a product's demand potential. When learning is relatively late in a product's lifecycle, but the upside potential is high, the manufacturer benefits from using a rigid contract. In contrast, flexible contracts are ideal when early demand signals are informative for this allows the opportunity to leverage this learning over the remainder of a product's lifecycle.

2.6 Information Asymmetry

We next consider settings where demand is uncertain, but when one of the parties in the channel has better private information about demand. First, consider the situation where the private knowledge resides with the manufacturer. Private knowledge in and of itself does not create problems for contracting. The problems for contracting arise because it may not be possible for the manufacturer to credibly reveal this information to the retailer. The private knowledge of demand for the manufacturer may

10. For instance, Hewlett-Packard learns about the ultimate demand potential for both personal computers and printers at about the same time (roughly speaking a few weeks after retail launch—HP managers believe that this is the point at which POS data becomes reliable). However, a few weeks is a long time in the personal computer's lifecycle whereas it is a relatively short time in a printer's lifecycle. Based upon HP's internal sales data, the average lifecycle for personal computers is 22 weeks and that for printers is about 18-22 months.

arise because of the market research they conduct in the process of new product development (e.g., concept testing or test marketing). The results of these investigations could be shared with the retailer. However, retailers generally are suspicious of this data because they believe that these data are partial to the manufacturer.[11]

Chu (1992) examines the implications of private information about demand in two different settings. In the first setting, he assumes that the informed manufacturer pre-commits to a level of pre-launch advertising expenditures and a wholesale price. Retail demand in this setting is specified as $Q_i = \alpha - \beta_i p + f(A)$, $i \in H, L$ where p is retail price, A is advertising expenditure and β_i which represents price sensitivity is private information.[12] The retailer decides whether to accept the offer or not. Conditional on acceptance, the retailer then sets the retail price for the product. It is assumed that the true demand is revealed to the retailer upon accepting the manufacturer's offer and prior to setting of the retail price. He shows that manufacturer with high-demand will signal this private information by over-advertising and over-pricing (relative to the complete information setting). These distortions while being expensive for the high-type manufacturer ensure that the low-type manufacturer will not mimic the strategy and the retailer's opportunity cost, F, for participation are satisfied. Chu then alters the order of play in the game by assuming that the retailer has the power to propose a contract which requires the manufacturer to pay a lump-sum fee (called a slotting allowance), S. Manufacturer accepts of rejects this offer. In the case of acceptance, the manufacturer then sets wholesale price and advertising and retailer then sets the retail price. Information about β_i is still private to the manufacturer. He shows that the slotting allowance, S, serves as a useful tool for the retailer to screen across the different possible types. In particular, by setting S equal to the high type manufacturer's full information profits the retailer ensures that only the high type accepts this contract. The slotting allowance therefore serves to screen across manufacturers and allow the retailer to effectively appropriate channel surplus. Comparing the signaling versus screening scenarios, he shows that the total channel profits through screening are higher if the effectiveness of advertising is sufficiently low. The intuition being that low advertising effectiveness results in a signaling equilibrium that requires raising wholesale price and advertising to levels that create a large loss in channel profits. The benefits of a slotting allowance is that all the money is kept in the channel. Chu and Messinger (1997) is similar. The retailer screens products by moving first and committing to auction off shelf space for n products. The makers of the n most profitable products bid allowances equal to producer profits from the $(n + 1)^{st}$ product.

Lariviere and Padmanabhan (1997) investigate the optimality of slotting allowance in a similar setting. The manufacturer has private information about demand that he would like to communicate to the retailer. The difference from the earlier approaches

11. For instance, manufacturer may only reveal positive results. The site selection for tests may be biased. Product pitches are always favorable but the failure rates for new products are very high. See Chu (1992) for more on this. See also Desai and Srinivasan (1995) for a model of signaling by an upstream party in a franchising channel.

12. A lower-price sensitivity translates to higher demand realization. Hence, the high-type manufacturer will have $\beta_H < \beta_L$.

is that the manufacturer is the Stackelberg leader and proposes the terms of trade. Additionally, retail demand is specified as $Q_i = \alpha_i - \beta p + f(e)$, $i \in H, L$ where p is retail price, e is retailer effort in merchandising the product and α_i which denotes demand potential for the new product is private information. They show that variables used for signaling by the manufacturer depends critically on the retailer's opportunity cost. If the retailer's opportunity costs are low, the manufacturer can credibly signal high demand through the wholesale price alone. A high wholesale price is differentially more expensive for a manufacturer endowed with a lower demand product. The high demand manufacturer therefore increases the wholesale price to the point at which the low demand manufacturer's sales fall to zero. A slotting allowance is not required because the retailer's opportunity costs are sufficiently low that the retailer can recover them despite the high wholesale price. This contract is no longer feasible when the retailer's opportunity costs are sufficiently high. The retailer would refuse the product even if he were certain of its demand because the prohibitive wholesale price does not allow him to recoup his costs. The high demand manufacturer must consequently alter the terms of trade. In equilibrium, she reduces the wholesale price but not enough for the retailer to recoup his costs fully. The shortfall is made up by a slotting allowance. The wholesale price reduction makes entering the market attractive for the low demand manufacturer unless the slotting allowance dissipates her potential earnings. In equilibrium, it does this exactly. The slotting allowance thus plays a dual role; it allows the high demand manufacturer to signal her information credibly and simultaneously assume part of the retailer's cost of stocking a new product.

2.7 Bargaining

In the literature discussed above, the implicit assumption is that the manufacturer has the power to make take-it-or-leave-it offers to the retailer. With the rise of the large and powerful retailers such as Toys-R-Us and Wal-Mart there has been a continuing academic debate on how the changing configurations of channel power has affected the management and the general functioning of distribution channels. Iyer and Villas-Boas (2003) is an attempt to understand the relative power in a distribution channel in a general and theoretically well-founded bargaining game. The overall logic of this paper stems from the fact that many distribution systems face problems of product non-specifiability: i.e., the parties involved cannot fully specify the product to be exchanged in a contract. Because of this, channel members can be opportunistic which has an impact on channel coordination. Opportunism in a vertical relationship is modeled in this paper through the possibility of renegotiation of an initial ex-ante contract. This captures the idea that a powerful party might renege on an initial agreement, even after the product is delivered, and demand extra payment. It is the presence of such opportunism that allows the bargaining process to have an impact on the market decisions (such as setting the retail price).

Iyer and Villas-Boas consider bargaining in a distribution channel consisting of a manufacturer who produces the product which cannot be fully specified in a contract and a retail intermediary who takes a market action (e.g., setting the retail price) and sells the product to the consumer market. The retailer's action (price) is unobservable and the manufacturer cannot fix it in a contract. Specifically, the coordination problems investigated in Iyer and Villas-Boas are predicated on the following factors: product non-specifiability, demand uncertainty and unobservability of retail behavior. In the standard case discussed in the literature when the product is specifiable, it is well known that a two-part tariff can coordinate the channel and maximize total channel profits (see Moorthy 1988). This is the case even when demand uncertainty and unobservability of retail behavior are present in the channel. However, they show that (if any renegotiation is costless) the presence of non-specifiability of the product can lead to the two-part tariff not being an equilibrium contract even in the simplest possible channel structure involving one manufacturer and one retailer. This is because the fixed fee in the two-part tariff does not affect the opportunistic behavior on the part of the manufacturer and will therefore not be accepted by the retailer. Rather, bargaining takes place on a simple wholesale price and it affects the market outcome (i.e., retail prices). Thus, trading on a simple wholesale price, and not on the more complex two-part tariff, is an equilibrium outcome. This, in a broad sense, ends up legitimizing the stream of distribution channels research which is based upon the simple wholesale price (double marginalization) approach.

Next the relative bargaining power impacts upon channel coordination. Interestingly, the paper finds that greater relative power of the retailer coordinates the channel in markets where the retailer effort is important. Thus, the presence of a powerful retailer might, in fact, be beneficial for the channel as a whole (and in some cases, beneficial to all channel members). On the other hand, "excess" manufacturer power can increase double marginalization in the channel and reduce coordination. Iyer and Villas-Boas show that a manufacturer is not always better off when its relative power increases.[13] If the level of manufacturer power is already high, any further increase ends up hurting the manufacturer because it drives the retailer to charge too high prices and shrink the channel pie. This result provides a contrast to the idea in the behavioral and descriptive literature that greater power always means greater benefits for a channel member. From an ex-ante and strategic perspective high levels of power can actually hurt the manufacturer.

3. COMPETITION IN THE DOWNSTREAM MARKET

We have studied in detail the contractual problems which arise between a single manufacturer and a single retailer. We now move on to the case of multiple retailers and retail competition. A new type of externality crops up when several retailers are involved, i.e., externalities across retailers and this section is devoted to the study of such problems. Probably the best way to motivate the study of a multiple retailer model is to examine an important paper by Mathewson and Winter (1984). Mathewson and Winter develop a model of spatial retail competition. Each retailer has her own geographical area and some degree of local monopoly power. Retailers inform consumers about the product by advertising locally and a consumer must be informed

13. See also Van Meighem (1999) for a somewhat similar point made in the context of sub-contracting.

of the product's existence before buying. The proportion of consumers informed is a function of the retailer's advertising spending A. It is assumed that manufacturer's cannot take over the local advertising function. There is also a spillover of each retailer's advertising message into neighboring areas which are outside the advertising outlet's market area. Mathewson and Winter look for the minimum sufficient set of restraints which will yield the same efficient outcome that maximizes total channel profit. In the presence of retail competition, the paper identifies 3 basic types of externalities: i) the vertical externality or the double marginalization effect which was present even for the one-manufacturer-one-retailer case. ii) a "horizontal non-pecuniary" externality which occurs due to "spillover" of advertising messages, i.e., a proportion of advertising messages from one outlet increases the demand at other outlets. This resulting problem is also called the "free-riding" problem and iii) a "horizontal pecuniary" externality, i.e., an increase in the price at one outlet increases the demand at other outlets through positive cross-elasticity of demand.

It should be evident that the last two externalities mentioned above are due to the presence of multiple retailers. In the Mathewson and Winter model the result of the vertical externality is that the retail price p is set too high and the level of advertising A is set too low. The horizontal non-pecuniary externality leads to too low an A being provided. This is because with spillovers retailers do not appropriate the full benefit of their advertising. Finally, the horizontal pecuniary externality results in retail price being too low. The net effect in this model therefore is that A is too low and the effect on retail price p is indeterminate.

Let us now examine the results of the model. If there are no spillovers and retailers operate under Nash conjectures, then a franchise fee plus exclusive territories is a sufficient set of instruments. This is because under exclusive territory (ET) the horizontal pecuniary externality is eliminated. The manufacturer can then set $w = c$ to eliminate the vertical externality and extract all excess profits at the retail level using 'e' the franchise fee. Exclusive territories also prevents "hold up" problems: once the manufacturer extracts profits through a franchise fee, the manufacturer may have an incentive to franchise another retailer nearby. Thus, if the retailer is not protected by an ET right she is always susceptible to a "hold up" since the franchise fee is paid up front. Another alternative method of solving the horizontal free-riding externality is to use RPM (a price floor). The manufacturer can set $w = c$ which takes care of the vertical externality. Then, the retail price can be fixed at $p = p^\star$ to take care of the horizontal pecuniary externality. Next, if there are advertising spillovers then setting $w = c$ and using an RPM $p = p^\star$ no longer works because even with the above the horizontal externality remains and too little advertising will be provided. Mathewson and Winter suggest that with spillovers the manufacturer should set $w < c$ and increase the retail margin so as to provide an incentive for each retailer to provide A. The excess retail profits can always be extracted using the fixed fee F.

The next important issue given multiple retailers is the phenomenon of retail differentiation. Differentiation among retailers in price and non-price service factors is a central feature of markets ranging from automobiles and appliances to gasoline and is especially observed in the co-existence of high-service retailers and lower-price discount retailers. Therefore, how a manufacturer should manage retail differentiation is an important channel management question. Yet, the approach in the existing literature after Mathewson and Winter on retail competition has been to examine coordination under the implicit (symmetric) contracting assumption that offering uniform contracts to all the retailers in the market is always optimal (for example, Perry and Porter 1985 and Winter 1993). A paper by Iyer (1998) brings this assumption into question and asks when (if ever) is it optimal for the manufacturer to use the channel contract to induce retail differentiation (even if the retailers were ex-ante identical in their cost and other characteristics). The paper identifies the type of channel contracts that can endogenously induce symmetry as opposed to differentiation among retailers. Thus retail differentiation in price and service is itself endogenous to the coordination of the channel in that some market conditions will induce differentiation, while others will induce symmetry in retailer actions.

To identify these conditions, Iyer (1998) develops a framework that highlights the role of two basic types of consumer heterogeneity in retail markets. Consumers are heterogeneous in their locations (as in the spatial models of horizontal differentiation) and in their valuations for retail services (as in the models of vertical differentiation). The paper shows that offering uniform contracts to all the retailers is sufficient only in markets with substantial locational differentiation relative to the differences in the valuation for service. Channel coordination in these markets simply requires mechanisms which ensure that retailer interests are aligned so that they compete by offering a mix of price and service that is desirable from the manufacturer's point of view. However, in markets with small locational differentiation and substantial diversity in consumer valuations, the manufacturer's problem is not just to align retailer interests, but to also use the channel contract to endogenously induce the optimal level of retail differentiation in price and service. This helps the manufacturer to better cater to the diversity in consumer willingness to pay and to prevent the excessive competition that the retailers would otherwise have indulged in. In this paper, this is achieved through the use of a menu of contracts as in the theory of mechanism design. Menu-based contracts induce differentiated retailer behavior despite the fact that the retailers are not "forced" into accepting different terms of trade. This aspect can be useful in shielding manufacturers from litigation under the Robinson-Patman act.[14]

Next Iyer (1998) shows that for relatively high-ticket items retailers tend to be excessively biased towards competing in the provision of retail services. The correlation between consumer willingness to pay for service and travel costs implies that for high-ticket products, the competing retailers will focus on the more service sensitive customers at the cost of ignoring the price sensitive consumers in the market. The

14. One can think of the asymmetric equilibrium of this paper as one in which consumers in the market are screened by the manufacturer by two different service-quality offerings made through the use of two retailers. Villas-Boas (1998) examines the related problem of a multi-product manufacturer who offers two products of different qualities to the consumer market but through a single retailer. This paper shows that accounting for the retailer makes the manufacturer increase the differences in the qualities of the products. Another related problem is one where the asymmetry in the retail market is exogenously specified as in dual distribution models (See for example Bell et. al (2002) and Coughlan and Soberman (2002).

manufacturer is therefore likely to encourage greater price competition among the retailers. In contrast, for low-ticket items the manufacturer prefers to reduce price competition and encourage greater provision of services. This provides an endogenous rationale for the use of price ceilings versus floors.

Padmanabhan and Png (1997) analyze the role of returns policies in influencing the basis of downstream competition. Kreps and Scheinkman (1983) showed that, where competing sellers first order stocks then compete on prices, the resulting equilibrium is similar to that when sellers compete directly on quantities (Cournot competition). Absent a returns policy, the retailers compete in a manner so as to sell all the retail stock. To the extent that the retailers are constrained by the stocks, these constraints limit the intensity of price competition. Padmanabhan and Png (1997) show, however, that, if the manufacturer accepts returns, then the retailers will not be constrained by their stocks. Equivalently, a returns policy effectively nullifies the constraint of previously-ordered stocks. Because retailers are not constrained by stocks, they compete only on price. Singh and Vives (1984) show that price (Bertrand) competition is more intense than quantity (Cournot) competition. This means that retail prices are lower and retailers' sales are higher. Because, the manufacturer's wholesale price in this model is the same with and without returns this implies that the manufacturer's profits will be higher with the provision of returns policies. Furthermore, the provision of returns policies reduces the retailer's profit as well.[15] Essentially, when retailing is competitive, the manufacturer needs to coordinate two retail-level variables—the two retail prices. Hence, the single instrument of the wholesale price is not adequate and a returns policy, by providing a second instrument, raises the manufacturer's profit. Broadening the model to incorporate uncertainty about demand potential, they show that the optimality of a returns policy (from the manufacturer's perspective) depends on the marginal cost, c, and uncertainty. In particular, the difference in profits between a returns policy and no returns policy is decreasing in the marginal cost and decreasing in uncertainty (e.g., spread between high and low demand holding the probability of these events constant).

Another interesting issue that arises in the context of multiple retailers is the issue of contract observability. Consider the case that retailers are unable to observe the contract that a manufacturer has negotiated with rival retailers in the same market. McAfee and Schwartz (1994) study this problem and analyze bilateral contracting between a manufacturer and retailer without commitment regarding the contracts that might be stuck between the manufacturer and other retailers. Each retailer then fears that the manufacturer might opportunistically renegotiate with another retailer to increase bilateral profit at that retailer's expense. In such a case, if the manufacturer were able to credibly commit at the outset to all contracts, then this would eliminate opportunism. One might think that a natural response to this problem is some form of nondiscrimination commitment such as a most favored customer contract. For example, the manufacturer can commit to make past customers retroactively eligible

15. Note that the returns policy also helps to overcome the double marginalization problem.

for any future discounts. McAfee and Schwartz's main result is that given two-part tariffs the use of most favored customer commitments can be ineffective. The intuition is that once a retailer accepts a lower marginal wholesale price for a higher fixed fee, a second retailer may well prefer a higher wholesale price and a lower fixed fee even if the nondiscrimination contract is available to it ex-post. Thus nondiscrimination clauses can have little bite even with symmetric retailers.[16]

4. COMPETITION IN THE UPSTREAM MARKET

In this section we briefly focus on the additional problems which arise when there are multiple manufacturers each one selling their own product. This introduces upstream competition which results in possible externality across brands. Channel contracts in this case can have an effect on competition between brands.

One way of reducing interbrand competition among manufacturers is by using a common retailer or a common agency. Bernheim and Whinston (1985) link interbrand competition to a negative externality across manufacturers. They show that one way to internalize this externality is to vest all the decision making power in a single down stream retailer who is a residual claimant. This retailer will make the correct pricing decisions that maximize channel profits.

McGuire and Staelin (1983) examine a situation where two competing manufacturer each sell through an exclusive retailer (who then compete in the end product market). Let demand at retailer i, $i = 1, 2$ be represented as $Q_i = 1 - p_i + \theta p_2$. For the case of exclusive dealing and assuming that the manufacturer is the stackelberg leader, the prices at the retail stage are obtained as

$$p_i = \frac{1}{2-\theta} + \frac{2}{(2+\theta)(2-\theta)} w_i + \frac{\theta}{(2+\theta)(2-\theta)} w_{3-i}. \tag{15}$$

Manufacturer's take these retailer strategies in their respective optimization problems. The optimal wholesale prices are

$$w_i = c + \frac{2+\theta}{4-\theta-2\theta^2}(1 - c(1-\theta)). \tag{16}$$

It is easy to obtain the manufacturer's profit expressions by plugging these back into their profit functions. It is possible to follow a similar exercise to obtain the manufacturer's profits under alternative configurations as well (e.g., both manufacturer's are vertically integrated, one is integrated and other uses an independent retailer, etc.). McGuire and Staelin (1983) show that the total channel profits with an independent retailer are higher than with a vertically integrated manufacturer if and only if $\theta \geq 0.43$. Additionally, each manufacturer's profit is higher when both use independent

16. For some other interesting papers on contract unobservability in vertical relationships see Hart and Tirole (1990) and O'Brien and Shaffer (1992).

channel than when both are vertically integrated if $\theta \geq 0.71$. Comparing the equilibrium channel structures, they show that both manufacturers choosing vertical integration is a Nash equilibrium strategy for all values of θ. In addition, each manufacturer choosing an independent channel is also a Nash equilibrium on the channels profit criterion for $\theta \geq 0.77$ and on the manufacturer profit criterion for $\theta \geq 0.93$. Moorthy (1988) shows that the intuition for their result relies on the whether the responses at the manufacturer or retailer level are strategic substitutes or strategic complements and whether the two products are demand substitutes or complements. When a manufacturer switches from vertical integration to decentralization, he raises his equilibrium price due to double marginalization. But this also raises the competing manufacturer's retail price. The first effect has a negative impact on channel profits while the second effect has a positive impact when the products are demand substitutes. The raising of the competing manufacturer equilibrium retail price is not always the case nor is it always beneficial. This happens only if there is strategic complementarity between the decisions of the two manufacturers: when one firm raises its price, the competitor must also find it optimal to do so. For example, if there are economies of scale in retailing costs, then the two firms may be strategic substitutes at the retailing level and switching to the decentralization might actually induce the other manufacturer to lower price. That will decrease channel profits if the two products are demand substitutes (exactly the opposite case of the McGuire and Staelin model).

Coughlan and Wernerfelt (1989) and Bonanno and Vickers (1988) have considered the effect of allowing two-part tariffs in a channel with upstream competition between manufacturers who use exclusive retailers. With two-part tariffs the manufacturers can use the fixed fee to extract retail profits. Therefore they can choose any wholesale price (even below marginal costs). The consequence of this is that if two-part tariffs were allowed, decentralization will always be the equilibrium outcome in the McGuire and Staelin type of channel regardless of the strategic relationship between the manufacturers—strategic substitutability or complementarity.

Another interesting and related paper is Lal and Villas-Boas (1996) which analyzes retail price promotions and manufacturer trade deals in a channel with manufacturers who distribute through exclusive retailers. They find that the introduction of retailers leads to smaller frequency and depth of consumer price promotions. They also find that trade promotions are more frequent than consumer promotions. Lal and Villas-Boas also analyze a model where a manufacturer would choose exclusive retailing as an equilibrium strategy.[17]

The mitigation of price competition is also the theme of Shaffer (1991) who examines slotting fees as a facilitating practice that allows retailers to diminish price competition. The idea being that manufacturers competing to obtain shelf-space at the retail level charge a high wholesale price and then repatriate their profits via a slotting allowance to the retailer. Assuming observability of wholesale prices, retailer i by

17. See also Lal and Villas-Boas (1998) for a related analysis of price promotions and trade deals with multi-product retailers.

committing to a manufacturer's offer of a high wholesale price and related slotting allowance provides an incentive to retailer *j* to raise her price which in turn drives up the retailer *i*'s profits. A similar effect holds for retailer *j*. Hence, retail prices as well as retailer profits would be higher when manufacturers are able to use slotting allowances to obtain shelf-space.

Exclusive dealing is another practice which affects competition between manufacturers. It is a contract under which a buyer promises to buy his requirements of one or more products only from a particular seller. Interestingly exclusive dealing and exclusive territories are often observed together. This is because the exclusive territory contract is often used as an up front payment by the manufacturer to entice the retailer to carry only the manufacturer's product. This is true, for example, in the case of fast-food franchises where the exclusive dealing contract involves a lot of relationship-specific investments made by the franchisee. There is a possibility for post-contractual opportunism on the part of the franchisor wherein he can establish a new franchisee in a nearby location after the first franchisee sinks in her investment. Exclusive territories protects the retail outlet from such hazards.

5. SUMMARY AND FUTURE RESEARCH ISSUES

We have reviewed in this chapter the growing literature on strategic interactions amongst institutions in a distribution channel. We started from the simplest setting of a manufacturer transacting with a single retailer in a deterministic demand setting and proceeded to broaden this setting by examining research dealing with multiple agents within a layer in the channel and/or multiple inputs contributing to demand realization in the final goods market. Extending the benchmark setting one step at a time allowed us to highlight: (a) the efficacy of different mechanisms for channel coordination, and, (b) the consequences of price as well as non-price mechanisms on the strategic choices of the agents in the channel. Our understanding of channel interactions and relationships has improved considerably due to the research efforts discussed here. However, the richness of distribution channels implies that there are considerable opportunities for systematic and sustained research effort. We outline some of the fruitful areas for exploration in the following paragraphs.

The growth in e-commerce and its implications for channel management has been picked up by the academic research community. The initial work in this area focused on understanding the implications of online retailing for retail pricing and market behavior (e.g., Lal and Sarvary 1999, Brown and Goolsbee 2000, and Iyer and Pazgal 2003). More recently, researchers have begun to explore the implications of these internet institutions for strategic conduct of the channel institutions and their interactions (e.g., Chen, Iyer and Padmanabhan 2002). There are many fruitful avenues for exploration in this domain. For instance, the price-discovery feature of online retailing enables consumers to improve their bargaining position with manufacturers and/or retailers. How does this shape the conduct of the channel institutions and the governance mechanisms? How does the evolution of e-commerce impact the emergence of different types of intermediaries and what are their implications for channel performance.

We discussed research (Rey and Stiglitz 1988 and McGuire and Staelin 1983) involving multiple manufacturers. However there is a gap in the understanding of problems which arise when multiple manufacturers deal with many downstream retailers. The formal analysis of markets like grocery or hardware where competing retailers sell products of multiple manufacturers is important. In such cases, apart from considerations of interbrand or upstream competition, researchers also need to examine issues related to optimal retailer selection and the retail contracts which competing manufacturers would adopt. A particularly interesting issue will be to examine the conditions under which manufacturers choose exclusive versus common retailers. We believe that considering a channel with both manufacturer and retailer competition using a bargaining approach would be fruitful. Another emerging phenomenon here is the increasing usage of own-brands by retailers. This refers to retailers featuring their own store-brands (e.g., Safeway Select) along with the manufacturer brands in their shop-floors. Initial research in this area has attempted to understand why retailer's might do this and how they should go about marketing these brands (e.g., Sayman, Hoch and Raju 2002). The challenges of this phenomenon for competitive strategies of retailers and manufacturers, and, interactions among these players is of considerable interest to academics and practitioners alike (see for example, Corstjens and Lal 2000, Cotterill, Putsis and Dhar 2001).

A large proportion of the current models of channel interactions utilize a single period (i.e., static) setting. However, the empirical reality is that interactions among channel members are characterized for the most part by long term relationships. This implies that the interactions among the channel entities are likely involve extensive negotiations, bargaining and renegotiation. Elaborating the static model to dynamic settings is needed. This would allow us to examine the role of vertical restraints and their efficacy in contexts that allow for unfolding of uncertainty, information revelation, bargaining and renegotiation.

This review primarily examined the large analytical literature on channel contracts and their effects. However, precious little empirical work has been done in this area. This is quite surprising because there are many instances where there are disagreements over the role of even the most basic types of channel contracts. Empirical validation of the predicted role and efficacy of the various restraints is a rich area for future research. In this review we found several situations where different packages of channel contracts were minimally sufficient (i.e., gave the same efficient outcome for the manufacturer). For example, in several cases in the literature one finds that RPM or a two-part tariff or quantity forcing are equivalent ways of solving the vertical externality. However as a empirical researcher one would also be interested in the performance of these restraints in reality. It is possible that enforcement of some restraints such as exclusive territories could involve significant transaction costs. Moreover market characteristics and retailer characteristics could greatly affect the desirability of a restraint. For example, a market characterized by a high degree of consumer search would increase the incentives for free riding and consequently the cost of monitoring.

Other interesting questions include whether empirical evidence can be identified to help discriminate between the various possible uses of channel contracts as predicted by theory. Is there, for example, empirical evidence which will help us discriminate between whether a specific restraint like exclusive dealing was used for channel efficiency reasons or for preventing entry of a competing product. Another important empirical question relates to whether risk-sharing really is an objective of retailer contracts? If so, what is the nature of these contracts? How does this alter contracts for small retailers versus the large retail corporations.

REFERENCES

Alderson, W. (1965), *Marketing Behavior and Executive Action*, Richard D. Irwin, Inc., Homewood, IL.

Arrow, K. J., S. Karlin and H. Scarf (1958), Studies in the Mathematical Theory of Inventory and Production, Stanford University Press.

Bell D.R., V. Padmanabhan and Y. Wang (2002), "The Effect of Partial Forward Integration on Retailer Behavior: An Explanation for Colocated Stores," Working Paper, Wharton School of Business, University of Pennsylvania.

Bernheim, B. D. and M. D. Whinston (1985), "Common Marketing Agency as a Device for Facilitating Collusion," *Rand Journal of Economics*," 16, 269–281.

Bonano, G. and J. Vickers (1988), "Vertical Separation," *Journal of Industrial Economics*, 36, 257–266.

Brown, J. and A. Goolsbee (2000), "Does the Internet Make Markets More Competitive: Evidence from the Life Insurance Industry," *Journal of Political Economy*, 110, 3, 481–507.

Bucklin, Louis P. (1969), "A Theory of Distribution Channel Structure," IBER Special Publications, Berkeley, CA.

Cachon, G. P. (2002), "Supply Chain Coordination with Contracts," in S. Graves and T. de Kok (Eds.), *Handbook in Operations Research and Management Science: Supply Chain Management*, North Holland Publishing (forthcoming).

Cachon, G. P. and M. Lariviere (2002), "Supply Chain Coordination with Revenue Sharing Contracts: Strengths and Limitations," Working Paper, Wharton School of Business, University of Pennsylvania.

Caves, R. E. and W. F. Murphy (1976), "Franchising: Firms, Markets and Intangible Assets," *Southern Economic Journal*, 42, 572–586.

Clark, A. J. and H. E. Scarf (1960), "Optimal Policies for a Multi-Echelon Inventory Problem," *Management Science*, 6, 475-490.

Chen, Y., G. Iyer and V. Padmanabhan (2002), "Referral Infomediaries," *Marketing Science*, 21, 4, 412–434.

Chu, W. (1992), "Demand Signaling and Screening in Channels of Distribution," *Marketing Science*, 11, 4 (Fall), 327–347

Chu, W., and P. Messinger (1997), "Information and Channel Profits," *Journal of Retailing*, Vol. 73, No. 4, 487–499.

Cooper, R. and T. W. Ross (1985), "Product Warranties and Double Moral Hazard," *Rand Journal of Economics*, 103–113.

Corstjens, M. and R. Lal (2000), "Building Store Loyalty Through Store Brands," *Journal of Marketing Research*, 37, 3, 281–291.

Cotterill, R. W., W. P. Putsis and R. Dhar (2001), "Do Models of Vertical Strategic Interaction for National and Store Brands Meet the Market Test," *Journal of Retailing*, 77, 1, 83–109.

Coughlan, A., E. Anderson, L. W. Stern and A. I. El-Ansary (2001), *Marketing Channels,* Prentice Hall Publishers.

Coughlan A. and D. Soberman (2002), "The Strategic Role of Outlet Mall Retailing," Working paper, Insead.

Coughlan, A. and B. Wernerfelt (1989), "On Credible Delegation By Oligopolists: A Discussion of Distribution Channel Management," *Management Science*, 35, 226–239.

Desai P. and K. Srinivasan (1995), "Demand Signaling Under Unobservable Effort in Franchising," *Management Science*, 10, 1608–23.

Donohue, K. L. (2000), "Efficient Supply Contracts for Fashion Goods with Forecast Updating and Two Production Modes," *Management Science*, 46, 11, 1397–1411.

Emmons, H. and S. Gilbert (1998), "Returns Policies in Pricing and Inventory Decisions for Catalogue Goods," *Management Science*, 44, 2, 276–283.

Goodman, D. A. and K. W. Moody (1970), "Determining Optimal Promotion Quantities," *Journal of Marketing*, 34, 31–39.

Grossman, S. and O. Hart (1986), "The Costs and Benefits of Ownership: A Theory of Lateral and Vertical Integration," *Journal of Political Economy*, 94, 691–719.

Hart O. D. and J. Tirole (1990), "Vertical Integration and Market Foreclosure," *Brookings Papers on Economic Activity*, special issue, 205–276.

Holmstrom, B. (1982), "Moral Hazard in Teams," *Bell Journal of Economics*, 13, 324–340.

Iyer, G. (1998), "Coordinating Channels under Price and Non-price Competition," *Marketing Science*, 17, 4, 338–355.

Iyer, G. and A. Pazgal (2003), "Internet Shopping Agents: Virtual Co-location and Competition," *Marketing Science*, 22, 1, 85–106.

Iyer, G. and V. Padmanabhan (2002), "Dynamic Uncertainty and Supply Chain Contracting," Working Paper, Haas School of Business, University of California, Berkeley.

Iyer, G. and J. M. Villas-Boas (2003), "A Bargaining Theory of Distribution Channels," *Journal of Marketing Research*, vol. XL, February, 80–100.

Jeuland, A. and S. Shugan (1983), "Managing Channel Profits," *Marketing Science*, 2, 239–272.

Jeuland, A. and S. Shugan (1988), "Reply to: Managing Channel Profits: Comment," *Marketing Science*, 7, 1, 103–106.

Kandel, E. (1996), "The Right to Return," *Journal of Law and Economics*, 329–356.

Kreps, D. M. and J. A. Scheinkman (1983), "Quantity Precommitment and Bertrand Competition Yield Cournot Outcomes," *Bell Journal of Economics*, 14, 326–337.

Lal, R. (1990), "Improving Channel Coordination Through Franchising," *Marketing Science*, 9, 4 (Fall), 299–318.

Lal, R. and J. M. Villas-Boas (1996), "Exclusive Dealing and Price Promotions," *Journal of Business*, 69, 2, 159–172.

Lal, R. and J. M. Villas-Boas (1998), "Price Promotions and Trade Deals with Multiproduct Retailers," *Management Science*, 44, 7, 935–949.

Lal, R. and M. Sarvary (1999), "When and How is the Internet Likely to Decrease Price Competition," *Marketing Science*, 18, 485–503.

Lariviere, M. A. (1999), "Supply Chain Contracting and Coordination with Stochastic Demand," in S. Tayur, R. Ganeshan and M. Magazine (Eds.), *Quantitative Models for Supply Chain Management,* Kluwer Academic Press, 233–268.

Lariviere, M. A. and V. Padmanabhan (1997), "Slotting Allowances and New Product Introductions," *Marketing Science*, 16, 2, 112–128.

Lilien, G., P. Kotler and K. S. Moorthy (1992), Marketing Models, Prentice-Hall Publishers.

Marvel, H. P. and J. Peck (1995), "Demand Uncertainty and Returns Policies," *International Economics Review,* 36, 3, 691–714.

Mathewson, G. F. and R. A. Winter (1984), "An Economic Theory of Vertical Restraints," *Rand Journal of Economics*, 15, 27–38.

Mathewson, G. F. and R. A. Winter (1985), "The Economics of Franchise Contracts," *Journal of Law and Economics*, 28, 471–501.

Mathewson, G. F. and R. A. Winter (1986), "The Economics of Vertical Restraints in Distribution," in F. Mathewson and J. Stiglitz (Eds.), *New Developments in the Analysis of Market Structures*, Cambridge: MIT Press.

McAfee, R.P. and M. Schwartz (1996), "Opportunism in Multilateral Vertical Contracting: Nondiscrimination, Exclusivity, and Uniformity," *American Economic Review*, 84, 1, 210–230.

McGuire, T. and R. Staelin (1983), "An Industry Equilibrium Analysis of Downstream Vertical Integration," *Marketing Science*, 2, 161–192.

Milgrom, P. and John Roberts (1986), "Price and Advertising Signals of Product Quality," *Journal of Political Economy*, 94, 4, 796–821.

Moorthy, K. S. (1987), "Managing Channel Profits: Comment," *Marketing Science*, 6, 4, 375–379.

Moorthy, K. S. (1988), "Strategic Decentralization in Channels," *Marketing Science*, 7, 4, 335–356.

Nahmias, S. (1993), *Production and Operations Analysis*, Irwin Press, Boston, MA.

Narayanan, V. G. and A. Raman (1997), "Contracting for Inventory in a Distribution Channel with Stochastic Demand and Substitute Products," Working Paper, Harvard Business School.

O' Brien, D.P. and G. Shaffer (1992), "Vertical Control and Bilateral Contracts," *Rand Journal of Economics*, 23, 3, 299–308.

Padmanabhan, V. and I. P. L. Png (1997), "Manufacturer's Returns Policies and Retail Competition," *Marketing Science*, 16, 1, 81–94.

Padmanabhan, V. and R. C. Rao (1993), "Warranty Policy and Extended Service Contracts: Theory and An Application to Automobiles," *Marketing Science*, 12, 3, 230–248.

Pasternack, B. A. (1980), "Filling Out the Doughnut: The Single Period Inventory Model in Corporate Pricing Policy," *Interfaces*, 10, 5, 96–100.

Pasternack, B. A. (1985), "Optimal Pricing and Return Policies for Perishable Commodities," *Marketing Science*, 4, 2, 166–176.
Perry, M. K. and R. H. Porter (1985), "Can Resale Price Maintenance and Franchise Fees Correct Sub-optimal Levels of Retail Services," *Journal of Industrial Organization*, 8, 115–141.
Petruzzi, N. and M. Dada (1999), "Pricing and the Newsvendor Problem: A Review with Extensions," *Operations Research*, 47, 183–194.
Priest, T. G. (1981), "A Theory of Product Warranties," *Yale Law Journal*, 90, 127–152.
Rao R.C. and S. Srinivasan (1995), "Why are Royalty Rates Higher in Service-Type Franchises," *Journal of Economics and Management Strategy*, 4,1, 7–31.
Rey, P. and J. Tirole (1986), "Vertical Restraints from a Principal-Agent Viewpoint," in L. Pellegrini and S. Reddy (Eds.), *Marketing Channels: Relationships and Performance*, Lexington Books, 3–30.
Rey, P. and J. Stiglitz (1988), "The Role of Exclusive Territories," Mimeo.
Rubin, P. (1978), "The Theory of the Firm and Structure of the Franchising Contract," *Journal of Law and Economics*, 223–234.
Sayman, S., S. J. Hoch and J. S. Raju (2002), "Positioning of Store Brands," *Marketing Science*, 21, 4, 378–397.
Shaffer, G. (1991), "Slotting Allowance and Resale Price Maintenance: A Comparison of Facilitating Practices," *RAND Journal of Economics*, 22, 1 (Spring), 120–135.
Singh, N. and X. Vives (1984), "Price and Qantity Competition in a Differentiated Duopoly," *Rand Journal of Economics*, 15, 4, 546–554.
Silver, E., D. Pyke and R. Peterson (1998), *Inventory Management and Production Planning and Scheduling*, John Wiley and Sons, New York, NY.
Spence, A. M. (1973), "Job Market Signaling," *Quarterly Journal of Economics*, 87, 355–374.
Spengler, J. J. (1950), "Vertical Integration and Antitrust Policy," *Journal of Political Economy*, 58, 347–352.
Stern, L. W. and B. A. Weitz (1997), "The revolution in distribution: challenges and opportunities," *Long Range Planning*, 30, 6, 823–829.
Sullivan, M. (1992), "What Caused Slotting Allowances?," Working paper, University of Chicago.
Tirole, J. (1988), *The Theory of Industrial Organization*, Cambridge: MIT Press.
Tsay, A. T. and N. Agarwal (2002), "Modeling Conflict and Coordination in Multi-Channel Distribution Systems: A Review," in D. Simchi-Levi, D. Wu and M. Shen (Eds.), *Supply Chain Analysis in the eBusiness Era*, Kluwer Academic Publishers.
Van Meighem, J. A. (1999), "Coordinating Investment, Production and Sub-contracting," *Management Science*, 54, 7, 954–971.
Villas-Boas, J. M. (1998), "Product Line Design for a Distribution Channel," *Marketing Science*, 17, 2, 156–169.
Winter, R. A. (1993), "Vertical Control and Price Versus Non-price Competition," *Quarterly Journal of Economics*, 1 (February), 61–78.

5. QUANTITY DISCOUNTS FOR SUPPLY CHAIN COORDINATION†

S. CHAN CHOI, LEI LEI, and QIANG WANG

ABSTRACT

Coordinating activities among members of a supply chain/distribution channel has been a subject of great attention in the past two decades. Among various methods to coordinate independent channel members, a number of studies in the literature suggest quantity discount as a mechanism to achieve incentive-compatible coordination between a seller/manufacturer and his buyer/retailer. The rationale behind this coordination mechanism is that a quantity discount schedule can be designed to align the players' interests with the maximum channel gain.

This chapter first reviews two separate streams of quantity discount models that have been developed in the literature. The operations management literature views quantity discount as a way to minimize the system-wide cost of operation. The manufacturer can offer quantity discount such that the retailer finds it optimal to order a larger quantity that minimizes the total channel's operating cost. On the other hand, the marketing literature employs quantity discount to induce the retailer to lower the retail price than the level she would choose otherwise. The increased market demand more than compensates the reduced margin so that the total channel profit is maximized. In both models, the channel members' profit goals are aligned with the total channel profit, and the resulting channel coordination creates efficiency gains that can be shared between the members.

† This research was partially funded by the Rutgers Business School Research Resources Committee Grant. The authors thank Professor Amiya Chakravarty for valuable comments and suggestions on earlier versions of this chapter, and Carter Daniel of the Rutgers Business School for his help in improving its readability.

Then we review models that combine these two sources of efficiency gains. When the two models are combined, there is a synergy between the two effects. However, the resulting models tend to be difficult to solve analytically. Finally, we introduce a quantity discount model that also includes a third-party logistics partner. Due to the model complexity, we examine the model using a couple of popular demand functions and show that it is also feasible to coordinate three supply chain partners using quantity discount. We conclude the paper with a discussion of further research directions.

1. INTRODUCTION

Improved integration of activities across multiple companies sharing components of a supply chain is a concern of increasing interest and importance (Shapiro 2001, p.7). A large number of studies in supply chain management have focused on normative optimization models with various assumptions on institutional arrangement and operational details. When two or more independent companies are involved in a supply chain process, a "solution" that integrates their activities for a system-level optimality requires that (a) the modeler has a complete knowledge of the partners' costs and capacities, and (b) all involved channel partners implement the business process as prescribed by the optimal solution.

The first requirement becomes increasingly realistic as the companies integrate their information technology on various operational levels (e.g., Wal-Mart and its suppliers; Procter and Gamble and its retailers). However, the second requirement would be difficult to enforce unless (i) there is a binding agreement among the supply chain partners, or (ii) the prescribed "solution" is *incentive compatible*. Several studies implicitly assume the former mechanism to derive optimal decisions on production schedule, inventory management, and *retail price* (Feichtinger and Hartl 1985; Jorgensen 1986). Eliashberg and Steinberg's (1987) model also includes *wholesale price* decision, in addition to the retail price, product delivery, and inventory policies. However, the game-theoretic models in these studies involve the manufacturer's profit maximization problem subject to the retailer's optimality condition, rather than the joint profit maximization via coordinated efforts between the channel members.

On the other hand, more sustainable channel coordination can be achieved by an incentive compatible mechanism, in which the action of a channel member who acts on a self-interest motivation is aligned with the objectives of the other members in the channel. It is a more attractive mechanism to achieve coordination among the supply chain (or, equivalently, the distribution channel) members for the benefit of the whole system.[1] This paper reviews such models of channel coordination that employ quantity discount in transfer price as a mechanism for incentive compatibility.

In Sections 2 through 5, we provide a critical review of the literature on quantity discount models. Section 2 provides an overview of the existing literature and classifies the existing models on several dimensions. Section 3 presents quantity discount models

1. In this paper, we use the terms "supply chain" and "distribution channel" interchangeably. Also, we focus our study on the models that consider two players in a supply chain, a seller and a buyer (or, interchangeably, a manufacturer and a retailer).

based on transaction efficiency, and Section 4 presents models based on the elastic demand argument. Models that combine these two factors are reviewed in Section 5. Section 6 presents an extension that also includes a transportation partner and shows that these quantity discount rationales are still valid in a more complex setting. We conclude the paper in Section 7 by suggesting further research areas.[2]

2. AN OVERVIEW OF LITERATURE

According to Dolan (1987), there are three popular forms of quantity discount schedules:

- A two-part tariff: a fixed charge F (either per period or per order), plus a uniform price p for all units
- A two-block (incremental) tariff: a per-unit price p_1 up to an order quantity $\hat{Q}$, then per-unit price p_2 for all units greater than $\hat{Q}$; an n-block tariff is defined similarly
- All-unit quantity discount: unit price p_1 up to a certain quantity level $\hat{Q}$, and when it is exceeded, another (lower) unit price p_2 is applied to all units

Dolan shows that a two-block tariff is a more general form: Any [per-order] two-part tariff can be written as an equivalent two-block tariff, but the converse is not true in general. In this paper, we focus on analytical models of quantity discount pricing schedules that are intended to coordinate operations among supply chain partners. In particular, we focus on a two-echelon supply chain system in which there are a seller (or a manufacturer) and a buyer (a distributor or retailer).[3]

A number of studies in the operations research and marketing literature have suggested quantity discounts as a mechanism to achieve coordination between the seller and buyer. The basic rationale behind the coordination mechanism is that a quantity discount schedule can be designed such that the objectives of the seller and buyer are incentive compatible with the maximum channel gain. This alignment of interests is attributed to two major factors working independently or combined: (a) *transaction efficiency* stemming from the system-wide saving of transaction costs including inventory holding, set up, and ordering costs (e.g., Monahan 1984; Chakravarty 1984, 1989; Lal and Staelin 1984; Dada and Srikanth 1987; Kim and Hwang 1989), and (b) *increased market demand* due to a lower optimal retail price that results from a lower wholesale price (e.g., Jeuland and Shugan 1983; McGuire and Staelin 1986). More recent studies attempt to combine these two quantitative discount models (e.g., Weng 1995; Chen, Federgruen and Zheng 2001).

2. Although this paper focuses on the use of quantity discount as a tool for coordinating member actions of a supply channel, it needs to be mentioned that there are two other major purposes of quantity discount pricing found in the economics literature: perfect price discrimination against a single customer (Buchanen 1953; Garbor 1955; Moorthy 1987) and partial price discrimination against a set of heterogeneous customers (Oi 1981).

3. Without loss of generality, it can also be a group of homogeneous buyers. In this paper, we use the term "buyer" and "retailer" interchangeably.

In addition to these basic factors, there are other dimensions that can characterize the quantity discount models:

- Objective function: Each channel member's objective can be modeled by either profit maximization or cost minimization. When the retail price is fixed, these two objectives are equivalent.
- Discount form: The discount can be offered as a single-break discrete schedule or as a continuous function of order quantity. Quantity discounts in the transaction efficiency models are based on per-order quantity, while those in the demand-based models are based on the cumulative demand during the planning period.
- Inventory holding cost: The cost can be defined as either unit-based or value-based. The value-based holding cost assumes that a large part of the cost is attributed to the capital cost.
- Demand function: The demand can be either fixed or a decreasing function of retail price. In the latter case, the retail price is also part of the decision variable.
- Number of buyers: A quantity discount schedule can be designed for a single buyer or multiple buyers. Note that homogeneous buyers can be treated as a single buyer. Heterogeneous-buyer models are significantly more complex.
- Product form: A product can be a nonperishable or perishable. Most studies assume nonperishable goods. With a perishable product, it has been shown that price discount needs to be steeper.

Table 1 classifies quantity discount models using these dimensions. Throughout this paper, the following notations will be used:

D = total yearly number of units demanded at the retail level
Q = order size by the buyer
$Q^\star$ = optimal order quantity (EOQ) in the absence of coordination
$Q^{\star\star}$ = optimal order quantity with channel coordination
H_s, H_b = the seller's and the buyer's yearly inventory holding costs per unit
S_s, S_b = the seller's and the buyer's order processing costs (per order)
P_0 = initial wholesale price before discount in the absence of coordination
$p^{\star\star}$ = optimal wholesale price after discount with channel coordination
Π_s, Π_b = profits of the seller and the buyer
C_s, C_b = total operating costs of the seller and the buyer.

3. QUANTITY DISCOUNT SCHEDULES BASED ON TRANSACTION EFFICIENCY

Working independently, Monahan (1984), Chakravarty (1984), and Lal and Staelin (1984) have shown that channel coordination between the seller and buyer can be achieved by a quantity discount schedule designed to provide incentive compatibility between the two agents' respective objectives: profit maximization for the seller and cost minimization for the buyer. For such a quantity discount to work, the schedule needs to be designed to induce the buyer to change his ordering policy so that (a) the buyer is no worse off, and (b) the seller's profit is increased, even if the total demand does not change (i.e., Pareto optimality).

Table 1. Quantity discount model classification

(a) Transaction efficiency models

Profit or cost	Discount form	Inventory holding cost assumption	
		Value-based	Unit-based
Profit Maximization	Continuous Function	Kim and Hwang (1989)	Lal and Staelin (1984)
			Kohli and Park (1989)
	Discrete Function	Lee and Rosenblatt (1986)	
		Chakravarty and Martin (1991)	
Cost Minimization	Continuous Function	Chakravarty (1984; 1989)	Dada and Srikanth (87)
		Monahan (1984)	Munson and Rosenblatt (2001)
	Discrete Function	Banerjee (1986)	Chakravarty and Martin (1988)

- Perishable item: Chakravarty (1989); all others model with nonperishable
- Stochastic demand: Chakravarty and Martin (1991); all others use with deterministic demand
- Multiple buyers: Chakravarty (1984,1989), Chakravarty and Martin (1988); all others model single buyer

(b) Demand-based models

	Continuous quantity-price function	Two-part tariff
Single Retailer	Jeuland and Shugan (1983)	McGuire and Staelin (1986)
Multiple Retailer		Ingene and Parry (1995)

- All models, by nature, deal with profit maximization and stochastic demand without inventory-related component. In addition, all are for nonperishable goods.

(c) Mixed models

Single Retailer	Weng (1995)
Multiple Retailer	Chen, Federgruen and Zheng (2001)

- Both models have profit maximization, stochastic demand, and continuous discount schedules with fixed tariff. In addition, both models assume unit-based inventory holding cost for nonperishable goods.

First, we start with five major assumptions that simplify the analytical approach:

- The buyer's inventory policy can be described as the simplest EOQ model with deterministic demand, no stock outs, no backlogs, and deterministic lead times.
- The seller has a complete knowledge on the holding and ordering costs governing the buyer's inventory policy
- There are no competitive price reactions to the seller's discount policy
- The demand is fixed at a uniform rate (in Section 4, we relax this assumption)
- There is a single buyer (or a group of homogeneous buyers; this assumption is relaxed in several models)

In the rest of this section, we review quantity discount models based on transaction efficiency only. First, we provide the rationale for quantity discount in Section 3.1. A graphical interpretation of a quantity discount is provided in Section 3.2, which shows that the Pareto optimal discount schedule is not unique when the inventory cost is unit-based. Section 3.3 presents several mechanisms to divide the total joint gain between the seller and the buyer. Assuming unit-based inventory holding cost, Sections 3.4 and 3.5 derive all-unit and incremental quantity discount schedules, respectively. We show that the optimal order quantity is independent from the discounted price. Section 3.6 shows that the independence no longer holds when inventory cost is value-based. It turns out that the channel-optimal solution is the one that gives the entire surplus to the buyer.

3.1 A rationale for transaction efficiency-based quantity discounts

3.1.1 The buyer's basic EOQ

When there is no channel coordination effort, the decision problem in the supply channel can be viewed as a Stackelberg game with the seller as the leader. With the knowledge of the buyer's reaction as a function of the seller's decision variable, the seller can optimize his own profit as the Stackelberg leader. The buyer's problem is to choose an order quantity that minimizes the total cost consisting of purchasing price, order-processing cost, and inventory cost:

$$C_b(Q) = pD + S_b(D/Q) + H_b(Q/2). \tag{1}$$

The buyer's optimal order quantity is the well-known EOQ model:

$$Q^\star = (2S_b D/H_b)^{1/2}, \tag{2}$$

which is independent of the purchasing cost. The resulting total cost for the buyer is

$$C_b(Q^\star) = pD + (2DS_bH_b)^{1/2}. \tag{3}$$

Note that in this formulation, the inventory holding cost is assumed to be per *unit* cost. Thus, this EOQ formula does not depend on the wholesale price. When the inventory cost depends on the *value* of the product, however, the EOQ is a function of the wholesale price p, (i.e., $Q^\star = (2S_b D/H_b p)^{1/2}$). The value-based inventory model becomes substantially complicated, and we present such models in Section 3.6.

3.1.2 The seller's problem

The seller's problem is to maximize his profit by influencing the buyer's order quantity. Instead of accepting the buyer's myopic EOQ Equation (2), the seller can design the wholesale price as a function of the order quantity, thereby providing an incentive to the buyer to choose a different order quantity. When the demand is assumed fixed, the seller's profit maximization problem can be simplified as a cost minimization problem. The seller's total cost consists of his order processing cost less the reduced inventory cost due to the buyer's order size of Q:

$$C_s(Q) = S_s(D/Q) - H_s(Q/2). \tag{4}$$

Note that in this formulation, the seller is assumed to carry a "large" fixed amount of inventory that does not depend on the buyer's order quantity. Thus, whenever the buyer orders, the seller's inventory holding cost is reduced by that amount. In Section 5, we relax this assumption to the case where the seller either orders or manufactures the product in a lot size that is an integer multiple of Q.[4]

Without coordination between the seller and buyer, the buyer's optimal order quantity is $Q = Q^\star$, and the corresponding seller's total cost is

$$\begin{aligned} C_s(Q^\star) &= (S_sD)/(2S_bD/H_b)^{1/2} - H_s(2S_bD/H_b)^{1/2}/2 \\ &= (H_bS_s - H_sS_b)[D/(2H_bS_b)]^{1/2}, \end{aligned} \tag{5}$$

and the resulting joint channel cost is

$$\begin{aligned} J(Q^\star) &= C_b(Q^\star) + C_s(Q^\star) \\ &= pD + (H_b(2S_b + S_s) - H_sS_b)[D/(2H_bS_b)]^{1/2}. \end{aligned} \tag{6}$$

On the other hand, when the channel members coordinate to minimize the joint cost

$$J(Q) = pD + \frac{D}{Q}(S_b + S_s) + (H_b - H_s)\frac{Q}{2}, \tag{7}$$

the jointly-optimal order quantity can be derived from the first order condition of Equation (7) as

$$Q^{\star\star} = [2(S_b + S_s)D/(H_b - H_s)]^{1/2}, \tag{8}$$

and the resulting joint channel cost at $Q^{\star\star}$:

$$J(Q^{\star\star}) = pD + [2D(H_b - H_s)(S_b + S_s)]^{1/2}. \tag{9}$$

Note that Equation (8) requires $H_b > H_s$, and comparing it with Equation (2), we have $Q^{\star\star} > Q^\star$.[5] That is, the joint-optimal order quantity that minimizes the total

4. Under this relaxed assumption, a redefined H_s can be used in Equation (4) with a positive sign. (See Section 5.)
5. It is a widely accepted assumption that the buyer's unit holding cost is greater than that of the seller's.

channel cost needs to be greater than that of the uncoordinated case. Since $J(Q^{\star\star}) \leq J(Q^{\star})$, $C_b(Q^{\star\star}) + C_s(Q^{\star\star}) \leq C_b(Q^{\star}) + C_s(Q^{\star})$. Thus, $C_b(Q^{\star\star}) - C_b(Q^{\star}) \leq C_s(Q^{\star}) - C_s(Q^{\star\star})$, which implies that when the channel members coordinate to minimize the joint cost, the increased order quantity decreases the seller's cost more than the increase in the buyer's cost. By passing part of the cost saving to the buyer to cover the increased buyer cost, the seller can motivate the buyer to order $Q^{\star\star}$ and still benefit from it. This can be achieved by offering the buyer a quantity discount. Figure 1 shows how the quantity discount can motivate the buyer to increase his order quantity by shifting the total unit cost curve to the right.

3.2 A graphical interpretation of quantity discount

It is interesting to note from Equation (8) that the joint cost-minimizing order quantity is independent of the wholesale price.[6] It means that there is a range of wholesale prices that can be offered to motivate the buyer to order $Q^{\star\star}$, producing the same efficiency gain. Any discount schedule can induce the buyer to order $Q^{\star\star}$ as long as it is designed such that the buyer's total purchasing cost is minimized at $Q^{\star\star}$. Dada and Srikanth (1987) provide a graphical analysis that defines the wholesale price range and its profit sharing implications.

Consider the maximum wholesale price the buyer would be interested in paying for an order quantity greater than his current EOQ, $Q^{\star}$. The status quo total cost for the buyer is $C_b(Q^{\star}, P_0)$, which increases when the order quantity deviates from $Q^{\star}$. Since we know that $Q^{\star\star} \geq Q^{\star}$ from the above discussion, we are interested only in a larger order quantity than $Q^{\star}$. In order to motivate the buyer to order any quantity $Q \geq Q^{\star}$, the wholesale price needs to be reduced to compensate the buyer's operating cost increase. The buyer would be interested in ordering a larger quantity Q at a maximum price $p(Q)$ only if $C_b(Q, p(Q)) \leq C_b(Q^{\star}, P_0)$. Using Equation (1), we obtain

$$p(Q) \leq \frac{C_b(Q^{\star}, P_0)}{D} - \frac{S_b}{Q} - \frac{H_b Q}{2D} \equiv P_{\max}(Q), \tag{10}$$

where $P_{\max}(Q)$ is the maximum average price for the buyer to order a quantity Q greater than $Q^{\star}$. Substituting the buyer's cost at $(Q^{\star}, P_0)$ with Equation (3), it is straightforward to show that, when $Q \geq Q^{\star}$, $P_{\max}(Q)$ is a decreasing, concave function of Q.

For the seller's problem, let us redefine the seller's cost (4) to be the "augmented cost" in Lal and Staelin's (1984) context by including the sales revenue term:

$$C_s(Q) = S_s(D/Q) - H_s(Q/2) - pD. \tag{11}$$

6. This is due to the assumption of per unit-based inventory cost. In Section 3.6, we shall discuss results in which the inventory cost is value-based.

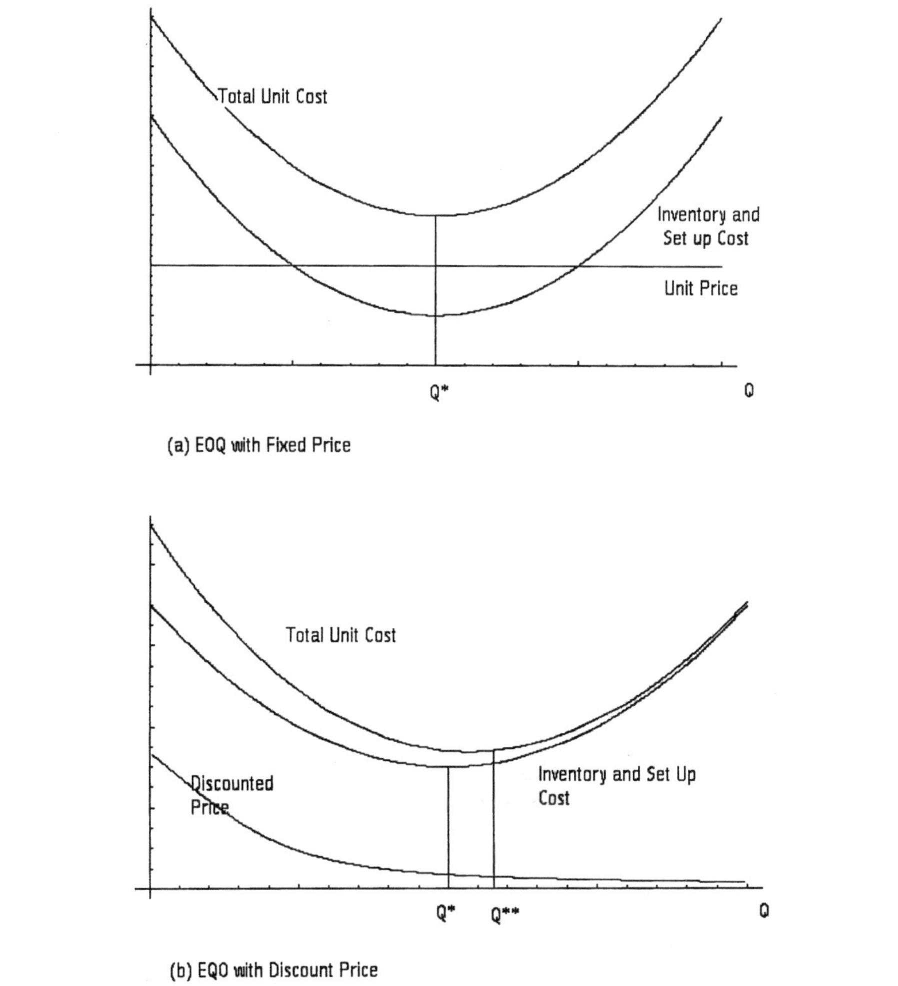

Figure 1. Cost curves for the buyer.

The seller would be interested in offering a quantity discount price $p(Q)$ only if $C_s(Q, p(Q)) \leq C_s(Q^\star, P_0)$, which is equivalent to

$$p(Q) \geq \frac{S_s}{Q} - \frac{H_s Q}{2D} - \frac{C_s(Q^\star, P_0)}{D} \equiv P_{\min}(Q), \tag{12}$$

where $P_{\min}(Q)$ is the minimum average price the seller is willing to charge to motivate the buyer to order a quantity Q greater than $Q^\star$. It is straightforward to show that,

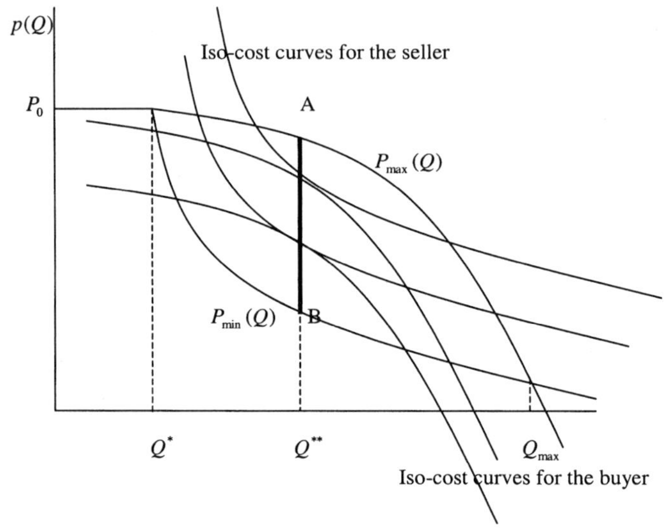

Figure 2. The feasible region and pareto optimal price-quantity frontier.

P_{min} (Q) is a decreasing, convex function of Q. Figure 2 shows generic shapes of these curves. The two curves intersect at $Q = Q^\star$ and at $Q = Q_{max}$. It can be shown that the slope of P_{min} (Q) is steeper than that of P_{max} (Q) at $Q = Q^\star$. These two curves represent the bottom line iso-cost curves of the buyer and the seller, respectively. At any given order quantity such that $Q^\star < Q < Q_{max}$, both buyer and seller are better off if the wholesale price is within the boundary: $P_{min}(Q) < p(Q) < P_{max}(Q)$. Thus, any order quantity and wholesale price pair within this region is a feasible price discount schedule. This region is called the *feasible region.*

We now show that the joint cost is minimized when the difference between P_{max} and P_{min} is maximum. By subtracting Equation (12) from (10) and multiplying both sides by D, we obtain the system-wide cost saving:

$$D(P_{max}(Q) - P_{min}(Q)) = C_s(Q^\star) + C_b(Q^\star) - (S_b + S_s)\frac{D}{Q} - (H_b - H_s)\frac{Q}{2}. \tag{13}$$

Noticing that $J(Q^\star) = C_s(Q^\star) + C_b(Q^\star)$ and that $J(Q) = \frac{D}{Q}(S_b + S_s) + (H_b - H_s)\frac{Q}{2}$, (which is obtained by adding equations (1) and (11)), we have

$$J(Q) = J(Q^\star) - D(P_{max}(Q) - P_{min}(Q)). \tag{14}$$

That is, at an order quantity larger than $Q^\star$, the joint cost decreases by an amount proportional to the difference between the two boundary prices, (P_{max} (Q) $-$ P_{min}

(Q)). Thus, maximizing the price differential is equivalent to minimizing the system cost, and by definition it occurs at $Q^{\star\star}$.

That $Q^{\star\star}$ is independent of the wholesale price can also be seen graphically using iso-cost curves (see Figure 2). Within the feasible region, the optimal price-quantity pair $(p^{\star\star}, Q^{\star\star})$ occurs on the Pareto efficient frontier, which is the path of tangency (the thick line in the figure) between the two iso-cost curves. Thus, the system-optimal price is not unique: Any wholesale price $P_{\min} < p^{\star\star} < P_{\max}$ applied to the order quantity $Q^{\star\star}$ minimizes the joint cost *and* benefits both members.

The wholesale price, however, determines how the joint cost saving is distributed between the seller and buyer. Let G_s and G_b denote the seller and buyer's share of the total cost saving, respectively. At a given order quantity $Q^{\star} < Q < Q_{\max}$ and a wholesale price $P_{\min}(Q) \leq p \leq P_{\max}(Q)$, the joint gain is divided between the members as follows:

$$G_b(Q, p) = D(P_{\max}(Q) - p), \qquad (15)$$
$$G_s(Q, p)) = D(p - P_{\min}(Q)).$$

Intuitively, a higher wholesale price (but under $P_{\max}$) will bring a larger share of the total gain to the seller, and vice versa. Sections 3.4, 3.5, and 3.6 present various discount schedules that determine the wholesale price as a function of Q. But before that, we briefly discuss how the two channel members can partition the total cost saving.

3.3 Mechanisms for dividing the joint gain

The division of the total gain from the joint EOQ depends on the assumption of the game between the two players. When the seller plays a role as the Stackelberg leader, he can keep virtually all the gain by passing onto the buyer only a small incentive "just enough" to let him choose $Q^{\star\star}$ instead of $Q^{\star}$. The wholesale price in this case is just below $P_{\max}$. In the buyer-Stackelberg game, on the other hand, the wholesale price is just above $P_{\min}$. However, this logic applies only in the case where the inventory holding cost is unit-based. When the holding cost is value-based, the Pareto efficient frontier is not a vertical line (see Section 3.6), and the system-optimal order quantity depends on the wholesale price (Chakravarty 1984; Banerjee 1986; Chakravarty and Martin 1988, 1991; Kim and Hwang 1989).

Kohli and Park (1989) suggest the use of a bargaining solution to derive the partitioning of the joint cost saving by setting the wholesale price $p^{\star\star}$ to be charged at $Q^{\star\star}$. Nash's bargaining solution predicts that the price is set such that the product of the utilities of the buyer's and seller's respective gains be maximized. For example, when both channel members are risk neutral, the product of the utilities

$$u(G_s)u(G_b) = D^2(P_{\max} - p)(p - P_{\min})$$

is maximized at $p^{\star\star} = (P_{\max} - P_{\min})/2$, the mid-point between the two limit prices. If the buyer is more risk averse, then bargaining price will be higher, and the seller will take a larger share of the joint gain. Kalai and Smorodinski's (1975) bargaining model predicts that the solution occurs in the utility space $\{u_b(G_b), u_s(G_s)\}$ where the line connecting the "ideal point" $\{u_b(D(P_{max} - P_{min})), u_s(D(P_{max} - P_{min}))\}$ and the no-bargaining outcome $\{0, 0\}$ intersects the Pareto frontier (i.e., the line connecting $\{0, u_s(D(P_{max} - P_{min}))\}$ and $\{u_b(D(P_{max} - P_{min})), 0\}$. For risk neutral players, this results in the same solution as that of Nash.

While these two bargaining models do not incorporate the negotiation power of the agents, Eliashberg's (1986) arbitration model predicts that the solution will occur at the intersection of the Pareto frontier and an allocation trajectory. The allocation trajectory is inferred from the aggregation weights that reflect the relative power of negotiation of the buyer and seller. Eliashberg's model results in a larger share of the efficiency gain going to the party that has larger bargaining power. (See Kohli and Park (1989) for illustrations of these bargaining solutions.)

3.4 All-unit quantity discount schedules

Once the jointly optimal order quantity and the corresponding transfer price are determined, the next step is to make the solution (p^{**}, Q^{**}) more attractive to the buyer than any other price-quantity pair. Such discount schedules can be designed in the form of an all-unit quantity discount such that the average unit price curve touches the buyer's iso-cost curve p^{**} at Q^{**} and lies above it elsewhere. The simplest way to achieve it is to offer a single-step schedule: Charge the discount price p^{**} only when the buyer orders Q^{**} (or greater), and charge P_0 otherwise [Schedule $(p^{**} \mid Q^{**})$]. Any multiple-step variations can be designed as long as the average price lies above the iso-cost curve, except at (p^{**}, Q^{**}).[7]

On the other hand, a simple *continuous* discount schedule can be designed with a price break point at Q^* (Dada and Srikanth 1987). Once the seller decides to keep a proportion λ $(0 \leq \lambda \leq 1)$ of the joint gain, the following quantity discount [Schedule $p_\lambda(Q)$] can be used:

$$
\begin{aligned}
p_\lambda(Q) &= \lambda\, P_{max}(Q) + (1 - \lambda)P_{min}(Q), \text{ when } Q > Q^*, \text{ and} \\
&= P_0, \text{ when } Q \leq Q^*.
\end{aligned}
\tag{16}
$$

Note that $p_\lambda(Q^{**}) = p^{**}$. With this discount schedule, the buyer's cost minimization problem is aligned with the joint cost minimization, which in turn is aligned with the seller's cost minimization. To see this, note that substituting $p_\lambda(Q)$ for $p(Q)$ in Equation (15) results in

$$
\begin{aligned}
G_b(Q, p_\lambda(Q)) &= (1 - \lambda)D(P_{max}(Q) - P_{min}(Q)), \\
G_s(Q, p_\lambda(Q)) &= \lambda D(P_{max}(Q) - P_{min}(Q)).
\end{aligned}
\tag{17}
$$

7. Lal and Staelin (1984) provide one of the earlier quantity discount schedules that use an exponential approximation. Their schedule is somewhat discontinuous from the current presentation and is summarized in Appendix A.

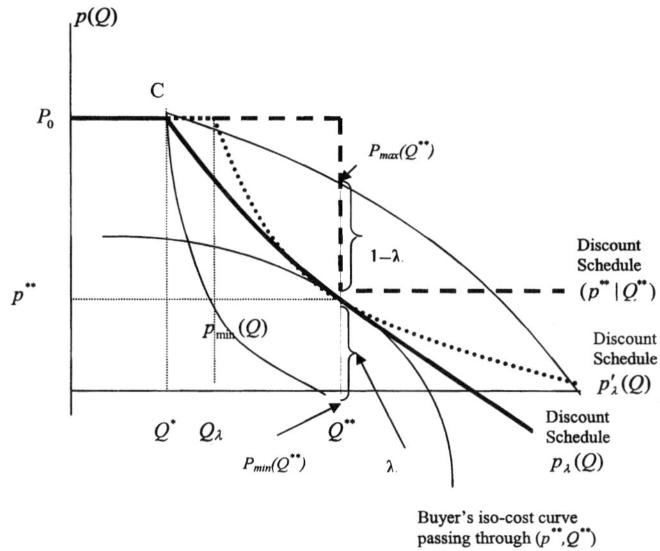

Figure 3. All-unit quantity discount schedules.

Thus, the buyer's objective of maximizing his share of the gain G_b is equivalent to maximizing the joint gain, providing incentive compatible objectives. The seller transfers to the buyer $(1 - \lambda)$ of the total saving realized at order quantity $Q^{\star\star}$.

A modification of schedule $p_\lambda(Q)$ can provide a stronger motivation for the buyer to order $Q^{\star\star}$. When the wholesale price is set at $P_{\min}(Q^{\star\star})$, all joint gain goes to the buyer, motivating the buyer to order $Q^{\star\star}$. However, the seller can take back a part of the gain by adding a per-unit fixed margin to $P_{\min}(Q^{\star\star})$ such that at $Q^{\star\star}$ the seller takes a proportion λ of the joint gain. The corresponding quantity discount schedule [Schedule $p'_\lambda(Q)$] is as follows:

$$\begin{aligned} p'_\lambda(Q) &= P_0, \text{ when } Q \leq Q_\lambda, \text{ and} \\ &= P_{\min}(Q) + \lambda\,(P_{\max}(Q^{\star\star}) - P_{\min}(Q^{\star\star})), \text{ for } Q > Q_\lambda. \end{aligned} \tag{18}$$

The price curve of $p'_\lambda(Q)$ is above $P_{\min}(Q)$ by the fixed margin. The price break point Q_λ can be obtained by equating the two equations in (18), and is the positive root of

$$H_s Q^2/2 \text{ p} + [S_s D/Q^\star - H_s Q^\star/2 - \lambda D(P_{\max}(Q^{\star\star}) - P_{\min}(Q^{\star\star}))]Q - S_s D = 0.$$

Figure 3 presents these three forms of all-unit quantity discount schedules and shows that $p'_\lambda(Q) > p_\lambda(Q)$ except at $Q^{\star\star}$, where all three price curves touch the iso-cost curve. This results in a steeper cost increase when the buyer's order deviates from the channel-optimal order quantity $Q^{\star\star}$, thus providing a stronger motivation for the buyer to choose $Q^{\star\star}$.

3.5 Incremental quantity discount schedules

Note that Schedules $p_\lambda(Q)$ and $p'_\lambda(Q)$ require the computation of the P_{max} and P_{min} as functions of Q, which may not be very attractive in practice. A similar incentive compatibility can be designed in the form of an incremental quantity discount that does not require the knowledge of such functions. A simple schedule can be designed such that the seller charges a discounted unit price of rP_0, $(0 \leq r \leq 1)$ for the quantity over the break point $\hat{Q}$ [Schedule $(\hat{Q}, r)$].

With this schedule, the total wholesale price for a quantity $Q > \hat{Q}$ is $P_0(\hat{Q} + r(Q - \hat{Q}))$, and the average unit price is $P_0(r + (1 - r)\hat{Q}/Q)$. Substituting this average unit price into the buyer's problem (1), we have

$$C_b(Q) = P_0(r + (1 - r)\hat{Q}/Q)D + S_b(D/Q) + H_b(Q/2). \tag{19}$$

Its minimum point occurs at

$$\tilde{Q} = \sqrt{\{2D\hat{Q}P_0(1 - r) + 2DS_b\}/H_b} = \sqrt{2D\hat{Q}P_0(1 - r)/H_b + Q^{\star 2}},$$

where $Q^\star$ is the buyer's myopic EOQ. The seller designs the incremental quantity discount schedule by solving $Q^{\star\star} = \tilde{Q}$ and $p^{\star\star} = P_0(r + (1 - r)\hat{Q}/Q^{\star\star})$:

$$r = (H_b Q^{\star 2} + 2p^{\star\star} Q^{\star\star} D - H_b Q^{\star\star 2})/(2P_0 Q^{\star\star} D), \tag{20}$$
$$\hat{Q} = Q^{\star\star} (p^{\star\star} - rP_0)/P_0(1 - r).$$

Any desired price-quantity position within the feasible region thus can be achieved by employing an incremental quantity discount schedule Equation (20). Its graphical property can be shown to be similar to that of the $p'_\lambda(Q)$ schedule in Figure 4, which follows from the fact that any incremental discount can be converted into an all-unit discount by using the average price function. Note that all these quantity discount schedules can be classified into two-block tariffs as discussed in Section 2.

3.6 Quantity discount under value-based inventory holding cost

One of the key assumptions in the above models was that the inventory holding cost is independent of the value of inventories. This simplification made the analytic results more tractable and provided a nice property of the Pareto efficient frontier being achieved at a single value of $Q^{\star\star}$ that is independent of the discounted price.

However, when the inventory holding cost is value-based, the problem becomes substantially more complicated. The buyer's total cost function now becomes

$$C_b(Q) = pD + S_b(D/Q) + pH_b(Q/2), \tag{21}$$

and the buyer's EOQ and the resulting cost are $Q^\star = (2S_bD/pH_b)^{1/2}$ and $C_b(Q^\star) = pD + (2DS_bpH_b)^{1/2}$, respectively. Note that the EOQ is now a function of the wholesale price. Due to the resulting functional complexity, Monahan (1984) uses an approximate function in deriving the optimal joint EOQ: $Q^{\star\star} = KQ^\star$ where $K = [(S_s/S_b) + 1]^{1/2}$. This ratio makes intuitive sense in that, as the seller's set-up cost is higher, the seller wants the buyer to order in a larger quantity. Interestingly, when the two set up costs are the same, the jointly optimal order quantity is at least 40% larger than the buyer's myopic EOQ. However, the approximation causes $Q^{\star\star}$ to be independent of the wholesale price as in the previously discussed models.

With the value-based inventory holding cost, it is shown that $Q^{\star\star}$ and $p^{\star\star}$ are not independent, and that the system-optimal discount solution can be unique. Facing the complexity, several authors formulate the problem as constrained optimization programs and provide numerical algorithms to compute the optimal discount schedule under various inventory/production assumptions (Banerjee 1986; Chakravarty and Martin 1991; Lee and Rosenblatt 1986). The numerical approach allows for more complex scenarios such as perishable products (Chakravarty and Martin 1989) and the buyer's order-grouping in multiple-item transactions (Chakravarty 1984).

On the other hand, closed-form quantity discount solutions were obtained by Chakravarty and Martin (1988) and Kim and Hwang (1989). Suppose, via a bargain or another gain-partitioning mechanism, the seller decides to take a portion λ $(0 \leq \lambda \leq 1)$ of the total gain. For simplicity, assume that the seller's order policy is a lot-to-lot ordering, thereby eliminating the seller's inventory holding cost. The seller's profit is $\Pi_s(p, Q) = pD - S_sD/Q$, and the buyer's cost is $C_b(p, Q) = S_bD/Q + H_bpQ/2 + pD$. Now the joint cost minimization problem becomes

$$\text{Min } J(p,Q) \equiv C_b(p,Q) - \Pi_s(p,Q) = S_bD/Q + H_bpQ/2 + S_sD/Q \tag{22}$$

$$\text{s.t. } (1 - \lambda)\{\Pi_s(p,Q) - \Pi_s(P_0,Q^\star)\} = \lambda\,\{C_b(P_0,Q^\star) - C_b(p,Q)\} \tag{23}$$

From (23), we get

$$p = 2\{\lambda QC_b(P_0,Q^\star) + (1 - \lambda)Q\Pi_s(P_0,Q^\star) + (1 - \lambda)S_sD - \lambda S_bD\}/\{Q(\lambda H_bQ + 2D)\}. \tag{24}$$

Substituting p in Equation (22) with (24), the joint cost is a function of only Q. Kim and Hwang (1989) show that this joint cost function is strongly unimodal in Q, and its minimum occurs at

$$Q^{\star\star} = (-Y + \sqrt{Y^2 - 4XZ})/(2X), \tag{25}$$

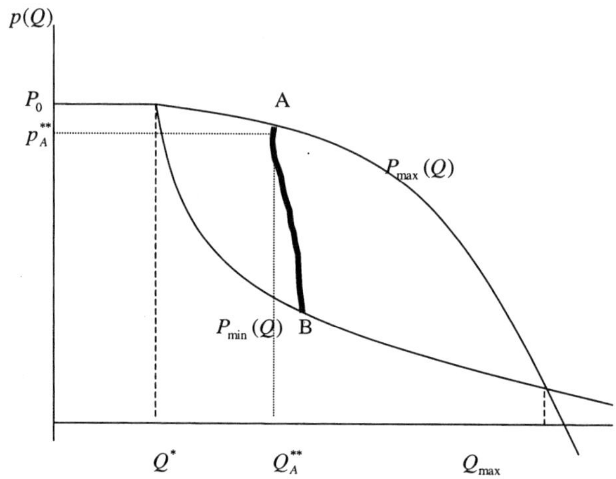

Figure 4. Quantity discount with value-based inventory holding cost.

where $X = 2H_b\{\lambda C(P_0, Q^\star) + (1 - \lambda)\Pi(P_0, Q^\star)\} - \lambda H_2^2 S_s$, $Y = -4\lambda H_b(S_b + S_s)D$, and $Z = -4(S_b + S_s)D^2$. Once $Q^{\star\star}$ is obtained, substituting it in Equation (24) results in the corresponding discounted wholesale price $p^{\star\star}$.

We saw in Section 3.2 that, when the inventory holding cost is unit-based, $Q^{\star\star}$ is fixed regardless of the value of λ, which in turn determines $p^{\star\star}$. In the current case, however, $Q^{\star\star}$ is a function of λ—the optimal order quantity depends on how the gain is split between the seller and buyer. Figure 4 shows the locus of the Pareto-efficient solution that is not a vertical line. Point A is the solution when $\lambda = 1$, and point B when $\lambda = 0$. Independent numerical examinations by Chakravarty and Martin (1988), Kim and Hwang (1989), and Dada and Srikanth (1987) show that the joint cost $J(p, Q)$ is minimized at point B, where the buyer takes all the gain. It implies that all other division of the gain is less efficient due to the constraint on λ (in Equation (23)). This opens a possibility for a fixed per-order charge while maintaining the quantity discount at the system-optimal level, which is left for future research.

Once the optimal solution pair ($p^{\star\star}$, $Q^{\star\star}$) is identified, incentive compatible discount schedules can be designed in the same way as in Sections 3.4 and 3.5. Deriving all-unit discount schedules is straightforward: A discrete step-discount schedule can offer a discounted price $p^{\star\star}$ at the break point $Q^{\star\star}$. Continuous all-unit discount schedules analogous to $p_\lambda(Q)$ and $p'_\lambda(Q)$ can be derived once $P_{\max}(Q)$ and $P_{\min}(Q)$ functions are identified.[8]

Also, an analogous procedure to that of Section 3.5 can be used to derive an incremental quantity discount schedule $(\hat{Q}, r)$ that charges a discounted price rP_0 for the quantity exceeding breaking point $\hat{Q}$. To find the discount rate and the price-breaking

8. These functions can be derived from Equation (23) by setting $\lambda = 1$ and $\lambda = 0$, respectively.

point, the buyer's cost function (19) is modified to reflect the impact of price on the inventory holding cost:

$$C_b(Q) = P_0(r + (1 - r)\hat{Q}/Q)D + S_b(D/Q) + H_b(rP_0Q + (1 - r)P_0\hat{Q})/2. \quad (26)$$

Its minimum point occurs at

$$\hat{Q} = \sqrt{\{2(S_b + (1 - r)P_0\hat{Q})D\}/(rP_0H_b)} = \sqrt{(2(1 - r)D\hat{Q} + Q^{\star 2}H_b)/(rH_b)}.$$

The seller designs the incremental quantity discount schedule by solving an equation system $Q^{\star\star} = \hat{Q}$ and $p^{\star\star} = P_0(r + (1 - r)\hat{Q}/Q^{\star\star})$:

$$r = (H_bQ^{\star 2} + 2p^{\star\star}Q^{\star\star}D)/(2Q^{\star\star}D + H_bQ^{\star\star 2}), \quad (27)$$
$$\hat{Q} = Q^{\star\star}(p^{\star\star} - rP_0)/P_0(1 - r).$$

When value-based inventory holding cost is assumed, therefore, the jointly optimal order quantity depends on how the gain is divided between the buyer and seller. This dependency makes the problem substantially more complicated. Although closed-form discount schedules were obtained in simple cases, more complicated scenarios would require, at least for now, using numerical procedures to compute the optimal price-quantity pair and a discount schedule. For instance, although Chakravarty and Martin (1988) were able to obtain a similar closed-form solution in a single buyer case, they had to resort to a numerical computation for a multiple-buyer scenario. In the latter case, individual buyer-level discount policies may not be possible under a single value of λ. Chakravarty and Martin (1988) argue that (a) the resulting price discrimination may have legal issues and (b) the set of individual buyers' iso-cost curves may not have a feasible region. To bypass these issues, they suggest grouping of buyers for a common order interval (or equivalently, order quantity). The grouping method is based on the shortest-path algorithm that minimizes the joint cost under a negotiated value of λ.

4. QUANTITY DISCOUNT SCHEDULES BASED ON DEMAND ELASTICITY

In the above quantity discount schedules, all channel members benefit from the coordination due to the efficiency gain created by reduced transaction costs. This efficiency gain is achieved even without considering the impact of the reduced wholesale price on retail price, which in turn may affect the demand. The marketing channel literature suggests that the optimal reaction of a retailer facing a decrease in the wholesale price is to pass some of the savings to the final consumers. The resulting increase in demand more than compensates for the decreased retail price. This section considers this second source of gain as a motivation for channel coordination.[9]

When there is no coordination between the retailer and manufacturer, the retail price chosen by the retailer in a Nash or a Stackelberg game is not Pareto-efficient. The maximum channel profit can be achieved by integration or perfect coordination (Jeuland and Shugan 1983). The coordination problem is to design the wholesale price (or transfer price) schedule such that it is in the retailer's best interest to choose a retail price that also maximizes the total channel profit. Here again, quantity discount can be used to design such a price schedule. In this section, we present models solely based on demand elasticity. More complicated models that combine transaction efficiency *and* demand elasticity will be presented in the next section. Section 4.1 discusses the base model that provides a rationale for channel coordination via quantity discount. More refined quantity discount models are presented in Section 4.2. The model is extended to the case of multiple retailers in Section 4.3.

4.1 Base models under demand elasticity

In a simple channel model with a manufacturer and a retailer facing a downward-slopping demand, it is straightforward to show that the joint profit-maximizing retail price is lower than the Nash or Stackelberg equilibrium retail price. Using a simple linear demand function, Table 2 shows that the joint channel profit is the highest and that the price is the lowest in the integrated channel.[10] Thus, the integrated case serves as the benchmark for coordination. This implies that the manufacturer can use quantity discount in order to induce the retailer to lower the retail price down to the channel-optimal level.

In addition to the notation used in Section 2, we define several new variables:

$D(\mathrm{x})$ = yearly demand as a function of retail price x ($dD/dx < 0$)
$p^\star$ = optimal wholesale price in the absence of coordination (same as P_0 above)
$x^\star$ = retail price in the absence of coordination
$x^{\star\star}$ = retail price with channel coordination
c_s, c_b = per unit variable costs of the manufacturer and the retailer
F_s, F_b = yearly fixed costs of the manufacturer and the retailer
g_s, g_b = margin per unit of the manufacturer and the retailer

Note that we do not explicitly consider the operating costs in this section. The retailer's and the manufacturer's profit functions are specified respectively as follows:

$$\Pi_b = g_b D(x) - F_b, \tag{28}$$
$$\Pi_s = g_s D(x) - F_s,$$

9. From this section, we will call the members the "manufacturer" and the "retailer" to emphasize the latter's role of setting the retail price.
10. Within the proper parameter range, we have $\prod_j^{Integ} \geq \prod_j^{Nash} \geq \prod_j^{Stk}$ and $x_j^{Integ} \leq x_j^{Nash} \leq x_j^{Stk}$.

Table 2. Comparison of retail prices and profits among three game scenarios[a]

	Nash	Stackelberg[b]	Integrated
$\Pi_s + \Pi_b$	$2(1 - b(c_b + c_s))^2/(9b)$	$3(1 - b(c_b + c_s))^2/(16b)$	$(1 - b(c_b + c_s))^2/(4b)$
x	$(2 + b(c_b + c_s))/(3b)$	$(3 + b(c_b + c_s))/(4b)$	$(1 + b(c_b + c_s))/(2b)$

[a] Assumed a simple linear demand function, $D(x) = 1 - bx$, and zero fixed costs.
[b] Either seller-leader or buyer-leader Stackelberg game

where demand $D(x)$ is a decreasing function of retail price, which is the sum of all variable costs and margins:

$$g_s + g_b + c_s + c_b = x. \tag{29}$$

In this model, the manufacturer's and the retailer's decisions have symmetrical effects on each other's profits.

In the absence of joint ownership, the two parties need to find a coordination mechanism to achieve the highest joint profit. Among the mechanisms suggested by Jeuland and Shugan (1983) are (a) simple contracts to fix respective margins at the joint optimal level, (b) implicit understanding (Jeuland and Shugan 1985), and (c) profit sharing using a wholesale price adjustment. In the first two mechanisms, each player has an incentive to deviate from the agreed upon margins and benefit at least in the short-term. However, a profit sharing mechanism can be made incentive compatible.

Such incentive compatible profit sharing can be achieved if the manufacturer and the retailer's profits are aligned with the joint profit with respect to the respective decision variables (i.e., margins). Suppose the manufacturer receives a fraction $k_1 (0 \leq k_1 \leq 1)$ of the total channel profit plus a fixed amount k_2. The manufacturer and the retailer profit functions are given by

$$\Pi_s = k_1[(g_s + g_b)D(x)] + k_2 - F_s, \tag{30}$$
$$\Pi_b = (1 - k_1)[(g_s + g_b)D(x)] - k_2 - F_b.$$

Note that both profits are linearly proportional to the joint profit, so that the first order conditions for maximizing Π_s, Π_b, and $(\Pi_s + \Pi_b)$ all correspond to

$$(g_s + g_b)\frac{dD(x)}{dx} + D(x) = 0. \tag{31}$$

A realistic mechanism for implementing this profit sharing is to employ a quantity discount pricing between the manufacturer and retailer. The manufacturer can offer a quantity discount, which is designed to induce the retailer to choose the jointly optimal retail price. Since the jointly optimal retail price is lower than that of an uncoordinated case, the manufacturer can lower the wholesale price so that the retailer can pass through a part of it to the retail price.

Now we derive a quantity discount schedule that can achieve this goal. The manufacturer's profit under coordination in (30) can be rewritten as

$$\Pi_s = k_1(x - c_s - c_b)D(x) + k_2 - F_s. \quad (32)$$

If this coordinated target profit level is to be achieved via a wholesale price schedule, Equation (32) needs to be equated with the manufacturer's profit function in (28):

$$\Pi_s \equiv (p - c_s)D(x) - F_s = k_1(x - c_s - c_b)D(x) + k_2 - F_s. \quad (33)$$

Solving (33) for wholesale price p, we obtain

$$p(D) = k_1 x(D) + (1 - k_1)c_s - k_1 c_b + k_2/D, \quad (34)$$

which represents the wholesale price as a function of demand. One can check the retailer's incentive compatibility by substituting the wholesale price $p(D)$ into the retailer's profit function in (28) to obtain the retailer's target profit in (30). That is, with the wholesale price (34), the retailer's profit is linearly aligned with the joint channel profit. An examination of the wholesale price (34) reveals that the wholesale price decreases as the demand increases: i.e., $dp(D)/dD < 0$. Thus, the wholesale price function represents a quantity discount. This quantity discount schedule has three components: a revenue sharing mechanism (the first term), a cost sharing mechanism (the next two terms), and a fixed payment or franchise fee, k_2, per the planning period (the last term). Its implementation does not require a complete knowledge of the demand function, but requires the knowledge of both variable costs.

For a division of the Pareto-optimal joint profit between the manufacturer and the retailer, Jeuland and Shugan (1983) suggest a bargaining solution in the determination of k_1 and k_2. As discussed in Section 3.3, two common bargaining solutions lead into an equal division of the additional profits. Eliashberg's (1986) arbitration model can also be applied when there is an asymmetry in the bargaining power.

4.2 Using two-part tariffs

The quantity discount schedule (34) may require substantial administrative costs due to its complexity. McGuire and Staelin (1986) provide a slight modification of the schedule that simplifies the components. Their solution is to set the wholesale price to be

$$p(D) = c_s + k_2/D \quad (35)$$

and to charge a fixed franchise fee $M(D) = k_1\{(x(D) - p(D) - c_b)D + k_2\}$. The total retailer's payment to the manufacturer is

$$p(D)D + M(D) = k_1(x(D) - c_s - c_b)D + c_s D + k_2, \quad (36)$$

which results in the average per unit transfer price being the same as in the Jeuland and Shugan's schedule. Moreover, the retailer's profit function is now linear in the total channel profit:

$$\Pi_b = (x - p(D) - c_b)D - F_b - M(D) \tag{37}$$
$$= (1 - k_1)(p - c_s - c_b)D - F_b - k_2'$$

which is exactly the same as the retailer's target profit in (30).

Moorthy (1987) also shows that Jeuland-Shugan's quantity discount schedule, which consists of a fixed franchise fee plus per-unit price, is not the only method. In Jeuland-Shugan's pricing (34), the total price the retailer pays is $p(D)D$, and thus his marginal *transfer price* is

$$\frac{\partial p(D)D}{\partial D} = k_1\left(\frac{\partial x(D)}{\partial D}D + x(D) - c_s - c_b\right) + c_s, \tag{38}$$

which is independent of the fixed fee k_2. Accordingly, the retailer's *total marginal cost* is the transfer price plus his cost: $k_1(\frac{\partial x(D)}{\partial D}D + x(D) - c_s - c_b) + c_s + c_b$. To determine optimal retail price, the retailer will seek to equate this marginal cost with his marginal revenue:

$$k_1\left(\frac{\partial x(D)}{\partial D}D + x(D) - c_s - c_b\right) + c_s + c_b = \frac{\partial x(D)}{\partial D}D + x(D). \tag{39}$$

We now show that this retailer's optimality condition is the same as that of the joint optimality condition. The first order condition for maximum joint channel profit, $\Pi_j = (x(D) - c_s - c_b)D - F_s - F_b$, is $\frac{\partial x(D)}{\partial D}D + x(D) - c_s - c_b = 0$. Under this condition, it is easy to see that the condition (39) is satisfied; i.e., the retailer's profit is maximized at the same retail price that maximizes the channel profit. This implies that the fixed payment part k_2 is not needed for channel coordination. A revised parameter $k_1' = k_1 + k_2/(x - c_s - c_b)D(x)$, which generates a straight linear quantity discount schedule, is equivalent to the two-parameter quantity discount schedule.[11]

However, Moorthy (1987) shows it does not have to be a quantity *discount* to achieve channel coordination. Any pricing schedule, including a quantity surcharge, that can equate the retailer's marginal cost with his marginal revenue (i.e., $D\frac{\partial x(D)}{\partial D} + x(D) = c_s + c_b$) at the channel-optimal demand level $D^{\star\star}$ will work, as long as the marginal cost is strictly below his marginal revenue at all demand levels $D < D^{\star\star}$ (i.e., the marginal cost curve "cuts" the marginal revenue curve from below

11. k_1' can be obtained by equating the resulting profit functions as follows: $k_1'(p - c_s - c_b)D(p) - F_s = k_1(p - c_s - c_b) D(p) + k_2 - F_s$ for the seller and $(1 - k_1')(p - c_s - c_b)D(p) - F_b = (1 - k_1)(p - c_s - c_b) D(p) - k_2 - F_b$ for the buyer.

at $D^{\star\star}$). The simplest such pricing is to charge the retailer a constant price of c_s (the manufacturer's marginal cost), and to extract the desired profit in the form of a fixed payment or a franchise fee, $k_1(x^{\star\star} - c_s - c_b)D(x^{\star\star}) + k_2$. McGuire and Staelin's (1986) quantity discount schedule (35) is a special case of this more general form, where $k_1 = 0$.

4.3 Multiple retailers

While most models presented above are concerned with a single retailer or a group of homogeneous retailers, manufacturers in the modern economy typically distribute through many retailers. Thus a more realistic channel coordination model needs to allow more than one retailer. Ingene and Parry (1995) and Chen, Federgruen, and Zheng (2001) respectively present quantity discount schedules that can coordinate the channel members' decisions for maximum joint profit. Chen et al.'s model also combines the EOQ and elastic demand approaches, and will be discussed in the next section.[12]

Ingene and Parry (1995) consider a model with N independent retailers with different cost and demand structures, where N is endogenously determined. The retailers are assumed to have exclusive territories so that actions taken by a retailer have no effect on the other retailers' demands. Let x_i and $D_i(x_i)$ respectively denote the retail price and the corresponding demand function of the i-th retailer. We present three scenarios of pricing by the manufacturer: (a) integrated retail operation, (b) autonomous retail divisions, and (c) independent retailers.

Again, the integrated channel provides the benchmark profit:

$$\Pi_I = \sum_{i=1}^{N_I} (x_i - c_i - c_s) D_i(x_i), \tag{40}$$

where N_I is the number of retailers that are profitable, and c_i is the unit variable cost of the i-th retailer (for notational simplicity, subscript b is omitted here). Since all retail prices are centrally determined, maximizing Equation (40) with respect to x_i results in N_I first order conditions:

$$\frac{d\Pi_I}{dx_i} = D_i(x_i) + (x_i - c_i - c_s)\frac{dD_i(x_i)}{dx_i} = 0. \tag{41}$$

Define the price elasticity at retail outlet i as

$$\eta_i = -\frac{x_i}{D_i}\frac{dD_i(x_i)}{dx_i} > 0. \tag{42}$$

A recursive solution for Equation (41) can be obtained as:

12. Chakravarty and Martin's (1988) multiple-buyer model is based on transaction efficiency and was discussed in the previous section.

$$x_i^{**} = \frac{(c_i + c_s)\eta_i^{**}}{\eta_i^{**} - 1} \text{ subject to } x_i^{**} > 0. \quad (43)$$

One necessary condition for this solution is that the price elasticity at the optimal solution needs to be greater than one.

Next we determine the number of retail outlets in the integrated channel, N_I. Let Π_i^{**} denote retailer i's profit at this integrated optimum, and suppose the retailer index i is ordered according to their profits. The profit-maximizing central planner will retain only those retail outlets that provide N_I positive profit contributions such that

$$\Pi_1^{**} > \Pi_2^{**} > \Pi_3^{**} > \ldots > \Pi_{N_I}^{**} \geq 0 > \Pi_{N_I+1}^{**} \ldots \quad (44)$$

In the autonomous retail division scenario, the firm can use transfer pricing for channel coordination. The manufacturing division will charge T_i per unit to retailer i, who in turn will maximize his own profit with respect to the retail price:

$$\Pi_i = (x_i(T_i) - c_i - T_i)D(x_i(T_i)). \quad (45)$$

The optimal retail price is derived from its first order condition:

$$x_i^{*} = \frac{(c_i + T_i)\eta_i^{*}}{\eta_I^{**} - 1}. \quad (46)$$

Comparing Equation (46) with (43), we find that the only transfer price that makes the retail price equal to the benchmark price ($x_i^* = x_i^{**}$) is $T_i = c_s$, $\forall i$. At this price, each retailer's marginal revenue $x_i + (D_i(x_i)/D_i'(x_i))$ equals to his and the manufacturer's combined marginal cost $(c_i + c_s)$. This is consistent with Moorthy's (1987) price schedule (35), but the fixed fee is not relevant for the total channel profit goal.

In the independent retailer scenario, the manufacturer can also charge all retailers the same wholesale price, which equals to his marginal unit cost, c_s. This wholesale price motivates each retailer to choose the system-optimal retail prices. However, in order to profit from the transaction, the manufacturer needs to levy each retailer a periodic fixed fee $k_{2,i}$ as in (35). Thus, the average wholesale price for retailer i is:

$$p_i = c_s + \frac{k_{2,i}}{D_i}. \quad (47)$$

It is obvious that this wholesale price is an individualized two-part tariff schedule.

However, Ingene and Parry (1995) argue that charging different fixed fees for different retailers requires substantial information, and that a preferred pricing is to use a simple two-part tariff that has common fixed fee (i.e., $k_{2,i} = k_2$, $\forall i$). Although the fixed fee is not relevant to the retailer's pricing decision, it affects a retailer's decision whether to participate in the channel, since he needs a non-negative net profit (i.e., $\Pi_i^{\star\star} - k_2 \geq 0$) in order to participate.

From the manufacturer's point of view, increasing k_2 will decrease the number of retailers participating (N_c), which is an integer number, and this will lead into a profit function that is saw-tooth shaped. The total number of retailers to participate, however, is smaller than that of the integrated channel case (i.e., $N_I > N_c$), because the added fixed fee will make the marginal retailers' profits negative. This implies that there is an optimal level of fixed fee $k_2^{\star\star}$, which sets the optimal number of participating retailers.

However, when the same fixed fee is used, all retailers except $i = N_c$ will have a positive surplus profit. When the number of participating retailers is externally fixed at N_c, Ingene and Parry show that channel coordination may not be the best choice for the manufacturer's profit. A profit-maximizing manufacturer will set the wholesale price $p^\star$ that is different from $p^{\star\star} = c_s$, and the corresponding fixed fee $k_2^\star$ such that he extracts more of the positive profit from the retailers. Such pricing schedule $(p^\star, k_2^\star)$ can provide the manufacturer a profit level that is greater than or equal to the channel-coordinating pricing schedule $(p^{\star\star}, k_2^{\star\star})$. This is a direct contrast from the single retailer case above, in which channel coordination can benefit both the manufacturer and retailer.[13]

5. COMBINED QUANTITY DISCOUNT MODELS

In the previous two sections, we have seen two separate sources of total channel gain when the channel members coordinate their actions. In both cases, by providing a wholesale price quantity discount, the manufacturer can induce the retailer to choose a decision (order quantity or retail margin) that is optimal for the whole channel. However, there are some differences in these two forms of quantity discounts. In the transaction efficiency models, the quantity discount is based on the *quantity per order.* The yearly demand is assumed fixed, and the retail price is irrelevant. The discount types are all-unit and incremental quantity discount schedules, which do not have the fixed charge component. In the elastic demand model, on the other hand, the discount is based on the *yearly (or per planning horizon) demand.* The simplest discount schedule is the two-part tariff, in which the marginal wholesale price equals the manufacturer's cost and a periodic fixed fee is charged in order to meet the manufacturer's profit goal.

A natural extension is to combine these two sources in a single model. In a combined model, the quantity discount schedule is compounded by the fact that the manufacturer needs to induce the retailer not only to order the jointly optimal quantity but also to set the wholesale price so as to support the benchmark retail price. In this section, we summarize the models that consider both transaction efficiency and elastic demand in designing quantity discount schedule for channel coordination. First, we

13. However, as will be seen in the next section, when the manufacturer can apply different wholesale prices $w_i^\star$ to different retailers, channel coordination is still optimal.

present a model that considers one retailer in Section 5.1, followed by a model that deals with multiple retailers in Section 5.2.

5.1 One retailer case

When we combine the above two sources of gain, it is intuitive that the solution would combine an all-unit or incremental discount with a periodic fixed fee. Indeed, Weng (1995) shows that this combined quantity discount is sufficient for channel coordination. In his model, two slight modifications were made from those in the previous two sections: (a) the manufacturer has his own inventory replenishment policy (instead of maintaining "sufficient" inventories assumed in Section 3), and (b) the retailer has no variable cost other than the transaction cost. The inventory holding cost is assumed to be unit-based.

The manufacturer's yearly profit is represented as

$$\Pi_s(p) = (p - c)D(x) - S_sD(x)/Q - H_sQ/2. \tag{48}$$

The manufacturer's own replenishment cost S_s includes order processing and ordering (or setup, if he makes the product) costs per retailer's order. His inventory holding cost H_s also depends on whether he buys or makes the product.[14] Due to assumption (a), the manufacturer's inventory holding cost is a real *cost* that is a function of the retailer's order quantity Q. Thus, the inventory holding cost is *subtracted* from the profit.[15] Note that both costs are expressed in the unit of the *retailer's* order quantity.

The retailer maximizes his yearly profit with respect to two variables (the retail price and the order quantity):

$$\Pi_b(x,Q) = (x - p)D(x) - S_bD(x)/Q - H_bQ/2. \tag{49}$$

In Section 4, we notice that the retailer's variable cost plays little role in shaping the discount schedule, and assumption (b) allows for a simpler formulation for the retailer profit function without loss of generality.

When there is channel coordination, the joint profit function is

$$\Pi_j(x, Q) = \Pi_s(p) + \Pi_b(x, Q). \tag{50}$$

For a given retail price x, the joint operating cost is minimized by the following joint EOQ:

$$Q_j^{**}(x) = [2S_jD(x)/H_j]^{1/2}, \tag{51}$$

14. In the operations management literature, it is well known that the seller's optimal lot size is an integer multiple of the retailer's order quantity Q. Both S_s and H_s are functions of this multiple coefficient (m). (See Weng (1995) and Joglekar (1988).) We assume the seller always sets the optimal value of m.

15. In Section 3, due to the "sufficient" inventory assumption, the inventory holding cost was *subtracted* from the *total cost* by the amount of the retailer order (see equation 4).

where $S_j = S_s + S_b$ and $H_j = H_s + H_b$, and the resulting operating cost is $[2S_jH_jD(x)]^{1/2}$. In the integrated channel, the wholesale price terms are cancelled out in the joint profit function.

As before, let $x^{\star\star}$ denote the coordinated retail price, which is the optimal solution for the joint profit function (50). Note that, at $x^{\star\star}$, the demand $D(x^{\star\star})$ and joint optimal order quantity $Q_j^{**}(x^{\star\star})$ is uniquely defined. Now the manufacturer's coordination problem can be decomposed into two interrelated sub-problems. The first is to motivate the retailer to set the retail price at $x^{\star\star}$ instead of $x^{\star}$ that maximizes his profit (49). This can be achieved by adjusting the wholesale price such that the retailer's *marginal cost* is same as the joint marginal cost. Given $x^{\star\star}$, the second problem is to induce the retailer to choose $Q_j^{**}(x^{\star\star})$ instead of $Q_j^{*}(x^{\star\star})$ that minimizes his own operating cost. This can be achieved by setting the *average wholesale price* such that the joint gain is shared between the channel members according to a pre-determined ratio. These two problems correspond to those in Sections 4 and 3, respectively, and they are in fact Stackelberg games with the manufacturer as the leader.

We first consider the order quantity problem. As in Section 3.2, we can derive the feasible region in which both the manufacturer and retailer are no worse off. For any $(x, Q_j(x))$, the minimum wholesale price p that makes the manufacturer no worse off needs to satisfy $\Pi_s(p \mid Q_j(x)) = \Pi_s^{\star}$. Let $P_{\min}(x)$ denote such retail price, which is obtained as

$$P_{\min}(x) = c + \{\Pi_s^{\star}/D(x) + (S_s/S_j + H_s/H_j)[S_jH_j/2D(x)]^{1/2}\}. \tag{52}$$

Likewise, the maximum wholesale price that makes the retailer no worse off is obtained from $\Pi_b(x, Q_j(x)) = \Pi_b{\star}$:

$$P_{\max}(x) = x - \{\Pi_b^{\star}/D(x) + (S_b/S_j + H_b/H_j)[S_jH_j/2D(x)]^{1/2}\}. \tag{53}$$

Applying Equations (52) and (53) to (50), we have

$$\Pi_j(x \mid Q_j(x)) = (\Pi_s^{\star} + \Pi_b^{\star}) + (P_{\max}(x) - P_{\min}(x))D(x). \tag{54}$$

This is analogous to Equations (10) and (12), except that the minimum and maximum wholesale prices are now a function of the retail price instead of the order quantity. However, since the joint order quantity $Q_j(x)$ is uniquely defined as a function of the retail price (Equation (51)), the feasible region can be defined in the same way as in Figure 2. As in Section 3 (Equation (14)), Equation (54) suggests that the difference between the minimum and maximum prices represents the unit gain due to the decreased operating cost.

Suppose the coordinated retail price $x^{\star\star}$ is already in place as will be discussed below. The joint gain is maximized at the point where ($P_{\max}(x^{\star\star}) - P_{\min}(x^{\star\star})$) is the greatest, and the corresponding optimal order quantity $Q^{\star\star}(x^{\star\star})$ is independent of the wholesale price. The increased joint profit due to coordination can be partitioned in the same way as discussed in Section 3.3. Suppose the manufacturer wants to keep a proportion, λ, of the joint gain: $\lambda[P_{\max}(x^{\star\star}) - P_{\min}(x^{\star\star})]D(x^{\star\star})$. This implies that the unit wholesale price needs to be $p_j^{**} = \lambda P_{\max}(x^{\star\star}) + (1 - \lambda)P_{\min}(x^{\star\star})$. As in Section 3, either incremental or all-unit discount schedules that make the *average wholesale price* equal to p_j^{**} at $Q^{\star\star}$ $(x^{\star\star})$ and above the retailer's iso-cost curve elsewhere will coordinate the order quantity.

The next problem is to motivate the retailer to choose the corresponding joint optimal retail price $x^{\star\star}$. The retailer's profit under the joint EOQ, $Q_j^{**}(x^{\star\star})$, is

$$\Pi_b(x) = (x - p_j^{**})D(x) - [2S_bH_bD(x)]^{1/2}. \tag{55}$$

Note that the retailer's optimal retail price, $x^{\star}(p_j^{**})$, which maximizes his profit (55), does not have to be the same as the joint-optimal price $x^{\star\star}$. (See Appendix B for a concavity condition of Equation (55).) As shown in Table 2, the coordinated retail price is lower than an uncoordinated one: $x^{\star}(p_j^{**}) > x^{\star\star}$ (see Weng (1995) for a proof). The retailer, in an uncoordinated pricing setting, has an incentive to choose a higher retail price, thereby upsetting the underlying assumptions of the joint EOQ. In order to induce the retailer to choose $x^{\star\star}$, the manufacturer has to lower the *marginal wholesale price* than p_j^{**}, while maintaining the *average wholesale price* at p_j^{**} at order quantity $Q_j^{**}(x^{\star\star})$.

In Section 4, where the operating costs were absent, the simplest solution was to set the wholesale price at the manufacturer's marginal cost (c_s) with a fixed charge. However, the operating costs in Equation (48) increases the manufacturer's, and hence the system's, marginal cost. Suppose the manufacturer's marginal cost is a fraction $(1 - \beta, 0 < \beta \leq 1)$ of the target average wholesale price p_j^{**}. If the manufacturer makes the wholesale price the same as his marginal cost, $(1 - \beta)\, p_j^{**}$, then the retailer's marginal cost becomes the same as the system's marginal cost. By equating his marginal cost with marginal revenue, the retailer chooses a retail price that is jointly optimal: $x^{\star}((1 - \beta)p_j^{**}) = x^{\star\star}$.[16] The manufacturer can recover his target revenue in the form of a periodic fixed fee $\beta p_j^{**}D(x^{\star\star})$. This fixed fee is a fraction of the retailer's gross profit, making the effective *average wholesale price* equal to p_j^{**}.

In summary, when transaction efficiency and elastic demand models are combined, both kinds of quantity discount schedules are required: (a) an all-unit or incremental quantity discount to motivate the retailer to choose Q_j^{**}, and (b) a two-part pricing which further reduces the unit wholesale price to the manufacturer's marginal cost but still maintains the average wholesale price by charging a periodic fixed fee.

5.2 Multiple retailers

When there are nonidentical multiple retailers, Chen, Federgruen, and Zheng (2001) show that individualized quantity discount schedules can be developed. Unlike Ingene and Parry's (1995) model, the number of retailers, N, in Chen et al.'s model is exogenously given. They, too, show that per-order quantity discount alone cannot guarantee to achieve perfect coordination among the retailers with different demand rate, even when their variable costs are the same.

16. Weng (1995) argues that there is a unique value of β that satisfies this condition without providing an interpretation.

Chen et al.'s discount mechanism is similar to that of Weng's (1995) in that it combines the order quantity-related discount and the fixed fee. However, in the multiple-retailers model, these discounts and fees are individualized for each retailer's demand and cost structure. This is in contrast to Ingene and Parry (1995) who advocate same fees to be applied to all retailers for practical considerations.[17] The proposed pricing schedule for coordination is as follows:[18] Each retailer i, $i = 1,2, \ldots, N$ is

(a) charged a basic per unit cost equal to

$$c_s + \frac{H_s Q_s^{**}}{2D_i}; \text{ and}$$

(b) given a per unit discount of

$$\frac{H_s \min\,[Q_s^{**}, Q_i]}{2D_i},$$

where Q_s^{**} is the manufacturer's own optimal order quantity from the integrated channel model. In addition, each retailer is charged a periodic fixed fee of $k_{2,i}$, which is chosen in a contract or through bargaining. Adding these two components, we obtain the average wholesale price to retailer i becomes

$$p_i(Q_i) = c_s + \frac{H_s Q_s^{**}}{2D_i} - \frac{H_s \min\,[Q_s^{**}, Q_i]}{2D_i}. \tag{56}$$

Notice that this wholesale price is basically the average cost to the manufacturer to fulfill an order from a retailer with yearly demand of $D_i(p_i)$ ordering a quantity of Q_i each time. The last term represents a discount based on the manufacturer's reduction in inventory holding cost. With this wholesale price, retailer i's profit before the fixed fee becomes

$$\Pi_i(D_i, Q_i) = p_i(D_i)D_i - \left(c_s + c_i + \frac{H_s(Q_z^{**} - \min\,[Q_s^{**}, Q_i])}{2D_i}\right) \tag{57}$$
$$D_i - \frac{H_i Q_i}{2} - \frac{S_i D_i}{Q_i}.$$

17. As discussed in the previous section, Ingene and Parry (1995) show that the seller's profit based on coordination with a common fixed fee can be lower than that from the uncoordinated profit maximization. However, they also point out that individualized fixed fees will make the coordinated solution the global optimum.
18. Instead of using the order quantity variable (Q), Chen et al's formulation uses the order interval T such that $Q = DT$. Since they are equivalent, we use Q for consistency of presentation. In addition, we omit the "account management expenses" of Chen et al. without loss of generality. For simplicity, the fixed per order costs are all paid by the buyer.

On the other hand, when the whole channel is under a central planning (i.e., an integrated channel), the joint profit maximization problem is

$$\Pi_j(D,Q) = \sum_{i=1}^{N} \{(p_i(D_i) - c_s - c_i)D_i - S_iD_i/Q_i - H_s \max\,[Q_s, Q_i]/2 - (H_i - H_s)Q_i/2\} - S_s\sum_i D_i/Q_s$$
$$= \sum_{i=1}^{N} \Pi_i(D_i, Q_i, Q_s) - S_s\sum_i D_i/Q_s. \tag{58}$$

Note that each retailer's maximization of Equation (57) also maximizes the total channel profit (58), thus the discount schedule above is incentive compatible. As in the demand-based quantity discount schedules in Section 4, the fixed fees are used for redistributing the joint gain from the retailers to the manufacturer. Again, it needs to be emphasized that this schedule is not a unique one for channel coordination. For example, one can use other methods for setting the fixed fees, including a Nash bargaining solution or Eliashberg's (1986) arbitration solution.

6. A THREE-PARTNER MODEL: CHANNEL COORDINATION INVOLVING LOGISTICS PARTNERS

In this section, we introduce a new analytical result that coordinates channel operations and guides the development of discount policies for a supply chain involving a manufacturer, a retailer, and a third party logistics partner (i.e., a transporter). In particular, we extend the results of Weng (1995) to a more general case where a transporter is explicitly involved in the channel and where the total supply chain profitability is affected by the way the three partners coordinate. It turns out that it is the transporter's cost, not the transportation price he charges, that is relevant to the channel coordination problem. We first present the model and a general result, followed by a specific example that examines the property of optimal price-order quantity combinations.

6.1 The three-partner model setting

In addition to the notations used in the previous sections, let u denote the unit transportation price that the transporter charges the manufacturer. Assume that the manufacturer pays for the transportation cost. Thus, the manufacturer's yearly profit can be represented as

$$\Pi_s(p) = (p - c)D(x) - S_s\,D(x)/Q - H_sQ/2 - uD(x) \tag{59}$$
$$= (p - c - u)D(x) - S_sD(x)/Q - H_sQ/2.$$

Similarly, we can represent the retailer's and transporter's profits as

$$\Pi_b(x, Q) = (x - p)D(x) - S_bD(x)/Q - H_bQ/2 \tag{60}$$

$$\Pi_t(u) = uD(x) - T(p,Q)D(x)/Q = (u - T_2(p))D(x) - T_1(p)D(x)/Q. \tag{61}$$

where $T(p,Q) = T_1(p) + T_2(p)Q$, and $T_1(p)$, and $T_2(p)$, stand for the fixed cost per transportation batch and the unit transportation cost, respectively. When each partner acts for his own interests (i.e., the retailer determines ordering quantity, $Q^\star$, the manufacturer determines wholesale price, $p^\star$, to the retailer, etc.), the retailer's ordering quantity becomes his own EOQ or $Q_b = (2S_bD(x)/H_b)^{1/2}$ with a yearly profit

$$\Pi_b(x \mid Q^\star) = (x - p)D(x) - [2S_bH_bD(x)]^{1/2}. \tag{62}$$

Given a unit purchasing price p, the retailer's objective is to choose a unit retail price x that maximizes his own yearly profit. Let $x^\star(p)$ denote this retailer's optimal retail price.[19] With $x^\star(p)$, the retailer's EOQ becomes $Q_b(p) = (2S_bD(x^\star(p))/H_b)^{1/2}$, and the manufacturer's, and the transporter's, yearly profit becomes:

$$\Pi_s(p) = (p - c - u)D(x^\star(p)) - (S_s/S_b + H_s/H_b)[S_bH_bD(x^\star(p))/2]^{1/2}, \tag{63}$$

$$\Pi_t = (u - T_2(p))D(x^\star(p)) - T_1(p)[H_bD(x^\star(p))/2S_b]^{1/2}, \tag{64}$$

respectively. When uncoordinated, the three channel partners are assumed to play a three-player Nash game.[20]

The joint yearly profit function of the manufacturer, the transporter, and the retailer is defined as follows:

$$\begin{aligned}\Pi_J(x,p,Q) &= \Pi_s(p) + \Pi_t(u) + \Pi_b(x,Q)\\ &= (x - c - T_2(p))D(x) - (S_s + S_b + T_1(p))\\ &\quad D(x)/Q - (H_s + H_b)Q/2\\ &= (x - c')D(x) - S_JD(x)/Q - H_JQ/2\end{aligned} \tag{65}$$

where $c' = c + T_2(p)$, $S_J = S_s + S_b + T_1(p)$, $H_J = H_s + H_b$. At any given retail price x, the optimal joint EOQ is $Q_J = [2S_JD(x)/H_J]^{1/2}$, and the corresponding joint profit becomes

$$\Pi_J(x, p \mid Q_J(x)) = (x - c')D(x) - [2S_JH_JD(x)]^{1/2}. \tag{66}$$

19. $\Pi_b\ (x \mid Q_b)$ may or may not be concave in x, depending on the structure of $D(x)$. Nevertheless, we know that $x^\star\ (p)$ must exist, because as $x \to \infty$, $\Pi_b \to 0$, as $x \to p$, $\Pi_b < 0$, and when $p < x < \infty$, $\Pi_b > 0$.
20. With three partners, a Stackelberg game can be defined in several different ways, and they are very difficult to solve in our context.

This is equivalent to Equation (50) in the two-player channel, except that the cost parameters are redefined. By maximizing Equation (66) with respect to x, we can obtain the optimal retail price, $x^{\star\star}$, and the corresponding joint EOQ, $Q_J^{\star\star}$.[21]

LEMMA 1. Given the manufacturer's unit wholesale price $p^\star$, the retailer's unit retail price $x^\star(p^\star)$, and the transporter's operation cost per order $T_1(p) + T_2(p)Q$, the joint profit under the joint EOQ satisfies $\Pi_J(x^\star(p^\star), p^\star \mid Q_J(x^\star(p^\star))) \geq \Pi_s^\star + \Pi_t^\star + \Pi_b^\star$.

PROOF: (See Appendix C).

Derivable from the proof to Lemma 1, the following result can be shown to hold.

THEOREM 1. For any given $p > 0$, $x^\star(p) > p$, and the transportation partner's operation cost per order $T_1(p) + T_2(p)Q$, the supply chain yearly profitability under the joint EOQ satisfies $\Pi_J(x^\star(p), p \mid Q_J(x^\star(p))) \geq \Pi_s + \Pi_t + \Pi_b$.

Theorem 1 reveals a potential improvement on total profitability achievable by a joint optimization of partners' operational policies. Again, the joint gain can be distributed to the channel members as an incentive to coordinate. However, notice that the *transportation cost*, not the transportation price, is relevant in determining the joint gain. This is because the transportation price cancels out in the joint profit function, but the transportation cost increases the system's marginal cost. The retail price needs to be increased from our previous model to reflect this added cost. Due to the functional complexity, however, it is very difficult to analyze the property of the optimal price-quantity frontier. Instead, we examine its property using a specific demand function in the next section.

6.2 A specific example

Suppose that, for each shipment of quantity Q from the manufacturer to the retailer, the transporter's cost has the following structure:

$$T(p,Q) = T_1(P) + T_2(p)Q = a_1'\frac{p}{p_0} + a_2Q + a_3pQ = a_1p + (a_2 + a_3p)Q. \tag{67}$$

We consider a demand function used in Weng (1995): $D(x) = d/x^2$, where d is a large constant for market factor.[22] Replacing $D(x)$ with d/x^2 in Equation (62), the retailer's yearly profit becomes

$$\Pi_b(x \mid Q^\star) = \frac{d}{x} - \frac{dp}{x^2} - \frac{[2dS_bH_b]^{1/2}}{x}. \tag{68}$$

For the retailer to be willing to implement a joint optimal operation policy, Equation (68) needs to be no less than his uncoordinated profit: $\Pi_b(x \mid Q^{\star\star}) \leq \Pi_b(x \mid Q^\star)$. After a manipulation, we obtain the maximum wholesale price the manufacturer can charge:

21. The joint profit (66) is functionally equivalent to Equation (55), for which a concavity condition is presented in Appendix B.

22. Another demand function used by Weng, $D(x) = d/x^4$, also produces the same result. See Lei, Wang, and Fan (2002) for detailed analysis using these functions.

$$P_{\max} \le x - \frac{[2dS_bH_b]^{1/2}}{D(x)}.$$

Given the range of manufacturer's wholesale price, $p < P_{\max}$, the retailer's optimal retail price can be derived from his first order condition, $\frac{d\Pi_b}{dx} = \frac{1}{x^2}\left[\frac{2dp}{x} - (d - [2dS_bH_b]^{1/2})\right] = 0$, which produces the retailer's reaction function:

$$x^\star(p) = \frac{2p}{1 - [2S_bH_b/d]^{1/2}}. \tag{69}$$

Now we compare the coordinated joint profit with the sum of uncoordinated profits. Given the retailer's optimal retail price $x^\star(p)$, we have

$$\Pi_b + \Pi_s + \Pi_t = (x^\star - c - (a_2 + a_3p))\frac{d}{x^{\star 2}} - 2\left[1 + \left(\frac{S_s + a_1p}{S_b} + \frac{H_s}{H_b}\right)/2\right]\sqrt{\frac{dS_bH_b}{2x^{\star 2}}}.$$

$$\Pi_J = (x^\star - c - (a_2 + a_3p))\frac{d}{x^{\star 2}} - 2\left[1 + \frac{S_s + a_1\,p}{S_b} + \frac{(S_s + a_1p)H_s}{S_bH_b} + \frac{H_s}{H_b}\right]^{1/2}\sqrt{\frac{dS_bH_b}{2x^{\star 2}}}$$

Let $\Delta\Pi$ be the improvement on total profit jointly achieved by three partners. We have

$$\Delta\,\Pi = \Pi_J - \left(\Pi_b + \Pi_s + \Pi_t\right) = 2\left\{\left[1 + \left(\varphi(p) + \frac{H_s}{H_b}\right)/2\right] - \left[1 + \varphi(p) + \varphi(p)\frac{H_s}{H_b} + \frac{H_s}{H_b}\right]^{1/2}\right\}\sqrt{\frac{dS_bH_b}{2x^{\star 2}}} \tag{70}$$

where $\varphi(p) = \frac{S_s + a_1p}{S_b}$ and $x^\star$ is defined in Equation (69).

The quantity given by Equation (70) represents the improvement on joint profit that can be achieved through coordination. Now we show that, as manufacturer's wholesale price approaches either its lower c or upper bound $P_{\max}$, the improvement on supply chain profit by joint optimization approaches a constant:

$$p \rightarrow c,\ \Delta\Pi \rightarrow \frac{\left(\varphi(c) - \frac{H_s}{H_b}\right)^2 [d^{1/2} - (2S_bH_b)^{1/2}]\sqrt{S_bH_b/2}}{4c\left\{\left[1 + \left(\varphi(c) + \frac{H_s}{H_b}\right)/2\right] + \left[1 + \varphi(c) + \varphi(c)\frac{H_s}{H_b} + \frac{H_s}{H_b}\right]^{1/2}\right\}},$$

$$p \rightarrow P_{\max},\ \Delta\Pi \rightarrow \frac{\left(\varphi(P_{\max}) - \frac{H_s}{H_b}\right)^2 [d^{1/2} - (2S_bH_b)^{1/2}]\sqrt{S_bH_b/2}}{4P_{\max}\left\{\left[1 + \left(\varphi(P_{\max}) + \frac{H_s}{H_b}\right)/2\right] + \left[1 + \varphi(P_{\max}) + \varphi(P_{\max})\frac{H_s}{H_b} + \frac{H_s}{H_b}\right]^{1/2}\right\}}.$$

Furthermore, we have the following result (see Appendix C for the proof).

PROPOSITION 1. Consider a demand function $D(x) = d/x^2$. If $\left|\frac{S_s}{S_b} - \frac{H_s}{H_b}\right| > \frac{\alpha_1 p}{S_b}$, then the joint profit gain, $\Delta\Pi$, increases as p decreases.

In particular, if $H_s \approx H_b$ and $a_1 p \approx 0$, then $\Delta\Pi$ always increases as p decreases, as long as $S_s \neq S_b$. This implies that the most efficient wholesale price is $p^{**} = P_{\min}$, which is exactly the same result as that of two-partner models in Section 3.6. Any other division of the joint gain is less efficient unless there is a *per-order* fixed fee levied from the retailer while the minimum wholesale price is charged. Most existing models, however, solve for less efficient solutions for a given partition ratio of the joint gain.

Once the optimal *average wholesale price* $p^{\star\star}$ is determined to achieve a negotiated sharing of the joint gain, various all-unit or incremental quantity discount schedules can be designed in the same way as discussed in Section 3. However, it is not sufficient to induce the retailer to choose the target retail price $x^{\star\star}$. We know from Section 5 that, at wholesale price $p^{\star\star}$, the retailer's price reaction (69) will produce a higher retail price than the joint-optimal target retail price. In order to motivate the retailer to choose the joint-optimal retail price $x^{\star\star}$, the wholesale price can be set at the manufacturer's total cost, and a *periodic* fixed fee needs to be used to achieve the target manufacturer profit. Thus, the basic methodology of quantity discount still remains the same. However, manufacturer's total cost in this case also includes the transportation cost, so that the retailer's marginal cost is aligned with the system's marginal cost.

7. CONCLUSION AND FURTHER STUDIES

In this chapter, we have made a comprehensive review of analytical models on quantity discount pricing that coordinate the operations of channel partners. The review is based on a two-echelon supply chain (or a two-level channel) involving a manufacturer and a retailer. The quantity discount enables channel coordination between the members via (a) transaction efficiency and (b) elastic demand.

In the quantity discount models based on transaction efficiency, incentive compatible coordination between the channel members can be achieved by a price discount schedule that is designed to motivate the retailer to choose an order quantity that minimizes the total system cost. It is shown that either an all-unit discount or an incremental discount can provide such an incentive as long as the *average wholesale price* touches the retailer's iso-cost curve at the joint optimal order quantity level and above it elsewhere.

The elastic demand-based quantity discount models focuses on the impact of the wholesale price on the retailer's retail price decision. The main idea is to make the retailer's *marginal cost* equal to the system's marginal cost at the jointly optimal demand level. Thus the retailer's profit function is aligned with that of the system. Since the coordinated retail price is lower than the uncoordinated one, the manufacturer needs to lower the unit wholesale price. The lost revenue is recovered by charging a periodic fixed fee to achieve the desired profit (i.e., a two-part tariff). In combined models, which consider both transaction efficiency and demand elasticity, channel coordination requires combining an incremental or all-unit discount with a periodic fixed charge.

We also introduced a new model that considers the involvement of a third partner in a supply chain process—a third party logistics partner. Guidelines that are similar to the existing models were developed in designing quantity discount schedules. We also quantified the improvement on supply chain profitability from a joint optimization of the three partners using a special case and showed that the transporter's cost, not his price, is relevant for the coordination pricing.

Once theoretical foundations are laid for the use of quantity discount in coordinating channel decisions, the next step is to apply these models to practical problems. We note that mathematical programming is widely used to solve various types of practical problems in supply chain (Shapiro 2001). However, their "optimal solutions" which encompass decisions of multiple supply chain partners may not be truly optimal unless there is a mechanism to enforce the solution to other members. We believe that coordination among the members, which is achieved by incentive compatible goals, is the most stable one. Thus, it would be a fruitful future research topic to incorporate quantity discount into more practical mathematical programming models in order to derive optimal channel decisions that are incentive compatible.

APPENDIX A: A CONTINOUS DISCOUNT SCHEDULE (LAL AND STAELIN 1984)

Lal and Staelin suggest an incremental discount with m price breaks, in which the initial price is P_0 for any order $Q \leq Q_o$, P_1 for any quantity between Q_0 and Q_1, and so on down to price P_m, where $P_o > P_1 > \ldots > P_m$. This piece-wise linear discount schedule is approximated by a continuous discount schedule, $p(Q)$, which represents the average price. Their suggested function is:

$$p = P_0 \exp[-a(Q - Q^\star)] \quad \text{for } Q \geq Q^\star, \tag{A1}$$

$$p = P_0 \quad \text{for } Q \leq Q^\star,$$

where $a > 0$ is a constant and is a decision variable for the seller. The corresponding price-breaking point is set at $Q^\star$.

With this quantity discount price schedule (A1), the buyer's total cost function becomes:

$$C_b(Q) = pD + S_b\,(D/Q) + H_b\,(Q/2) \tag{A2}$$
$$= P_0 D \exp[\,-a(Q - Q^*)] + S_b(D/Q) + H_b(Q/2) \text{ for } Q \geq Q^*.$$

With the discount schedule (A1), the buyer can minimize his total cost by choosing an order quantity that satisfies the first order condition:

$$\frac{\partial C_b}{\partial Q} = -\,aDP_0\exp[-\,a[Q - Q^*)] - \frac{DS_b}{Q^2} + \frac{H_b}{2} = 0. \tag{A3}$$

Now the seller's problem is to choose the discount parameter $a = a^{\star}$ such that the buyer chooses $Q^{\star\star}$: i.e., find $a^{\star}$ that satisfies the following equation:

$$a^* DP_0\exp[-\,a^*(Q^{**} - Q^*)] = \frac{H_b}{2} - \frac{DS_b}{Q^{**2}}. \tag{A4}$$

This equation implies that the buyer's increased cost is cancelled out by the discount on the wholesale price.

While the resulting quantity discount schedule is sufficient to motivate the buyer to choose an order quantity, $Q^{\star\star}$, that minimizes the total channel cost, the price discount will reduce the seller's revenue at the same time reducing his operating costs. Thus, the seller needs to make sure that the cost reduction from the increased order quantity is sufficiently larger than the lost revenue. The seller's total cost after the price discount is

$$C_s(Q^{**}) = [P_0 - P_0 \exp[-a^*(Q^{**} - Q^*)]]D + DS_s/Q^{**} - H_sQ^{**}/2 \tag{A5}$$

Lal and Staelin show that the parameter conditions to ensure his total cost after the price discount is less than that before the discount, $C_s(Q^{\star\star}) < C_s(Q^{\star})$, are $D > 2Q^{\star\star}$ and $Q^{\star} > 0.364Q^{\star\star}$. These parameter conditions are interpreted as follows: (a) a buyer orders enough number of times in the first place to be able to reduce the number of orders per the time horizon, and (b) the required discount is not too large to cause a radical quantity change.

Once the seller's profitability condition is satisfied, an optimal value of $a^{\star}$ can be numerically found from the following equation, which is derived from (A4):

$$\frac{1}{DP_0}\left[\frac{H}{2} - \frac{DS_b}{Q^{**2}}\right] \exp\,[a^*(Q^{**} - Q^*)] - a^* = 0 \tag{A6}$$

This equation has two positive roots whenever order cost is such that $0.5S_b \leq S_s \leq S_b$ and inventory holding cost is such that $H_s < 0.8H_b$. Between the two roots of $a^{\star}$, the lower value provides the seller's best interest.

The solution to (A6) results in a wholesale price $p^{\star\star}$ (via equation (A1)) associated with the buyer's new optimal order quantity $Q^{\star\star}$ at which the total system cost is minimized. At this price, the entire system gain from the coordination goes to the seller, and the buyer's cost is no higher with the new order quantity than that without the price discount. When a different partitioning of the joint gain is desired for some reason, Lal and Staelin (1984) provide a variation of the discount schedule that also sets a price-breaking quantity different from $Q^{\star}$. They also provide a quantity discount schedule for multiple groups of heterogeneous buyers using multiple price-breaking points designed for respective buyer groups. Although a closed form solution could not be derived for this problem, they provided a numerical algorithm that produces an optimal unified pricing policy. The discount schedule starts with the buyer with the smallest demand with the first price-break point as described above, and then the average price is held fixed until the next price-break point at which the next step of discount schedule begins, and so forth. In this numerical algorithm, buyer group i's discount schedule depends on the previously determined values of a_{i-1} and its price-breaking point.

However, this analysis implicitly assumes zero marginal cost of the product, which is an unlikely scenario in reality. Note that the quantity discount schedule (A1) is a decreasing convex function in Q, and the limiting discounted wholesale price when $Q \rightarrow \infty$ is zero. That is, the schedule cannot rule out the possibility that the discounted wholesale price to induce the buyer to order $Q^{\star\star}$ can be below the cost. To prevent this drawback, the seller's cost minimization problem needs to be replaced with a profit maximization problem that also includes the variable cost.

In fact, Lee and Rosenblatt (1986) point out that one can construct examples such that the amount of price discount even exceeds the selling price of the product. Their formulation adds a set of explicit constraints that specify that the amount of the discount cannot be larger than the seller's original profit margin (pre-discount price less unit variable cost) plus a desired unit profit margin. The resulting constrained optimization problem is not amenable to an analytical approach, and the authors provide an algorithm to solve it numerically.

APPENDIX B: CONCAVITY OF PROFIT FUNCTION (55)

A sufficient condition for the existence of a unique optimal retail price $x^{\star}$ is that the buyer's profit function (55) is concave in x. It's concavity depends on the form of the demand function and the two cost parameters, S_b and H_b.

For notational simplicity, let $m = 2S_bH_b$, and use p for $p_j^{\star\star}$. Equation (55) can be rewritten as $\Pi_b(x) = (x - p)D(x) - [mD(x)]^{1/2}$. Taking its second order derivative, we have

$$\frac{\partial^2\Pi_b}{\partial x^2} = 2D' + \frac{m^2D'^2}{4(mD)^{3/2}} + (x - p)\ D'' - \frac{mD''}{2\sqrt{mD}}.$$

Taking a common denominator, $4D\sqrt{mD}$, which is positive, the numerator becomes

$$mD'^2 + 2D(4\sqrt{mD}D' - (m - 2(x - p)\sqrt{mD})D'').$$

The condition that this quantity is non-positive is

$$m = 2S_bH_b \leq \frac{16D^3 + 2D' + (x - p)D'')^2}{(D'^2 - 2DD'')^2}.$$

That is, when the buyer's order and inventory holding costs are sufficiently small, the buyer's profit function is concave in x, and there is a unique optimal value of $x^\star$.

APPENDIX C: PROOFS TO LEMMA 1 AND PROPOSITION 1

LEMMA 1. Given the seller's unit wholesale price $p^\star$, the retailer's unit retail price $x^\star(p^\star)$, and the transporter's operation cost per order $T_1(p) + T_2(p)Q$, the supply chain yearly profit under the joint EOQ satisfies $\Pi_J(x^\star(p^\star), p^\star \mid Q_J(x^\star(p^\star))) \geq \Pi_s^\star + \Pi_t^\star + \Pi_b^\star$.

PROOF. From Equations (62), (63) and (64), we have

$$\Pi_s^\star + \Pi_t^\star + \Pi_b^\star$$
$$= (x^\star(p^\star) - c - T_2(p^\star))D(x^\star(p^\star)) - 2\left[1 + \left(\frac{S_s + T_1(p^\star)}{S_b} + \frac{H_s}{H_b}\right)/2\right]$$
$$[S_bH_bD(x^\star(p^\star))/2]^{1/2}$$

Since

$$\Pi_J(x^\star(p^\star), p^\star \mid Q_J(x^\star(p^\star)))$$
$$= (x^\star(p^\star) - c - T_2(p^\star))D(x^\star(p^\star))$$
$$-2\left[1 + \frac{S_s + T_1(p^\star)}{S_b} + \frac{(S_s + T_1(p^\star)H_s}{S_bH_b} + \frac{H_s}{H_b}\right]^{1/2}[S_bH_bD(x^\star(p^\star))/2]^{1/2}$$

and $\left[1 + \left(\frac{S_s + T_1(p^\star)}{S_b} + \frac{H_s}{H_b}\right)/2\right]^2 - \left[1 + \frac{(S_s + T_1(p^\star)}{S_b}\right.$

$$\left. + \frac{S_s + T_1(p^\star)H_s}{S_bH_b} + \frac{H_s}{H_b}\right] \geq 0,$$

the claim $\Pi_J(x^\star(p^\star), p^\star \mid Q_J(x^\star(p^\star))) \geq \Pi_s^\star + \Pi_t^\star + \Pi_b^\star$ holds.

PROPOSITION 1. If $|\frac{S_s}{S_b} - \frac{H_s}{H_b}| > \frac{a_1 p}{S_b}$, then the improvement on supply chain profitability by joint optimization, $\Delta\Pi$, increases as p decreases.

PROOF. Given

$$\Delta\Pi = \frac{\left(\frac{S_s + \alpha_1 p}{S_b} - \frac{H_s}{H_b}\right)^2 \left[d^{1/2} - (2S_b H_b)^{1/2}\right]\sqrt{S_b H_b/2}}{4p\left\{\left[1 + \left(\frac{S_s + a_1 p}{S_b} + \frac{H_s}{H_b}\right)/2\right] + \left[1 + \frac{S_s + a_1 p}{S_b} + \frac{(S_s + a_1 p)H_s}{S_b H_b} + \frac{H_s}{H_b}\right]^{1/2}\right\}}.$$

Let $Y = \frac{\left(\frac{S_s + a_1 p}{S_b} - \frac{H_s}{H_b}\right)^2}{p}$, then $\frac{dY}{dp} = \frac{\left(\frac{a^1}{S_b}p\right)^2 - \left(\frac{S_s}{S_b} - \frac{H_s}{H_b}\right)^2}{p_2}$. If $|\frac{S_s}{S_b} - \frac{H_s}{H_b}| > \frac{a_1 p}{S_b}$,

$\frac{dY}{dP} < 0$, and Y increases when p decreases. Furthermore, since

$$\Delta\Pi = \frac{[d^{1/2} - (2S_b H_b)^{1/2}]\sqrt{S_b H_b/2}}{4\left\{\left[1 + \left(\frac{S_s + a_1 p}{S_b} + \frac{H_s}{H_b}\right)/2\right] + \left[1 + \frac{S_s + a_1 p}{S_b} + \frac{(S_s + a_1 p)H_s}{S_b H_b} + \frac{H_s}{H_b}\right]^{1/2}\right\}}.Y.$$

We know that as p decreases, Y increases, the denominator of $\Delta\Pi$ decreases, and thus $\Delta\Pi$ increases.

REFERENCES

Banerjee, Avijit (1986), "A joint Economic-lot-size model for purchaser and vendor," *Decision Sciences*, 17 (3), 292–311.

Buchanen, J. M. (1953), "The Theory of Monopolistic Quantity Discounts," *Review of Economic Studies*, 20, 3, 199–208.

Chakravarty, Amiya K. (1984) "Joint, Inventory Replenishment with Discounts Based on Invoice Value," *Management Science*, 30 (9), 1105–1112.

Chakravarty, Amiya K. and G.E. Martin (1988), "An Optimal Joint Buyer-Seller Discount Pricing Model," *Computers and Operations Research*, 15 (3), 271–281.

Chakravarty, Amiya K. and G.E. Martin (1989) "Discount Pricing Policies for Inventories Subject to Declining Demand," *Naval Research Logistics*, 36 (1), 89–102.

Chakravarty, Amiya K. and G.E. Martin (1991) "Operational Economies of a Process Positioning Determinant," *Computers and Operations Research*, 18 (6), 515–530.

Chen, Fangruo, Awi Federgruen, and Yu-Sheng Zheng (2001), "Coordination mechanisms for a distribution system with one supplier and multiple retailers," *Management Science*, 47, 5, 693–798.

Dada, Maqbool and K. N. Srikanth (1987), "Pricing Policies for Quantity Discounts," Management Science, 33, 10, 1247–1252.

Dolan, Robert J. (1987), "Quantity Discounts: Managerial Issues and Research Opportunities," *Marketing Science*, 6, 1, 1–22.

Eliashberg, Jehoshua (1986), "Arbitrating a Dispute: A Decision Analytic Approach," *Management Science*, 32, 8, 963–974.

Eliashberg, Jehoshua and Richard Steinberg (1987), "Marketing-production decisions in an industrial channel of distribution," *Management Science*, 33, 8, 981–1000.

Feichtinger, Gustav and Richard Hartl (1985), "Optimal Pricing and Production in an Inventory Model," *European Journal of Operational Research*, 19, 1, 45–56.
Garbor, A. (1955), "A Note on Block Tariffs," *Review of Economic Studies*, 23, 32–41.
Ingene, Charles A. and Mark E. Parry (1995), "Coordination and Manufacturer Profit Maximization: The Multiple Retailer Channel," *Journal of Retailing*, 71, 2, 129–151.
Jeuland, Abel, P. and Steven M. Shugan (1983), "Managing Channel Profits," *Marketing Science*, 2, 3, 239–272.
Jeuland, Abel, P. and Steven M. Shugan (1985), "Implicit Understanding in Channels of Distribution," *Management Science*, 31, 4, 435–460.
Joglekar, Prafulla N. (1988), "Comments on 'A Quantity Discount Pricing Model to Increase Vendor Profits,'" *Management Science*, 34, 11, 1391–1398.
Jorgensen, Steffen (1986), "Optimal Production, Purchasing and Pricing: A Differential Game Approach," *European Journal of Operational Research*, 24, 1, 64–76.
Kalai, Edud and Meir Smorodinsky (1975), "Other Solutions to Nash Bargaining Problem," *Econometrica*, 43, 513–518.
Kim, Kap Hwan and Hark Hwang (1989), "Simultaneous Improvement of Supplier's Profit and Buyer's Cost by Utilizing Quantity Discount," *Journal of the Operations Research Society*, 40, 3, 255–265.
Kohli, Rajeev and Heungsoo Park (1989), "A Cooperative Game Theory Model of Quality Discounts," *Management Science*, 35, 6, 693–707.
Lal, Rajiv and Richard Staelin (1984), "An Approach for Developing an Optimal Discount Policy," *Management Science*, 30, 12, 1524–1539.
Lee, Hau L. and Meir J. Rosenblatt (1986), "A Generalized Quantity Discount Pricing Model to Increase Supplier's Profits," *Management Science*, 32, 9, 1177–1185.
Lei, Lei, Qiang Wang and Chunxin Fan (2002), "Lower Bound on the Improvement of Supply Chain Profitability from Partial to Total Coordination," Working paper, Rutgers Center for Supply Chain Management, Rutgers Business School, Rutgers University.
McGuire, Timothy W. and Richard Staelin (1986), "Channel Efficiency, Incentive Compatibility, Transfer Pricing, and Market Structure: An Equilibrium Analysis of Channel Relationships," in *Research in Marketing: Distribution Channels and Institutions* (L.P. Bucklin and J.M. Carman eds.), V 8, Greenwich, CT: JAI Press, 181–223.
Monahan, James P. (1984), "A Quantity Discount Pricing Model to Increase Vendor Profits," *Management Science*, 30, 6, 720–726.
Moorthy, Sridhar (1987), "Managing Channel Profits: Comment," *Marketing Science*; 6, 4, 375–379.
Oi, Walter Y. (1971), "A Disneyland Dilemma: Two-Part Tariffs for a Mickey Mouse Monopoly," *The Quarterly Journal of Economics*, 85, 77–96.
Shapiro, Jeremy F. (2001), *Modeling the Supply Chain*, Pacific Grove, CA: Duxbury.
Weng, Z. Kevin (1995), "Channel Coordination and Quantity Discounts," *Management Science*, 41, 9, 1509–1522.

III. INTERFACES IN THE INTERNET SPACE

6. PRICING INTERNET SERVICE

RICHARD STEINBERG

ABSTRACT

Internet service needs to be appropriately priced, because Internet Service Providers want to remain competitive, and because pricing is increasingly seen by engineers as an essential tool for controlling network congestion. Thus, the pricing of Internet service lies at the interface of marketing and engineering. From this perspective we present a review of the major proposals for pricing Internet service. Topics include: Smart Market, Paris Metro Pricing, Max-Min Fairness, Proportional Fairness, CPE Pricing, and the Contract and Balancing Process. Background information on the history and technology of the Internet is also provided.

1. INTRODUCTION

How should Internet service be priced? The answer depends on who is pricing to whom. In this chapter, we will for the most part be concerned with how an Internet Service Provider (ISP) should price to its customers, the "end users." The methods we consider here can, for example, help AOL determine how it should price Internet service to your home, but also how much it should charge a business customer for a direct connection from the business's own networks to the Internet. We will also discuss Internet pricing from another—but closely related—point of view, viz., how the owner of a dedicated communication network should price bandwidth to businesses that host Internet applications. Internet service needs to be appropriately priced, both because Internet Service Providers want to remain competitive, and because pricing is increasingly seen by engineers as an essential tool for controlling network congestion. Thus, the pricing of Internet service lies at the interface of marketing and engineering. It is from this perspective that we present a review of the major proposals for pricing Internet service.

Our naïve opening question is quite general, so let us consider some specifics: Can Internet pricing be determined by an auction mechanism? Can that elusive concept, Quality of Service, be endogenously determined, leaving the ISP, so to speak, off the hook? Can Internet pricing be based on the philosophical concept of fairness? Can the individual packets that comprise Internet traffic be priced via a consumer demand function? And finally, can the pricing of bandwidth be based on a contractual agreement? In this chapter, we consider all these questions and the often ingenious methods that have been suggested for addressing them.

In the following section, we discuss what the Internet is, how it developed, and how it works. In Sections 3 through 7, we examine, respectively, Smart Market, Paris Metro Pricing, Max-Min Fairness, Proportional Fairness, and CPE Pricing. In Section 8 we look at a proposal for bandwidth pricing, the Contract and Balancing Process. Section 9 briefly covers charging. Section 10 concludes with related issues, including a mention of some other pricing proposals.

A review such as this is necessarily subjective, and other authors would have different lists of the "major" proposals. However, any credible review of Internet pricing must include two: Smart Market, because it was the first such proposal and has been enormously influential, and Proportional Fairness, because it is grounded in a deep and elegant theory that generalizes several earlier approaches. However, all the proposals discussed here are significant, and it is impossible to say which one, if any, will become the standard.

Finally, note that we do not consider the pricing of Internet *content*, i.e., digital information goods; for this topic, especially with regard to bundling, see Bakos and Brynjolfsson (2000). For a comprehensive overview of how the Internet is affecting marketing generally, see Barwise, Elberse and Hammond (2002).

2. THE INTERNET

The *Internet* is the publicly accessible network of computers that exchange data worldwide. However, this simple definition omits the Internet's most distinguishing trait. As Papadimitriou (2002) lucidly puts it:

> The most novel and defining characteristic of the Internet is its nature as an artifact that was not designed by a single entity, but emerged from the complex interaction of many economic agents (network operators, service providers, users, etc.), in various and varying degrees of collaboration and competition.

The truth of Papadimitriou's remark is evident from a recapitulation of the Internet's ontogeny.

The Internet appeared in embryo as the U.S. Defense Department's *Advanced Research Projects Agency Network*. ARPANET was designed to be a robust communication network that could survive a nuclear attack, even if one or more sites were destroyed. When ARPANET first went into operation in 1969, it was comprised of four host sites: UCLA, Stanford Research Institute, UC Santa Barbara, and the University of Utah. (Here *host* refers to any computer that has full access to other computers on the network.) By 1971, other educational and research institutions had joined ARPANET, so that it was now comprised of fifteen hosts. At this time, it was used primarily by a small number of computer scientists interested in developing networking technologies.

The number of hosts grew modestly over the next few years. In 1974, the word "internet" appeared in the title of a document by Vinton Cerf, a DARPA[1] scientist at Stanford Research Institute, and two Stanford University graduate students, Yogen Dalal and Carl Sunshine (Cerf, Dalal, and Sunshine 1974). They used the term as a short form for "internetwork," in the phrase "internetwork transmission control program." This refers to the method of transmitting files by breaking them down into small units of information called "packets" which are then sent over the network. The Transmission Control Program, they explained, "acts in many ways like a postal service since it provides a way for processes to exchange letters with each other" (Cerf, Dalal, and Sunshine, 1974, p. 3).

In 1978, the Transmission Control Program was split into the Transmission Control Protocol (TCP) and the Internet Protocol (IP), known jointly as TCP/IP, which remains to this day the standard communication language of the Internet. In 1980 and 1981 another—completely independent—network was founded, called CSNET (Computer Science NETwork), through grant funds from the National Science Foundation (NSF). By the end of 1981, ARPANET had two hundred hosts.

By 1983, all research networks were converted to TCP/IP and were inter-connected through ARPANET or CSNET. It was around this time that the capitalized term "Internet" came to mean a connected set of networks, specifically those using TCP/IP. By 1984, there were more than a thousand hosts.

In 1986, the National Science Foundation created a research network called NSFNET, which was significant for its high-speed backbone. (A *backbone* is a large transmission line for long-distance connection that carries data gathered from smaller lines that interconnect with it.) At the same time, the NSF supported the development of regional networks that could carry traffic from individual organizations, such as government agencies and universities, to the national backbone service. This led to the number of hosts multiplying from about two thousand early in the year, to more than five thousand by the end of the year.

In 1987, NSF commissioned Merit Network, Inc., MCI, IBM, and the State of Michigan to manage the NSFNET backbone project, and by year-end the number of hosts exceeded ten thousand. As more networks joined up, the Internet became truly international, and by 1989, there were over a hundred thousand hosts worldwide. In 1990, the ARPANET was retired, but NSFNET continued to grow. By now, the phrase "the Internet" had appeared in print (Wall and Schwarz 1990, Chapter 6, p. 260), and the number of hosts had passed three hundred thousand.

1. In 1973, ARPA had been renamed DARPA (D for "Defense").

Also in 1990, a computer scientist working at CERN (European Organization for Nuclear Research), Tim Berners-Lee, developed the World Wide Web as a tool for collaboration for the high-energy physics community. "The Web" was publicly released by CERN the following year. Also in 1991, the NSF removed commercial restrictions on the NSFNET. By 1992, the number of hosts had reached the million mark.

In 1995, NSFNET reverted back to a research network, and the main US backbone traffic was re-routed through various interconnected network providers. Thus, the Internet—in what is essentially its current form—was born. At the time, it had five million hosts. By one estimate, in July 2002 there were in excess of 165 million Internet hosts (Zakon 2003).

2.1 The web

Most people erroneously consider the Internet to be synonymous with the *World Wide Web*. The Web is essentially a method of accessing information via the Internet, where a Web *browser* downloads data from a Web *server*.

Every host computer on the Internet has associated with it a unique number known as an *IP address*. Through the use of browsers such as Internet Explorer and Netscape, the Web allows users to automatically search the Internet for the IP address of the computer that contains the requested file—e.g., a text document, graphic image, or a sound, video or multimedia file—and call it up for "viewing." Web pages are formatted in a system of symbols called HTML (HyperText Markup Language) that supports links to other files, i.e., access via a simple click of the mouse. The Web is just one of many ways that information can be disseminated over the Internet; among the others, email is probably the most familiar.

2.2 Network externalities and Metcalfe's law

The extraordinary rate of growth of the Internet can be explained by the commonplace that its value rises with the number of computers connected to it—this effect is called "network externalities"—which leads to a snowballing effect. By how much does the value rise? The most popular view is provided by *Metcalfe's Law*, which was proposed by George Gilder in a discursive column in Forbes ASAP (Gilder 1993). Gilder named the principle in honor of Robert Metcalfe who, as a member of the research staff of Xerox PARC (Palo Alto Research Center) in 1973, predicted the potential of a system for connecting hundreds of computers within a building using hardware running from machine to machine.[2] Gilder later provided a succinct formulation (Gilder 1999):

> Metcalfe's Law: The observation by Robert Metcalfe that the performance and value of a network rises by the square of the total power of the computers compatibly attached to it.

The idea behind this square law is very straightforward. Assume for simplicity that the power of each of the computers is roughly the same, which we set to be 1. Then

2. Metcalfe went on to design one of the more popular systems for doing this, called *Ethernet*.

if there are n computers attached to the network, the value of the network to each one of them is the number of computers, or n. Thus, the total value of the network is the number of computers times the value to each computer, or n^2.

Although Metcalfe's Law is frequently and enthusiastically cited, it also has its detractors. For example, Odlyzko (2000) dryly observes: "If this 'law' were indeed valid, there would be no need to worry about general connectivity. Two equal size networks would double the total value, and so the two carriers would have an irresistible economic incentive to link up, either through merger or some standard interconnection." Although he expands further on this idea, Odlyzko eventually concedes: "However, this law does reflect a fundamental truth, that the value of a point-to-point communication network increases faster than just in proportion to the number of users." And this is the only point we wish to observe here. But how exactly does the Internet work?

2.3 Packets, bytes, and bits

Although much of Internet traffic is carried over the same lines as telephone calls, the technology of the Internet—called *packet switching*—is very different from the technology of the public telephone network—called *circuit switching*. In a circuit switched network, a specific path to the destination is obtained for the message, and a fixed amount of capacity is reserved for the duration of the call, during which time no other information is sent along the same path.

By contrast, in a packet-switched network the file to be transmitted is first broken down into small units, called packets, which are affixed with a header containing the destination computer's IP address and other information. (The header also contains unassigned fields which can be used for future applications, e.g., congestion notification.) The packets are then sent through the network along individual paths. The choice of path is determined by routing algorithms which typically select shortest paths.

The storage requirements of packets are measured in *bytes*. A byte in text might correspond to a single character, such as a letter, number, or typographic symbol (e.g., 't', '9', or '@'), but in a visual image might correspond to a color or a pixel location. A byte is eight bits long, where a *bit* (i.e., *bi*nary digi*t*) is the smallest unit of data and has a value of 0 or 1.

The *data transfer rate,* or simply *rate,* at which data is sent through the network is measured in bits per second (bps)—or, more usually—megabits per second (Mbps). The rate depends on the load on the *network resources* involved. An example of a network resource would be a fiber optic cable between New York and London.

2.4 TCP/IP and Explicit Congestion Notification

TCP/IP (Transmission Control Protocol/Internet Protocol) is the basic communication language, or protocol, of the Internet. TCP enables two host computers to establish a connection, and manages the breaking down of the file into packets at the source (the sending computer) and the re-assembling of the message at the destination (the receiving computer). IP specifies the format of packets, as well as the addressing scheme, so that the packets get to the right destination. Although all packets from the same file will have the same destination IP address, in general each packet will have its own routing through the network. When the packets arrive at the destination computer, it is TCP on that machine that strips off the headers and reassembles the data into the original file. TCP on the receiving machine also sends "acknowledgement packets" back to the source computer to verify the integrity of the data received; if no acknowledgement packet comes back after a specified time interval, the packets that were not acknowledged are retransmitted by the sending machine.

TCP also has the responsibility for adapting the sending rate of the source computer to the capacity of the network. When a resource within the network becomes overloaded, one or more packets are lost. Loss of a packet is taken as an indication of congestion. Consequently, the destination informs the source, and the source immediately reduces its sending rate. The source then gradually increases its sending rate until it again detects congestion via the loss of packets, and the cycle begins anew.

Formally, *congestion* can be defined as the loss of network performance that occurs when the users of the network collectively demand more resources than the network can provide. Users of course need to know when the network is congested, but dropping packets is a crude way of notifying them: it is both wasteful (a dropped packet might have already consumed resources; it might also need to be resent) and untimely (until their packets are dropped, users are unaware of the existence of congestion).

These drawbacks have led to proposals for the introduction of congestion marking, whereby a packet encountering a long queue will have a specific bit in its header set to indicate "congestion experienced." The procedure is called *Explicit Congestion Notification*, or ECN. Users detecting ECN marks should respond by reducing their transmission rates. This will result in the network being able to share resources without having to drop packets, except in periods of exceptionally heavy use. Explicit Congestion Notification has now been made a "Proposed Standard" by the Internet Engineering Task Force (IETF), the body concerned with the evolution of the Internet architecture.[3]

2.5 Bandwidth and elastic traffic

Capacity in the Internet is usually defined in terms of *bandwidth*, the amount of data that can be transmitted per unit time, i.e., the maximum rate. A related concept is *throughput*, which refers to the amount of data transferred successfully from one place to another in a given amount of time.

Fiber optic cable, or simply *fiber*, refers to both the medium and the technology associated with the transmission of packets as light pulses along a glass or plastic fiber. Fiber has significantly more bandwidth than conventional copper wire and is in general not subject to electromagnetic interference. Most telephone company long-distance lines are now of optical fiber, and increasing share of Internet traffic is carried over fiber.

3. The ECN proposal is derived from the *binary feedback scheme* of Jain, Ramakrishnan and Chiu (1987).

In a network, a *node* is a connection point, either an end point—called simply an *end node*—or a redistribution point, for packet transmission. The management of traffic between nodes in a network so that the receiving device can handle all the incoming data is called *flow control.* This is particularly important where the sending device is capable of sending data much faster than the receiving device can receive it, a not uncommon situation. In cases were traffic can adjust to changes in delay and bandwidth while still meeting the needs of its applications, the traffic is called "elastic" (Shenker 1995; Kelly, Maulloo, and Tan 1998). A mathematical definition of elastic traffic is given in Section 5.

2.6 Quality of Service

Internet Quality of Service (QoS) refers to the set of notions surrounding Internet service characteristics. The Internet currently operates on a single service quality, viz., what is usually described as "best effort" service. This means that the network attempts to serve all users, transmitting packets on a first-come, first-served basis without making any explicit commitment to any user with regard to rate, guarantee of packet arrival, or any other measure of service quality. It is the users, not the network, who are expected to detect the existence of congestion and reduce their transmission speeds. Discussions of Internet QoS can be found in Odlyzko (1999c) and Park, Sitharam and Chen (2000).

2.7 Internet Service Providers

An *Internet Service Provider,* or ISP, is a company that provides access to the Internet. Typically, for a monthly fee, the ISP provides individuals with a software package, a username and password, and an access phone number. The user can then log on to the Internet and make unlimited use of a number of services such as access to the World Wide Web and email. Such a pricing system is called a *flat rate scheme.* In addition to providing Internet access, some ISPs have their own online independent content. ISPs also serve large companies, providing a direct connection from the company's own networks to the Internet.

In March 2002, the Internet Service Provider reported to have the largest share of global Internet usage (i.e., percentage of Internet users worldwide using a particular ISP) was America Online (AOL) at 13.58%, with its nearest competitor, Road Runner, having a share of only 2.76%. Road Runner is owned by America Online's parent company, AOL Time Warner. The third largest share was held by UUNET at 2.18% (WebSideStory 2002). Although these figures have undoubtedly changed since this study was conducted, the dominance of AOL in the ISP market at this point in time appears undisputed.

As of May 2003, AOL has four main pricing plans. Under the standard plan, customers pay a flat rate of $23.90 per month for unlimited usage. There is also an annual flat-rate plan payable in advance that works out to be $19.95 per month. Under the "limited usage" plan, customers pay $9.95 per month for five hours of usage, plus $2.95 for each additional hour. Finally, under the "light usage" plan customers pay $4.95 per month for three hours of usage, plus $2.50 for each additional hour. Each of these plans provides customers with AOL-specific content in addition to access to the Internet.[4]

3. SMART MARKET

Mackie-Mason and Varian (1995, 1996) contend that users should face prices that reflect the resource costs they generate. Specifically, the authors propose that, in any efficient pricing mechanism, the price a user pays should have three components: (1) a fixed connection fee; (2) a close-to-zero charge per packet when the network is not congested; and (3) a positive charge per packet when the network is congested, where the price varies so as to reflect the existing degree of network congestion.

In particular, they present their *Smart Market* scheme in which a per-packet charge is levied against the user application whenever the network is congested. By "smart," they are referring to the feature that packets are priced in real time to reflect the existing degree of network congestion. Under Smart Market, each packet has in its header a bid field that indicates how much its sender is willing to pay for transmission. The bids would be determined by input from three parties: the local administrator controlling access to the net, the actual user of the sending computer, and by the computer software itself.

Operating as a multi-unit Vickrey auction,[5] the network admits all packets with bid prices that exceed the current threshold value, which is determined by the marginal congestion cost imposed by the next additional packet. The users thus pay not the price they bid, but the market-clearing price, which is lower than the price of all admitted packets.

What is clever about this scheme is that users can be expected to bid their true values, the dominant strategy in the Vickrey auction (see Vickrey 1961, 1962).[6] Mackie-Mason and Varian provide some simple analytic models based on Smart Market pricing. They show that capacity should be expanded when the revenues from congestion fees exceed the cost of providing the capacity and, further, that this result holds in a competitive setting.[7] Further, they show that the competitive price will result in the optimal degree of congestion.

Ganesh, Laevens and Steinberg (2000) observe that the Smart Market proposal corresponds to the Bertrand model of competition (Tirole 1988), where each producer sets a price and is willing to supply any demand at that price, and consumers choose the quantity demanded based on the market price. In Smart Market, this corresponds to each user specifying a price he is willing to pay per packet, and the network ac-

4. These prices are for the U.S. market. AOL has similar pricing plans in other countries.

5. A Vickrey (1961) auction is a second-price sealed-bid auction for a single item. That is, the bidders simultaneously submit their bids for the item without knowledge of the bids of the other players, and the winner of the auction is the one with the highest bid, who is required to pay the second-highest bid price. This is generalized to auctions with m identical objects in Vickrey (1962).

6. Mackie-Mason and Varian mention in both of their papers cited above that the idea of using a second-price auction to allocate network resources goes back to Waldspurger et al. (1992), although in a different context.

7. The importance of evaluating the viability of a pricing proposal by considering it in a competitive setting cannot be overstated. See, for example, Section 5.

cepting packets from a collection of highest bidders. However, Ganesh et al. claim that Smart Market, while efficient from a theoretical point of view, is impractical to implement. They provide two reasons for this. First, there is the considerable difficulty of conducting repeated auctions at the speeds at which the Internet operates. Second, there is a fairness issue connected with comparing a bid on a packet that has recently entered the network with a bid on a packet that has been waiting for an extended period of time. Smart Market does not make an allowance for this.

What is notable in the Smart Market scheme is that it does not promise the customer a specific level of QoS—which might, in fact, be difficult to measure and maintain—but simply guarantees a priority level for packet transmission. A packet with a high bid will gain access sooner than one with a low bid, but the technology of the Internet means that delivery time cannot be guaranteed. This concept of abandoning the chimera of a QoS guarantee, and instead employing a more tangible and objective measure as a proxy for service level, is itself a significant contribution of the Smart Market proposal. We will see this approach used in other proposals we discuss; Paris Metro Pricing takes this idea the furthest.

4. PARIS METRO PRICING

Once upon a time the Paris Metro employed an unusual pricing scheme. Users were offered a choice of travelling First or Second Class, where the only difference between classes was the price charged. First Class carriages, being more expensive, typically had fewer passengers and thus were less congested. Hence, users with a strong aversion to congestion were willing to pay the higher First Class fare. Whenever First Class carriages became too popular, some users decided it was not worth the extra cost and switched to travelling Second, thus reducing congestion in First Class and restoring the quality differential. In a paper entitled "A modest proposal for preventing Internet congestion," Andrew Odlyzko (1997) made the inspired suggestion to apply this idea to the Internet:

> Following in the footsteps of Jonathan Swift,[8] I propose to turn a perceived burden into a solution, and rely on usage-sensitive pricing to control congestion, bypassing most of the complexity of other solutions. This should allow for simpler networks that are easier to design and deploy and operate faster.

Odlyzko calls his proposal *Paris Metro Pricing, or* PMP.[9]

Under PMP, the Internet Service Provider first partitions its network into several separate networks, each having a fixed fraction of the capacity of the original network, and then applies different charges to each. The ISP offers no guarantees of service quality; however, on average, higher-priced networks will be less congested. Users

8. Here Odlyzko cites Swift's 1729 satire, "A Modest Proposal for Preventing the Children of Poor People in Ireland from Being A Burden to their Parents or Country, and for Making them Beneficial to the Public."
9. He has further discussion on PMP in his follow-up papers, Odlyzko (1999a, 1999b).

will sort themselves according to their aversion to congestion and the prices charged. Odlyzko's attractively simple proposal raises two issues, discussed below.

4.1 The acceptability of PMP

Was the Paris Metro a special case, or would a pricing scheme with endogenously-determined quality be acceptable to consumers in other contexts? The answer is that there are, in fact, a number of other examples of PMP in practice—many from India. For example, De Palma and Leruth (1989) observed a two-price system in effect for the busses of New Delhi. The busses were all identical except for the price charged. Displayed on the front of each bus was a sign that read either "1R*e*" or "2R*s*," where—as might be expected—the 1 rupee busses were considerably more crowded.

In his original paper, Odlyzko (1997) warns that the analogy between the Paris Metro and the Internet should not be overdrawn. He points out that, on the Paris Metro, all passengers arrived at the destination at the same time; the different prices only paid for the expected differential in discomfort caused by congestion (e.g., probability of getting a seat). With PMP applied to the Internet, packets on a lower-priced network would have a higher probability of being dropped and thus a higher probability of requiring re-sending; therefore, the file could be expected to arrive later than on a higher-priced network. The pricing for New Delhi buses clearly has much in common with that for the Paris Metro. If we assume that the arrival rates for the two classes of busses was about the same, then bus choice would not have made a significant difference in arrival time.

But there exist other examples of real-world PMP that involve a time element via queuing, and thus arguably form closer analogies to Odlyzko's proposal for Internet pricing. Chander and Leruth (1989) cite the case of a government hospital in New Delhi, where two different options were available for treatment of routine ailments, fee and free. Users could choose to pay the fee and expect a shorter queue, although the same set of doctors treat the non-paying patients. Gibbens, Mason and Steinberg (2000) report a similar arrangement for medical care in the United Kingdom where, as an alternative to receiving free (at the point of delivery) medical care via the National Health Service, people can choose to pay for private treatment at what they will expect will be a greatly reduced waiting time.

In December 1995, a four-lane toll highway called the *91 Express Lanes* opened in the median of a 10 mile section of the Riverside Freeway (State Route 91) in California (Transportation Research Board 2002). This was the first privately-financed toll road in the U.S. in over fifty years. According to the California Private Transportation Company, "the private sector would take the risk and the State would get congestion relief at no cost to taxpayers."[10]

Perhaps the most unusual instance of PMP in practice is at Tirupati, a pilgrimage

10. http://www.91expresslanes.com/. The congestion toll in effect as of May 2003 ranges from as high as \$3.60, for eastbound traffic, 7:00 am to 8:00 am, Monday through Thursday, and \$4.75, for westbound traffic, 5:00 pm to 6:00 pm, Monday through Friday; and as low as \$1.00, 11:00 pm to 3:00 am, every day in both directions.

center in Andhra Pradesh, India where devotees go for *darshan* ("viewing of the deity"). Devotees can have free darshan, called *Sarvadarsanam* ("darshan for all"), but can alternatively choose to pay a fee, called *Special Darshan*. The motivation for paying is explained on the official Tirupati website: "Pilgrims who use this queue will have a shorter waiting time." In fact, Special Darshan is available in two categories, at 40.00 rupees per head and at 50.00 rupees per head. The timings are the same for Sarvadarsanam and the Special Darshan, and the queues merge at the inner sanctum. Remarkably, the system handles more than fifty thousand pilgrims per day.[11]

Odlyzko's motivating idea can be summarized as follows: A simple pricing scheme can induce users to separate themselves into classes that provide differing qualities of service, and the division can be self-stabilizing. The above examples tend to support that belief for a wide variety of services. However, there is another issue connected with PMP that deserves close scrutiny.

4.2 The viability of PMP pricing under competition

The Paris Metro is run by a government monopoly, the RATP –Régie Autonome des Transports Parisiens. Therefore, for purposes of applicability to Internet pricing, we might ask what occurs when competition is introduced into Odlyzko's PMP model. Gibbens, Mason, and Steinberg (2000) address this question through the use of game theory.

Their model is specified as follows. There are two competing, profit-maximizing Internet Service Providers, each of which may offer either one or two service classes. In the case where an ISP chooses to offer two service classes, it forms them by partitioning its network into two separate networks and charging separate prices on each. Congestion on a network is determined in equilibrium by two factors: the fraction of the ISP's total capacity allocated to the network, and the number of users on the network. The immediate consequences are that a network with low capacity and many users will have high congestion. Further, quality is demand-dependent, determined (in part) by the equilibrium choices of the prices. Gibbens et al. ask: In equilibrium, how many service classes will the two ISPs choose to offer?

They begin with the following assumptions. Each user joins one and only one network, resulting in Q_i users on network i. Upon joining a network, a user receives utility with three components: (i) a positive benefit, V, independent of the network joined, (ii) a dis-benefit, dependent on the degree of congestion, $K(Q_i)$, and preference for (i.e., aversion to) congestion, θ, and (iii) a dis-benefit from having to pay a price, p_i, to the ISP. Further, users are assumed heterogeneous in their preference for congestion. Those users with elastic traffic receive little dis-benefit from congestion, and will have low values of θ; those with inelastic traffic will be very sensitive to congestion, and will have high values of θ. A user's utility in joining network i is $U(\theta, i) = V - \theta K_i(Q_i) - p_i$, and the costs are set to zero so that the profit of the network is $\Pi = p_i Q_i$.

11. For more information, see http://www.tirupati.org/.

The authors make two simplifying assumptions: one about the form of the congestion function and the other concerning the distribution of consumer disutility for congestion. First, congestion on a network is assumed to be proportional to the number of users divided by the capacity of the network. Second, consumer disutility for congestion, θ, is uniformly distributed.

Gibbens et al. find that the unique equilibrium outcome is that neither ISP subdivides its network, i.e., neither firm will employ Paris Metro Pricing; further, the two firms charge the same price. In their analysis, the authors assume that the capacities of the subnetworks are fixed and symmetric. However, they also conducted numerical analysis which indicates that their results are robust to the assumption of fixed, equal, and symmetrically split capacities; in all cases considered, the ISPs each maximize their own profits by charging a single price and offering one network. The authors provide the economic interpretation of their results: As more products are introduced, the cost of increased competition outweighs the benefits from greater segmentation of the market.

The existence, uniqueness and symmetry of the equilibrium gives rise to the question of the extent to which these results depend on the form of the congestion function and the distribution of consumer disutility for congestion. Haimanko and Steinberg (2000) offer the following partial result. Consider any congestion function and any consumer disutility cumulative density function with the only assumptions being that these functions are each continuously differentiable and strictly increasing. Then with fixed, equal capacities in the case where neither ISP subdivides its network: (1) there does not exist an asymmetric equilibrium, and (2) if an equilibrium does exist, then it is unique. This tends to indicate that the results of Gibbens et al. are robust to the form of the congestion function and the distribution of consumer disutility for congestion. However, further research in this direction might yield new insights.

The above results support the view that Paris Metro Pricing will not be viable in a competitive market. However, in order to simplify their model, Gibbens et al. assumed that there is a fixed number of firms in the market (i.e., two). The authors suggest that the process of free entry may be a mechanism by which a range of prices and qualities can arise in equilibrium. This would be a worthwhile area for further research on a proposal that has garnered a considerable amount of attention.

4.3 The true origins of Paris Metro Pricing

One may wonder how the Régie Autonome des Transports Parisiens arrived at this innovative pricing scheme. The answer is that it was an accident of history.

When the Paris Metro opened in July 1900, intercity rail service in France had differential classes of service, and so it was decided that the Metro would have differential classes of service as well. (Although there were at the time three classes of service on the intercity rail, it was felt that two classes would suffice for the Metro.) The difference in pricing between Metro classes was justified by the differences in materials used in the carriages, and hence in the comfort levels for passengers. Due to this price differential, First Class was less crowded.

From the 1930s onwards, the basis of the Metro's rolling stock was the Sprague-Thomson carriage, which had upholstered seats in First Class and wooden seats in Second. In the early 1980s, these carriages were replaced with modern equipment. With modern cars and materials, comfort became harmonized between the two classes. Thus, for a time, essentially the only difference between the two classes was the price and, consequently, the level of congestion. Voilà: PMP.

Claude Quin, a member of the Communist Party who had became president of RATP in 1981, advocated equating the rates so as to eliminate the class differences. This was ultimately implemented in August 1991. Although the change was putatively motivated by egalitarian reasons, it was felt that this would lead to greater utilization of train capacity as well.[12]

5. MAX-MIN FAIRNESS

In the presence of congestion, not all users can be allocated their desired rate, and the network faces the task of having to decide what rate to allocate to each user. Understandably, it might want to do this in a manner that is considered to be "fair." But is fairness truly an appropriate concept for communication networks? Kelly, Maulloo, and Tan (1998) make the argument that it is. They point out that, although fairness has traditionally been considered an *economic* issue, involving as it does comparisons of utility, any discussion of the performance of a rate control scheme *must* address the issue of fairness, since a given scheme might maximize the rate of information passing through the network—an important criterion—but at the same time deny access to some users.

That being said, it is all too obvious that fairness can be interpreted in a variety of ways. In the setting of communication networks, the first criterion to be widely accepted is due to Jaffe (1980, 1981), and has come to be known as *Max-Min Fairness*. Jaffe begins by citing one interpretation that requires that all users obtain equal throughputs. He points out, however, that in a network with different users using lines of different capacities (it seems this would be a rather common situation), it is unlikely that such a policy would be desirable. He argues that the general interpretation of fairness as "all users are treated equally" may not be desirable in practical networks, as one user may be more important and thus more deserving of a higher rate. The criterion he arrives at is an equilibrium condition:

> A set of rates is *max-min fair* if no rate can be increased without simultaneously decreasing another rate that is equal or smaller.

If the network has a single bottleneck resource, then under Max-Min Fairness each user either receives his desired allocation or an equal share of the bottleneck resource.

12. I am grateful to Henri Zubar of Régie Autonome des Transports Parisiens and Emile Quinet of ENPC, CERAS, Paris for providing historical information (private communication, September 2002).

(For more details on Max-Min Fairness in communication networks, see Chapter 6 of Bertsekas and Gallagher (1992).)

5.1 Mathematical formulation

It is not difficult to provide a precise mathematical statement of Max-Min Fairness. We begin with some definitions. Consider a network with a set J of resources, and let C_j be the capacity of resource $j \in J$. A *route* r is a nonempty subset of J, where R denotes the set of routes in the network. Each route r can be associated with a user of the network. When user r is allocated a rate x_r, let $U_r(x_r)$ denote the utility to user r.

We assume that traffic on the network is such that, for all r, utility function $U_r(\cdot)$ is increasing, strictly concave, and continuously differentiable on $[0, \infty)$. Under these assumptions, the network is said to experience *elastic traffic* (Shenker 1995). Further, we assume that utilities are additive. That is, the aggregate utility of the vector of rates $x = (x_r, r \in R)$ is the sum of the utilities, i.e., $\Sigma_{r \in R} U_r(x_r)$.

Define the matrix $A = (A_{jr}, j \in J, r \in R)$ by:

$$A_{jr} = \begin{cases} 1 & \text{if } j \in r \\ 0 & \text{if } j \notin r \end{cases}$$

Thus, A describes which resources lie on which routes. Define the grand capacity vector $C = (C_j(\cdot), j \in J)$.

A vector of rates $(x_r, r \in R)$ is said to be *feasible* if $x \geq 0$ and $Ax \leq C$. A feasible vector of rates is *max-min fair* if for any other feasible vector y:

$$\exists\, r : y_r - x_r > 0 \quad \Rightarrow \quad \exists\, x_s \leq x_r : y_s - x_s < 0.$$

The compactness and convexity of the feasible region imply that such a vector x_r exists and is unique (see Kelly 1997a).

5.2 Nash bargaining solution

In 1991, Mazumdar, Mason and Douligeris (1991) proposed that fairness issues for communication networks be placed in a rigorous game-theoretic framework. They embark on the following sequence of reasoning. First, games can be broadly classified into cooperative and non-cooperative. However, in non-cooperative games each user acts individually to optimize his own performance without regard to the performance of the others. This leads to a Nash equilibrium solution, which might turn out to be Pareto inefficient. Since Pareto inefficiency is undesirable, a cooperative game is the preferable framework. Finally, they argue that the *Nash bargaining solution* is a suitable candidate for a fair, optimal operation point in the sense that it satisfies certain axioms of fairness and is Pareto optimal.

What the Nash bargaining solution does is predict an outcome based only on information about each bargainer's preferences, as modelled by an expected utility function over the set of feasible agreements and the outcome that would result in the case of disagreement. For more details, see Nash (1950) and Roth (1979). Although Mazumdar et al. make some connections with Jaffe's work, they do not compare the Nash bargaining solution with Max-Min Fairness, and so leave the relationship between the two solution approaches unresolved.

In a lecture before the London Mathematical Society, Frank Kelly (1997b) observed that the concept of Max-Min Fairness can be grounded in the moral philosophy of John Rawls. Rawls had developed essentially the same concept, which he called "maximin," as a component of his "Theory of Justice" (1971, 1999). As Rawls explains it, this rule for "choice under uncertainty" is one that

> a person would choose for the design of a society in which his enemy is to assign him his place. The maximin rule tell us to rank alternatives by their worst possible outcomes: we are to adopt the alternative the worst outcome of which is superior to the worst outcomes of the others.

Although Max-Min Fairness has attractive properties as a method for rate control, it cannot be considered as an option for pricing Internet service as it is lacking one essential feature: a market mechanism.

6. PROPORTIONAL FAIRNESS

In an influential series of papers, Frank Kelly and his co-authors describe a pricing and rate control model for elastic traffic in which each user chooses the charge per unit time he is willing to pay. Thereafter, the network determines each user's data transfer rate according to a fairness criterion applied to the rate per unit charge. Kelly shows that his criterion, which he calls *Proportional Fairness*, achieves a system optimum when the users' choices of charges and the network's choices of allocated data transfer rates are in equilibrium. We outline the main results here. For proofs and further development, see Kelly (1997a) and Kelly, Maulloo and Tan (1998).

For a rate vector x, the *proportional change* with respect to a rate vector y_r is:

$$\frac{y_r - x_r}{x_r}$$

A feasible vector of rates $x = (x_r, r \in R)$ is called *proportionally fair* if there exists no other feasible rate vector y for which the sum of the proportional changes over all users is positive. In other words, a rate allocation is proportionally fair if no aggregate improvement is possible. In symbols, a rate vector x is proportionally fair if, for all feasible rate vectors y:

$$\sum_{r \in R} \frac{y_r - x_r}{x_r} \leq 0$$

The concept of proportional fairness can be generalized to *weighted* proportional fairness. In particular, we can use as weights each user's willingness-to-pay.

A feasible vector of rates is *proportionally fair per unit charge* if, for all feasible rate vectors y:

$$\sum_{r \in R} w_r \frac{y_r - x_r}{x_r} \leq 0$$

Thus, a feasible set of rates $\{x\}$ in a network is proportionally fair per unit charge if, for any other set of feasible rates $\{y\}$, the sum of the weighted proportional changes is negative or zero.

6.1 Three optimization problems

There are three optimization problems to consider: the System Problem, the User Problem, and the Network Problem. We make use of the same notation as was used in Section 5.1. We also define the grand utility function vector $U = (U_r(\cdot), r \in R)$.

The System Problem

The *system optimal* rates are given by the solution to the following problem:

SYSTEM(U, A, C)

$$\text{maximize} \sum_{r \in R} U_r(x_r)$$

$$\text{subject to } Ax \leq C$$

$$\text{over } x \geq 0$$

In words, the system problem is to maximize aggregate utility, subject to the capacity constraints. This may appear to be a straight-forward problem, but Kelly, et al. (1998) point out that, despite the putative tractability of optimizing a strictly concave function over a convex set, the network operator is unlikely to know the grand utility function vector U. This brings us to the second of the three problem formulations:

The User Problem

Suppose that each user r can choose an amount to pay per unit time, w_r, and receive in return a flow proportional to w_r. Here w stands for *weight* or—in particular—*willingness-to-pay*. Thus, the rate allocated to user r will be:

$$x_r = \frac{w_r}{\lambda_r}$$

where λ_r is the charge per unit flow by the network to user r, which is presumed to be known to user r. Then r seeks to solve the following problem:

$\text{USER}_r(U_r; \lambda_r)$

$$\text{maximize } U_r\left(\frac{w_r}{\lambda_r}\right) - w_r$$

$$\text{over } w_r \geq 0$$

That is, user r chooses an amount to pay per unit time, w_r, and receives in return a rate $x_r = w_r/\lambda_r$.

The Network Problem

As observed above, the network operator is unlikely to know the grand vector of user utility functions, $U = (U_r(\cdot), r \in R)$. However, it is not unreasonable to assume that he could be knowledgeable about the grand vector $W = (W_r, r \in R)$ of user willingness-to-pay. This brings us to the third optimization problem:

$\text{NETWORK}(A, C; w)$

$$\text{maximize } \sum_{r \in R} w_r \log x_r$$

$$\text{subject to } Ax \leq C$$

$$\text{over } x \geq 0$$

Thus, this problem is formulated as if the network maximizes a logarithmic utility function, but with the constants $(w_r, r \in R)$ chosen by the users. The choice of a logarithmic utility function is justified by the following theorem (Kelly 1997a), which beautifully interrelates these three problems:

PROBLEM DECOMPOSITION THEOREM. *There exist vectors* $\lambda = (\lambda_r, r \in R)$, $w = (w_r, r \in R)$ *and* $x = (x_r, r \in R)$ such that

i $w_r = \lambda_r x_r$ for $r \in R$

ii w_r *solves* $\text{USER}_r(U_r; \lambda_r)$

iii x *solves* $\text{NETWORK}(A, C; w)$

In addition, the vector x is the unique solution to the SYSTEM(U, A, C) *problem.*

Thus, the SYSTEM problem can be solved by simultaneously solving the NETWORK and USER problems.

6.2 Proportionally Fair Pricing

This above development leads to the following result (Kelly 1997a):

PROPORTIONAL FAIRNESS THEOREM. *A rate vector x solves* NETWORK(*A, C; w*) *if and only if it is proportionally fair per unit charge.*

In a series of lectures presented to the Netherlands operations research community in January 2000, Richard Weber demonstrated that a proportionally fair allocation will also be the Nash bargaining solution, and that a weighted proportionally fair vector is the Nash bargaining solution where the players have unequal bargaining power, thus reconciling the work of Mazumdar, Mason, and Douligeris (1991) with that of Jaffe (1980, 1981). Details are provided in Chapter 10 of Courcoubetis and Weber (2003). See also the earlier work of Key and McAuley (1999).

Gibbens and Kelly (1999) make use of the theoretical model developed in Kelly, Maulloo, Tan (1998) to develop an implementable pricing scheme that makes use of *packet marking*. The motivating idea is that it might be easier for the network to achieve an efficient allocation by conveying information on congestion to the users, rather than asking the users to supply information to the network. In addition, the network is relieved of the burden of eliciting information from users about their preferences.

Called *proportionally fair pricing*, the scheme is very simple in its operation. Time is broken down into discrete slots. The network marks a packet if it arrives in a time slot in which aggregate arrivals exceed capacity. Here marks reflect the fact that such packets imposed a cost, in terms of delay or loss on another packet. End users are informed as to whether their packet was marked, and they are free to choose how to adapt their sending rates. Of course, a natural way to implement packet marking would be through the use of the ECN bit (discussed in Section 2) to carry the mark.

Gibbens and Kelly suggest that for a network with "potentially uncooperative" end users, it may be necessary for each mark to be associated with a small fixed charge to the user, ensuring that the end user will have both the appropriate incentive and the necessary information in order to use the network efficiently. Proportionally Fair Pricing aims to combine the flexibility of Smart Market with the simplicity of Paris Metro Pricing; further, it is a pricing system that can evolve naturally from existing standards and proposals (e.g., TCP and ECN).

7. CPE PRICING

Rather than using one bit to mark packets, why not make use of several bits and actually *price* packets? Ganesh, Leavens and Steinberg (2000, 2001) do exactly this, allowing the network to assign a price to each packet rather than to simply mark them.[13] The price assigned by the network reflects the degree of congestion encoun-

13. In their analysis the price is taken to be a real number, but they suggest that in practice a small number of bits of price feedback should provide sufficient accuracy.

tered by the packet; users are informed of how much they were charged and thus will be motivated to make use of this information to adjust their sending rates. Since the prices reflect the "social cost" imposed on others, users will have the correct incentives to adapt to congestion.

Like Gibbens and Kelly (1999), Ganesh et al. consider time to be broken down into discrete slots, in which a single network resource is shared by N users. In each time slot t, user i transmits a quantity of packets $x_i(t)$. The unit price of packet transmission, $p(t)$, is determined as a function $\phi(x)$ of $x = x_1 + x_2 + \ldots + x_n$, the traffic arriving at the resource in that time slot. Thus:

$$p(t) = \phi(x(t)), \qquad \text{where } x(t) = \sum_{i=1}^{N} x_i(t).$$

Each user i derives a utility, $u_i(x_i(t))$, which is a non-decreasing function of the number of packets he sends in time slot t. His total utility is assumed to be the sum of his utilities over all N time slots. In time slot t, user i chooses $x_i(t)$ so as to maximize:

$$u_i(x_i(t)) - x_i(t)\,\phi(x(t))$$

This expression denotes the single-stage payoff to user i as a function of the actions of all the users.

Now if the utility functions of all the users are common knowledge, and if the price function is also common knowledge, then this is the well-known Cournot competition model (Tirole 1988), and a Nash equilibrium will be sought. Ganesh et al. remark that this approach does not suffer from the problems associated with the "Bertrand" approach of Smart Market pricing (discussed in Section 3).

Nevertheless, they find this model to be unsatisfactory. Specifically, they point out that there can be multiple equilibria, and that sufficient conditions for uniqueness of the Nash equilibrium are neither simple nor intuitive. They find the Nash equilibrium approach to be unrealistic in this setting, as it assumes that the users would be aware of the utility functions of all the users of the network; however it is unlikely that a user will even know the *number* of users on the network.

7.1 An expectation-based model

Ganesh et al. point out the only information that a user needs to know in order to choose his sending quantity is the price in that time slot; it is irrelevant as to how that price was arrived at from the auctions of all the users. Essentially, the only information that a user could be assumed to have is the history of the prices and of his own actions. This leads to the following alternative modelling framework.

Each user i forms his own estimate, $\hat{p}_i(t)$, of the price in time slot, t, by exponentially smoothing all the price information available to him, viz., all the prices up until time $t - 1$. He then chooses his transmission rate, $x_i(t)$, so as to maximize his payoff as a function of this expected price. Thus, he chooses $x_i(t)$ so as to maximize:

$$u_i(x_i(t)) - x_i(t)\hat{p}_i(t).$$

The authors now need to make three assumptions pertaining to, respectively, the form of the utility functions, the form of the price function, and the method of user price-estimate updating.

First, the utility functions are assumed to be of the form:

$$u_i(x) = w_i(x^\beta - 1)/\beta, \qquad 0 \leq \beta < \infty.$$

If $\beta = 0$, then the formula is interpreted to mean $u_i(x) = w \log x$. Ganesh et al. explain that logarithmic utility functions date back to Daniel Bernoulli's explanation of the St. Petersburg Paradox (Bernoulli 1738). They also point out that the application of logarithmic utility functions to communication networks has been considered by Kelly, Maulloo and Tan (1998), who interpret parameter w_i as a measure of user i's willingness to pay for bandwidth, and by Massoulié and Roberts (1999), who provide an engineering interpretation of the welfare maximization problem for certain values of β and observe that TCP rate adaptation approximates the behaviour of a user seeking to maximize such a utility function with $\beta = -1$.

Second, the price function is assumed to be iso-elastic:

$$\phi(x) = \left(\frac{x}{C}\right)^k,$$

where C is a scale parameter which is associated with the physical capacity of the resource, and $k \geq 1$ defines the steepness of the penalty for demand in excess of capacity. The use of the iso-elastic price function is to keep resource utilization close to C, but at the same time to prevent demand from exceeding the resource capacity. The form of the price function, which they name the *CPE price mechanism*, corresponds to a constant price elasticity demand function, which has empirical support in the context of consumer demand (Simon 1989).

Third, users are assumed to use exponential smoothing to update their respective estimates of the price:

$$\hat{p}_i(t) = \alpha_i\, p_i(t-1) + (1-\alpha_i)\,\hat{p}_i(t-1),$$

where α_i is the smoothing constant, $0 < \alpha_i < 1$.

There are two major results in Ganesh et al. (2000). One is that there is a price $q^\star$ which is "self-consistent" in the following sense. If all users had $q^\star$ as their price estimate and optimized their transmission rates accordingly, then the price would in fact be $q^\star$, and this $q^\star$ is unique. The other result is that the price estimates of individual users converge to $q^\star$ if each smoothing constant α_i is sufficiently small.

In Ganesh et al. (2001), the authors generalize the earlier results to allow for delays in the price information being provided to the users. They also present simulation results that support their theoretical results.

8. PRICING BANDWIDTH

Consider now the owner of a *dedicated* (i.e., private) IP network who intends to lease capacity and would like to know how to price it. Potential customers would include *Content Service Providers*, i.e., businesses that host network applications, such as Blockbuster Video, which delivers full length films to customers online as well as through its rental stores, and Streetmail, which delivers information traditionally available from local newspapers and community calendars. Other customers could include small telecommunications companies requiring additional capacity, as well as companies with special bandwidth needs, such as a financial services firms requiring, e.g., the capability to make video broadcasts to its analysts.[14] How should the network owner charge its customers for bandwidth? Let us examine how this problem is now handled.

Currently, large network owners trade capacity via a complex procedure involving long-term leases known as *Indefeasible Right of Use* (IRU). An IRU is effectively the temporary ownership of a portion of the capacity of a fiber optic cable, and is granted by the company, or consortium of companies, that built the cable. IRUs typically cover a period of twenty to twenty-five years. In some cases, the IRU forbids the customer to resell the capacity during the term of the lease. IRUs are usually paid for up front in a single cash payment.

The role of IRUs is clear. A telecommunications firm interested in extending its business into a new region could lease capacity in the region, allowing it to enter the market but bypass the costs associated with a buildout. A major player could thus essentially piece together a communication network without having to incur the enormous expense of actually building the network. In this way it would assure itself of having sufficient capacity for when it requires it.

IRUs originated in the days of the Bell System monopoly prior to the break-up of AT&T in 1984. IRUs allowed competitors to utilize the large undersea cables that had been laid by AT&T to build out their networks. The IRU concept had a resurgence in popularity around 1997, when a number of new backbone providers arrived and had a need to build their markets quickly.

However, IRUs have recently come under criticism, with the accusation that they can be used to inflate revenues. Legal problems aside, there are serious practical problems with IRUs. In the antediluvian days of the Bell System, traffic was circuit-switched and reasonably predictable. However, the flexibility of the newer packet-switched technology has lead to enormous growth in demand for capacity, which has become far more difficult to predict, and IRUs have diminishing utility.

8.1 Contract and balancing process

The ECN bit in the packet header has an appealing interpretation as a hypothetical congestion charge. When a transmission line in a network is well below capacity, there will be few if any marks generated and the appropriate charge should be low;

14. Cable & Wireless issued a press release announcing a three-year global communications contract with UBS Warburg (Salgado, Valder Couser 2002). Secure data networks was among the services mentioned.

when a line is near capacity, many marks will be generated and the charge should be high. It seems plausible, then, that congestion marks could be used for controlling the supply of capacity.

A natural approach would involve the owner of a network transmission line being paid by each user based on the number of marks the user generates on the line. However, this simplistic method would produce a perverse incentive for the owner to increase congestion; for example, by making side payments to some users to generate traffic to maintain a state of high congestion. Further, given the likelihood of rapid variation in the rates at which congestion marks would be generated—a necessary consequence of their role in balancing demand over very short time scales—there is the additional issue of how prices can be appropriately averaged.

Anderson, Kelly, and Steinberg (2002) propose a *Contract and Balancing Process* (CBP) that addresses both issues. With CBP, bandwidth pricing is handled by a contractual process based around the use of ECN marks. It operates as follows. The network owner specifies a period of time for which potential users can contract a portion of the transmission line capacity. At the start of this period, the users pay the network owner according to their respective contractual amounts. During the period, the number of ECN marks received by each user is tabulated. At the end of the period, the users engage in a balancing process *with the other users*, based on each user's individual usage as compared with his respective contractual amount. The proportion of ECN marks generated by each user is employed as a proxy for usage, so that users who generated a larger-than-contracted proportion of ECN marks will make payments to the other users, while those who generated a smaller-than-contracted proportion of ECN marks will receive payments from the other users.

Observe that this approach allows volatile prices to be appropriately averaged, and thus mediates between rapidly fluctuating congestion prices and the longer time scales over which the bandwidth contracts would be written. Also note that CBP eliminates the incentive for the network owner to increase congestion. Finally, note that high usage by a user in one period—which may create considerable cost to other users—cannot be cancelled out by him through low usage in another period (a period in which aggregate demand might in fact be very low).

8.2 An example

Consider the following illustrative example involving three parties, network operator, N, and two customers, User 1 and User 2. Network owner N has a fiber optic cable with a capacity 100 megabits per second (Mbps), and is offering use of the cable over a six month period. He has set the contractual rate at c dollars per megabit and the balancing change at γ dollars per mark. User 1 chooses to contract for 40 Mbps of capacity, while User 2 contracts for 60 Mbps.

At the beginning of the contractual period, Users 1 and 2 make upfront cash payments to the network owner of $40c$ dollar and $60c$ dollars, respectively. Suppose over the course of the six-month period the number of marks generated by User 1 is exactly 40% of the total marks generated. Then, at the end of the contractual period no further payments are made. If, however, User 1 generated more than 40% of the total marks, then in the balancing process User 1 will pay User 2 an amount equal to the balancing charge, γ, times the excess number of marks he generated. In symbols, User 1 will pay User 2:

$$\Delta_{1,2} = \gamma[z_1 - 0.40(z_1 + z_2)]$$

where z_1 = number of marks generated by User *i*. If, however, User 1 generated less than 40% of the marks, then the expression above will be negative, and User 1 will *receive* this amount from User 2. Observe that the balancing payments are made only among the users and do not involve the network owner; the only payments received by the network owner are those made to him at the beginning of the period.

8.3 Results

Anderson et al. (2002) presents three main results. Consider the CBP scheme for a line with $n \geq 2$ users, where each user *i* contracts for a specified capacity γ_i over the balancing period at an immediate cost to him of $C_i(y_i)$. A line capacity $Y=\Sigma_i y_i$ is then available for use by all *n* users over the period.

The first result is that each user will have a unique optimal choice of contract quantity, given any set of contract quantities by the others. Further, a bound on the contract price is provided, beyond which the cost of participating in a contract is sufficiently expensive that a user will choose not to contract for any amount, but instead will pay for usage entirely through the balancing process.

The second result is that if the users have the same marginal cost and all follow a price-taking policy, then there is a unique Nash equilibrium for the contract quantities. Further, the time-averaged expected price is the unit cost of capacity.

The third result is that, under certain conditions, the second result can be generalized from a line to a network. Note that this leaves open the question as to whether there exists a useful sufficient condition to identify a unique Nash equilibrium in the network. This would be a good starting place for future research.

9. CHARGING

In order to implement any of these proposals, there must be in place a workable billing and accounting system so that users can be properly charged. Mackie-Mason and Varian (1995, 1996) propose that the accounting system for Smart Market be handled by the routers that compare the packet bid price with the threshold price. They recommend that accounting information on every *n*th packet be sent to a dedicated accounting machine that would determine the equilibrium access price. (They suggest $n = 1000$.) However, Smart Market would require a new field in the IP header to carry the requisite price information.

An ISP currently using flat-rate pricing that wishes to convert to PMP will require a way to keep track of each user's choice of network (which of course could change over time). Proportional Fairness requires little by way of billing and accounting, since each user is charged by his willingness-to-pay. The major implementation issue for CPE Pricing is that, like Smart Market, it would require a new field in the IP header to carry the price information. For the Contract and Balancing Process, Anderson et al. (2002) cite Briscoe, Rizzo, Tassel, and Damianakis (2000), who argue that ECN marks could allow the dispersal of charging operations, including accounting and billing, to customer systems. Briscoe et al. also implemented a mechanism for charging operations to be dispersed to end systems. Implementation is clearly an understudied issue in the pricing of Internet service and is a fertile area for future research.

10. CONCLUSION

Throughout this chapter, we have seen that, in order to properly address the topic of pricing Internet service, one needs to understand the current operation of the Internet and anticipate future technological developments, as well as be cognizant of the underlying marketing issues, including competition, real-time pricing, market segmentation, quality of service, differentiated services, and consumer perceptions of fairness. Fairness appears to have special significance; in addition to Proportional Fairness most of the Internet pricing proposals can probably be re-stated in terms of some type of fairness criterion.

In recent work, Bolton, Warlop and Alba (2003) study "consumer perceptions of price (un)fairness," and recommend that future work in that area consider whether consumers perceive fairness not only through comparisons with past prices, competitor prices, and perceived costs—their study's reference points—but also with respect to prices paid by other consumers. The authors suggest that consumers may be particularly sensitive to prices paid by others in the presence of price discrimination. Clearly, research in this area of marketing may have a significant influence on how Internet service pricing is modelled in the future.

There are a number of other interesting pricing proposals; we have space to mention only a few. Clark (1997) develops a pricing method based on the observation that marginal costs are nonzero only during congestion. In Clark's scheme, called *Expected Capacity Pricing*, there is a contract between the users and owner of the network as to the expected capacity that the network will provide the user. Based on this contract, the incoming traffic is metered and each packet is tagged as being either an *in* packet, i.e., within the profile of expected capacity, or an *out* packet, not within the expected capacity profile. At any point of congestion, only "out" packets can be dropped (or be recipients of ECN marks). Thus charging is related not to actual traffic, only expected traffic, which should lead to operational simplicity, since measurements within the network are not required.

Gupta, Stahl, and Whinston (1996) model user preferences as depending on the service acquired as well as the expected waiting time. In their scheme, called *Priority Pricing*, an incoming user service request is associated with an instantaneous value for the service, and a delay cost for this value. The user request can be fulfilled through several alternatives, each associated with a price and an expected waiting time. The user seeks the lowest total cost for a service, where the cost includes the cost of delay. Thus, an optimal choice is reflected by a choice of an alternative in a particular priority class, dependent upon the user's instantaneous value and delay cost. A billing and accounting system would be especially easy to implement for Priority Pricing, as the Internet Protocol already provides for a priority field in the packet header.

Marbach (2002) presents a pricing scheme that employs priorities to provide differentiated quality of service, where the users are free to choose the priority of their own traffic, but are charged accordingly. The users are assumed to choose an allocation of priorities so as to each optimize their own net benefit. Marbach shows that under his *Static Priority Pricing*[15] there exists a unique equilibrium, and the resulting allocation is max-min fair. For a thorough discussion of other Internet pricing proposals, the reader is referred to Courcoubetis and Weber (2003).

In conclusion, Kelly, Maulloo and Tan (1998) have predicted that, in the future, we can expect networks to have intelligence embedded at their end points which will act on behalf of human users. The authors suggest that this development is likely to lessen the distinction between engineering and economic issues and increase the importance of an interdisciplinary view. Further discussion on this intriguing topic can be found in the introduction and conclusion sections of Kelly (1996), as well as the introduction to Kelly (2000).

REFERENCES

Anderson, E., F. Kelly, and R. Steinberg (2002). A contract and balancing mechanism for sharing capacity in a communication network. Working Paper LSEOR 02.50, London School of Economics, Department of Operational Research, July.

Bakos, Y. and E. Brynjolfsson (2000). Bundling and competition on the Internet: Aggregation strategies for information goods. *Marketing Science* 19 (1), 63–82.

Barwise, P., A. Elberse, and K. Hammond (2002). Marketing and the Internet. In: B.A. Weitz and R. Wensley (eds.), *Handbook of Marketing*. London: Sage, pp. 527–557.

Bernoulli, D. (1738). Specimen Theoriae Novae de Mensura Sortis [Exposition of a new theory on the measurement of risk] *Commentarii Academiae Scientiarum Imperialis Petropolitanae*, Tomus V [*Papers of the Imperial Academy of Sciences in Petersburg*, Vol. V], 175–192.

Bertsekas, D. and R. Gallagher (1992). *Data Networks*. Second Edition. Englewood Cliffs, New Jersey: Prentice-Hall.

Bolton, L. E., L. Warlop, and J. W. Alba (2003). Consumer perceptions of price (un)fairness. *Journal of Consumer Research* 29 (4), 474–491.

Briscoe, B., M. Rizzo, J. Tassel, and K. Damianakis (2000). Lightweight policing and charging for packet networks. *Third IEEE Conference on Open Architectures and Network Programming (OpenArch 2000)*, 77–87.

Cerf, C., Y. Dalal, and C. Sunshine (1974). Specification of internet transmission control program. Request for Comments 675, NIC [Network Information Center] 2, INWG [International Network Working Group] 72, Network Working Group, Stanford Research Institute, December.

Chander, P. and L. Leruth (1989). The optimal product mix for a monopolist in the presence of congestion effects: A model and some results. *Journal of Industrial Organization*, 7 (4), 437–449.

Clark, D. (1997). Internet cost allocation and pricing. In: L.W. McKnight and J.P. Bailey (eds.), *Internet Economics*. Cambridge, MA: MIT Press, pp. 215–252.

Courcoubetis, C. and R. R. Weber (2003). *Pricing Communication Networks: Economics, Technology and Modelling*. Chichester: John Wiley and Sons.

Ganesh, A., K. Laevens, and R. Steinberg (2000). Dynamics of congestion pricing. Microsoft Research Technical Report MSR-TR-2000-70 & Judge Institute Working Paper WP 16/2000, July 2000.

15. Although in his paper he did not give a name to his pricing scheme, Marbach suggests (private communication, April 2003) that "Static Priority Pricing" would be appropriate, as the prices are set *a priori* and do not change as the state of the network changes (in contrast to Gupta et al. (1996), where prices are adapted to track a socially optimal resource allocation).

Ganesh, A., K. Laevens, and R. Steinberg (2001). Congestion pricing and user adaptation. *IEEE INFOCOM 2001*, 959–965.
Gibbens, R. (2000). Control and pricing for communication networks. *Philosophical Transactions of the Royal Society of London*, Series A 358, 331–341.
Gibbens, R. J. and F. P. Kelly (1999). Resource pricing and the evolution of congestion control. *Automatica*, 35 (12), 1969–1985.
Gibbens, R., R. Mason, and R. Steinberg (2000). Internet service classes under competition. *IEEE Journal on Selected Areas in Communications* 18 (12), 2490–2498.
Gilder, G. (1993). Metcalfe's law and legacy. *Forbes ASAP*, 13 September. [http://www.gildertech.com/public/telecosm_series/metcalf.html]
Gilder, G. (1999). *The Telecosm Glossary: An Opinionated Lexicon*. Second Edition. Gilder Technology Group.
Gupta, A., D. O. Stahl, and A. B. Whinston (1996). Priority pricing of integrated services networks. In: L. W. McKnight and J. P. Bailey (eds.), *Internet Economics*. Cambridge, MA: MIT Press, pp. 323–352.
Haimanko, O. and R. Steinberg (2000). Price symmetry in a duopoly with congestion, CORE Discussion Paper, No. 2000/56, December.
Jaffe, J. M. (1980). A decentralized, "optimal," multiple user, flow control algorithm. *Proceedings of the Fifth International Conference on Computer Communications*, pp. 839–844.
Jaffe, J. M. (1981). Bottleneck flow control. *IEEE Transactions on Communications*. COM-29 (7), 954–962.
Jain, R., K. K. Ramakrishnan and D.M. Chiu (1987). Congestion avoidance in computer networks with a connectionless network layer. DEC Technical Report 506, August. Reprinted in *Innovations in Internetworking*. ARTECH House Inc., 1988.
Kelly, F. (1996). Modelling communication networks, present and future. *Philosophical Transactions of the Royal Society* A358 (2000) A354, 437–463.
Kelly, F. (1997a). Charging and rate control for elastic traffic. *European Transactions on Telecommunications*, 8, 33–37. [Revised version available at http://www.statslab.cam.ac.uk/~frank/elastic.html]
Kelly, F. P. (1997b). Pricing and rate control for communication networks. 1997 Naylor Prize Lecture, London Mathematical Society. November 21. [Available at http://www.statslab.cam.ac.uk/~frank/TALKS/LMS/]
Kelly, F. (2000). Models for a self-managed Internet. *Philosophical Transactions of the Royal Society* A358 (2000) 2335–2348.
Kelly, F. (2001). Mathematical modelling of the Internet. In B. Engquist and W. Schmid (eds.), *Mathematics Unlimited—2001 and Beyond*. Berlin: Springer-Verlag, 685–702.
Kelly, F. P., A. K. Maulloo, and D. K. H. Tan (1998). Rate control in communication networks: Shadow prices, proportional fairness and stability. *Journal of the Operational Research Society* 49, 237–252.
Key, P. B. and D. R. McAuley (1999). *IEE Proceedings Software* 146, 39–43.
MacKie-Mason, J. K. and H. R. Varian (1995). Pricing the Internet. In: B. Kahin and J. Keller (eds.), *Public Access to the Internet*. Cambridge, MA: MIT Press, pp. 269–314.
MacKie-Mason, J. K. and H. R. Varian (1996). Some economics of the Internet. In: W. Sichel and D.L. Alexander L. (eds.), *Networks, Infrastructure, and the New Task for Regulation*. Ann Arbor: The University of Michigan Press, pp. 107–136.
Marbach, P. (2002). Priority service and max-min fairness. *IEEE INFOCOM 2002*, 266–275.
Mark, R. (2002). AOL maintains strong lead among ISPs worldwide. Press Release, WebSideStory, March 7, http://www.websidestory.com/
Massoulié, L. and J. Roberts (1999). Bandwidth sharing: objectives and algorithms. *IEEE INFOCOM 1999*, 1395–1403.
Mazumdar, R., L. G. Mason, and C. Douligeris (1991). Fairness in network optimal flow control: optimality of product forms. *IEEE Transactions on Communications* 39 (5), 775–782.
Nash, J. (1950). The bargaining problem. *Econometrica* 18, 155–162.
Odlyzko, A. M. (1997). A modest proposal for preventing Internet congestion. Unpublished manuscript, AT&T Labs, September.
Odlyzko, A. M. (1999a). Paris Metro pricing for the Internet. *Proceedings ACM Conference on Electronic Commerce (EC'99)*, ACM, 1999, 140–147.
Odlyzko, A. M. (1999b). Paris Metro pricing: The minimalist differentiated services solution. *Proceedings 1999 Seventh International Workshop on Quality of Service (IWQoS '99)*, IEEE, 159–161.
Odlyzko, A. M. (1999c). The economics of the Internet: Utility, utilization, pricing, and Quality of Service. Technical Report 99-08, DIMACS, February. [Available at http://www.dtc.umn.edu/~odlyzko/doc/recent.html].

Odlyzko, A. M. (2000). The history of communications and its implications for the Internet. AT&T Labs Research, working paper, June. [Available at http://www.dtc.umn.edu/~odlyzko/doc/recent.html].

Owen, D. (1999). Paris rail abolishes first class travel. *Financial Times*, September 1, 1999, p. 2.

de Palma, A. and L. Leruth (1989). Congestion and game in capacity: A duopoly analysis in the presence of network externalities. *Annales d'Economic et de Statistique* 0 (15–16), 389–407.

Papadimitriou, C. H. (2002). Learning the Internet. In: J. Kivinen and R.H. Sloan (eds.), *Computational Learning Theory (Proceedings of the 15th Annual Conference on Computational Learning Theory, COLT. Lecture Notes in Computer Science* 2375, Springer, p. 396.

Park, K., M. Sitharam, and S. Chen (2000). Quality of service provision in noncooperative networks with diverse user requirements. *Decision Support Systems, Special Issue on Information and Computation Economies* 28, 101–122.

Rawls, J. (1971). *A Theory of Justice*. Cambridge, MA: Harvard University Press. Also published in a revised edition, 1999.

Roth, A. (1979). *Axiomatic Models of Bargaining*. Berlin: Springer-Verlag.

Salgado, T., H. Valder, and C. Couser (2002). UBS Warburg selects Cable & Wireless as global telecoms supplier, Press Release, Cable & Wireless, September 5.

Shenker, S. (1995). Fundamental design issues for the future Internet. *IEEE Journal on Selected Areas in Communication* 13 (7), 1176–1188.

Simon, H. (1989). *Price Management*. New York: North Holland.

Tirole, J. (1988). *The Theory of Industrial Organization*. Cambridge, MA: MIT Press.

Transportation Research Board (2002). 91 Express Lanes Orange County, California. *Innovative Finance for Surface Transportation*, National Cooperative Highway Research Program, May [http://www.innovativefinance.org/projects/highways/91.asp].

Vickrey, W. (1961). Counterspeculation, auctions, and competitive sealed tenders. *Journal of Finance* 16 (1), 8–37.

Vickrey, W. (1962). Auction and bidding games. In: *Recent Advances in Game Theory*. The Princeton University Conference, 15–27.

Waldspurger, C.A., Hogg, T., Huberman, B.A., Kephart, J.O., and Stornetta, W.S. (1992). Spawn: A distributed computational economy. *IEEE Transactions on Software Engineering* 18 (2), 103–117.

Wall, L. and R.L. Schwartz (1990). *Programming Perl*. O'Reilly & Associates, Inc. 1990.

WebSideStory (2002). AOL maintains strong lead among ISPs worldwide. Press Release, March 7.

Zakon, R. H. (2003). Hobbes' Internet timeline. v6.0, February [http://www.zakon.org/robert/internet/timeline/].

7. USAGE VOLUME AND VALUE SEGMENTATION IN BUSINESS INFORMATION SERVICES

T.C.A. BASHYAM and U.S. KARMARKAR

ABSTRACT

Business Information Services are intermediaries that collect, package and distribute engineering, medical, legal, and such professionally-oriented information. Users of these services have three options for searching for and accessing information. First, a package service that delivers a database to subscribers using data storage media like CD-ROMs. Second, an online (Internet) service. Finally, users also have the option of self-service, where they collect and collate information directly from its sources. Potential subscribers are characterized by their annual usage volume for such services, and by a *value parameter*, defined as the maximum value they expect to derive from each search. We model a duopoly where service providers with either a package or online service compete for subscribers who also have the option of self-service. Under a plausible ordering of relative costs for online versus package technologies, we demonstrate that package and online services can co-exist, with the package provider serving high volume subscribers with a given value parameter, while the online provider serves low volume users. When the average value parameter in the subscriber base increases, service providers are shown to increase their database sizes and subscription charges. As search costs for online access drop relative to package services, we show that the package provider is eliminated from the market.

1. INTRODUCTION

The rapid growth in the volume of business information has resulted in the emergence of information service intermediaries that collect, collate, package and distribute business information (Davis and Davidson 1991, Rayport and Sviokla 1995). Reuters and Bloomberg (financial information), Aspect Development (semiconductor parts catalog, a subsidiary of I2 Technologies), Jeppesen (aeronautical information) and Celera Genomics (biosciences information) are all examples of such services in their respective industries.

Such information service providers have two options for delivering their service. First, a package service, where the provider collects and assembles the information into a database, and distributes copies using CD-ROMS or tapes. Second, an Internet-based service where the provider delivers such information online from a hosted database. Potential subscribers could choose one of these options or may prefer self-service, where they collect and assemble information directly from source. These alternatives differ in their cost structure for providers using them, and in the perceived value for potential subscribers. Specifically, a package service, may offer lower search costs for subscribers compared to an online service, given the longer access times in an online service due to network delays. The package provider cannot meter usage and can only use a fixed subscription fee. An online provider can meter usage and charge a marginal fee per use. Additionally, that provider can also lower their cost of updating and maintaining their service by having information sources submit data in a standard format. This paper examines how providers using these technologies compete, given their inherent cost and benefit tradeoffs, in a market where subscribers opt for whichever option (including self-service) maximizes their value.

Potential subscribers to the services are characterized by two attributes. First, by their annual usage volume, measured by the number of searches they make for information. Second, by a *value parameter*, defined as the maximum value they might derive from each search, illustrated in the following examples. Consider a hardware design engineer trying to pick a new component for a laptop computer by searching a semiconductor parts catalog such as Aspect Development‘s service (Bashyam and Karmarkar 2000). Suppose, by exhaustively searching for a component for the new notebook computer, the designer engineer could increase its battery life, and therefore its selling price by $150. We could use $150 as the value parameter for this user. Another design engineer using the same parts catalog to exhaustively search for a new component for a new modem card (a commodity product) could perhaps increase its selling price by $5, a relatively low value parameter, given the commodity nature of the end product. If these two users were to represent the extremes in this subscriber population, then, we could reasonably assume that the value parameter is some distribution over the range [$5, $150] for all subscribers in this market. In general, the value parameter is high when the opportunity cost of not finding the right information is high.

Subscribers indexed by their annual usage volume and value parameter choose the benefit maximizing service delivery option, which may be self-service, online or package service. Service providers compete for subscribers using either a package or online service. In this paper, we model the market as a duopoly. Service providers choose their profit maximizing database size and pricing—these design parameters of the service are shown to be critical for subscribers' benefit from using the service.

Our results are as follows. In a duopoly with a package and online provider, we show that the lower search cost option (package or online) is chosen by high volume subscribers of any given value parameter. The higher search cost technology is chosen by low volume subscribers of any given value parameter. Self-service is chosen by subscribers whose usage volume is too low to warrant a third party service. We also demonstrate the impact of subscribers' value parameter and usage volume distribution, and their search costs in the packaged technology on database sizes, pricing and market shares for service providers in a duopoly. We show that dispersion in the value parameter among subscribers expands the feasible set of database sizes, and fixed and marginal charges for subscription. Thus, as the average value parameter in the subscriber base for the services increases, both database sizes and fixed charges increase for the online and packaged service providers in the duopoly. As the search cost for a subscriber in the packaged technology approaches that in the online service, both providers lose profits, due to intense price competition. However, the package provider may be forced out of the market first because of their higher costs of maintaining their service.

2. LITERATURE REVIEW

There is an enormous body of literature on non-cooperative games (see Bashyam 1996 and references therein), which has guided our modeling of competition in Business Information Services. Economic models of services are also relevant to this work. Oren and Smith (1981) study the pricing of telecommunication services in the presence of demand externality. In their model, potential subscribers are indexed by their volume of usage, which in turn is determined by the size of the subscriber base for the service. For a monopolist service provider, they investigate the effect of pricing strategies, such as a fixed charge or a three part tariff, on market penetration and profit. Oren, Smith and Wilson (1983) extend the pricing problem to address competitive service markets. Subscribers there are characterized by both their volume of usage and by their willingness to pay for a given volume. Service providers are assumed to be identical, therefore, they have the same pricing structure and share the market equally in the symmetric equilibrium derived. Unlike the present work, market segmentation is not an issue since service providers are completely identical. However, a subscriber's willingness to pay for usage in their model is very similar in its economic effects to their value parameter in a information service in this paper, since we show that the marginal benefit for a subscriber is shown to be a function of their value parameter.

The rest of the paper is organized as follows. In Section 3, we describe the profit maximization model for service providers that takes into account tradeoffs in the competing technologies, including the cost of creating and updating the service, search costs for subscribers and pricing mechanisms. To facilitate its development, we first characterize the benefit function for subscribers as a function of their value parameter and usage volume under self-service, online and package options. The self-service benefit function determines the minimum annual benefit that a third-party service provider must exceed to capture that subscriber. In Section 4, we derive results for market segmentation in a duopoly along the usage volume and search parameter dimensions—these results drive numerical simulation of the equilibrium in a duopoly in Section 6. In Section 5, we introduce the duopoly model of the market. Section 6 provides numerical equilibrium results for and insights from the duopoly model. A final section concludes this work.

3. THE MODEL

In this section, we develop a profit maximization model for service providers when they use either a package design or an online service to capture subscribers. The following table lists the notation used in this paper.

i	index ($i = a,b,s$) denotes technology choice. $i = a$ (package), b (online), s (self-service)
N_i	database size for a provider using technology i; $i = a$ (package), b (online), s (self-service). For self-service, N_s is the database size created by a subscriber.
$K_i(N_i)$	total costs for provider i using technology i; $i = a$ (package), b (online), s (self-service)
$C_i(N_i)$	search cost per search in a database of size N_i; $i = a$ (package), b (online), s (self-service)
D	value parameter per search for a subscriber; $D_{min} \leq D \leq D_{max}$
Q	annual search volume for a subscriber; $0 \leq Q \leq Q_{max}$
$V(N,D)$	intrinsic value per search for a subscriber with value parameter D in a database of size N
$U_i(N,D)$	$= V(N,D) - C_i(N)$; net benefit per search for a subscriber with value parameter D
$W_s^*(Q,D)$	total annual benefit under self service option for a subscriber that makes Q searches per year of value D per search
m	is the total number of subscribers
$f(D)$	is the proportion of subscribers that have value parameter D
$F(D)$	is the proportion of subscribers with value parameter less than D i.e. this the cumulative distribution of D
$g(Q\|D)$	is the proportion of subscribers with value parameter D who make Q searches per year.
$G(Q\|D)$	is the proportion of subscribers of value parameter D who make less than Q searches per year. i.e. this is the cumulative distribution of Q for a given D.
P_i	is the fixed charge (annualized) for purchasing service i, $i = a$ (package), b (online)
p_b	is the marginal charge per search in the online service
$W_a(N_a,P_a,Q,D)$	total annual benefit under package service option for a subscriber that makes Q searches per year of value D per search for a given combination of database size and price
$W_b(N_b,P_b,p_b,Q,D)$	total annual benefit under online service option for a subscriber that makes Q searches per year of value D per search for a given combination of database size and prices
Pr_i	annual profit for provider i, $i = a$ (package), b (online)
Ms_i	is the market share, or the fraction of users that buy the service i, $i = a$ (package), b (online)
Q_{MS_i}	For a given D, this represents the set of usage volumes where service i is preferred over all other service alternatives, $i = a$ (package), b (online)
$S(N_b,P_b,p_b)$	Is the total usage for all online subscribers, for a given choice of database size, fixed and marginal prices for the online provider

As a first step, we detail the costs for a service provider using either package or online technology. A provider incurs two types of costs: first, fixed costs that do not depend on how much information (measured by number of records) the provider accumulates and second, costs that vary depending on the size of the database a provider chooses to create. Fixed costs include the cost of search software development, creating indexes and hardware and software costs for building a data factory (Bashyam and Karmarkar 2000). Costs that depend on the size of the database created include the costs for initial data entry and ongoing updates of the information. We abstract these costs into a function $K_i(N_i)$, $K_i(0) = 0$, that represents the cost of building and maintaining a database of size N_i. In the semiconductor parts catalog business, Aspect Development (Bashyam and Karmarkar 2000) incurred an initial cost of $3 million for creating its service. Its ongoing cost of maintaining its service (measured as the cost of updating data per part in the database) is a trade secret. In Aspect's case, $K_i(N_i)$ = $3 million + cost per part * number of parts (N_i). The variable cost per record is a critical determinant of the provider's profitability and viability. The cost per record in the legal publishing industry, for example, is $4.70 per record for data entry, formatting and indexing (Boeri and Hensel 1995, Reisman 1995). In general, the data entry costs in $K_i(N_i)$ is higher for content that is highly technical (e.g. a semiconductor parts catalog), has high legal liabilities associated with incorrect information (e.g. aeronautical information) and where the source data is in diverse and incompatible formats (e.g. some paper, some digital).

Subscribers incur a search cost for each query in a service, represented by $C_i(N_i)$, $C_i(0) = 0$, $i = s, a$, and b denoting "self service", package and online service options respectively. In general, search costs depend on the database size (higher the size, larger the search cost), access technology (more for online because of network latencies) and quality of indexing. The functional form $C_i(N_i)$ allows us to model the effect of service delivery technology and database size. A reasonable way to measure the marginal cost per search is to first quantify the time spent by the average subscriber to find the information they are looking for in databases of different sizes (say, N_1, N_2, N_3 and so on). The product of the search time and the loaded cost for the subscriber per unit time (i.e. annual salary divided by standard working hours per year) can then be used as a measure of the search cost per search for the different database sizes used. Because of the latencies in an online network compared to a local database, we would expect that $C_a(N) < C_b(N)$, for a given database size N, assuming the quality of indexing is comparable between online and package technologies.

Subscribers are indexed by their value parameter D and their annual usage volume Q. The value parameter D is the maximum benefit a subscriber would derive per search in a service if they searched exhaustively for information. The following example gives meaning to these attributes for the semiconductor parts catalog industry. Assume that potential subscriber firms design products priced in the range $1500–$15,000, each of which may have approximately 30 discrete components, based on the ratio of the sizes of an average printed circuit board and the components therein. Suppose electronic design engineers in those firms look for one particular component for their designs, say an 8-bit comparator or a floppy disk controller. Judicious component selection for a product priced in the given range could reduce the cost of the product or increase its selling price in the market by, perhaps, $150 to $1,500, or roughly $5 to $50 (=1500/30) per component depending on the product being de-

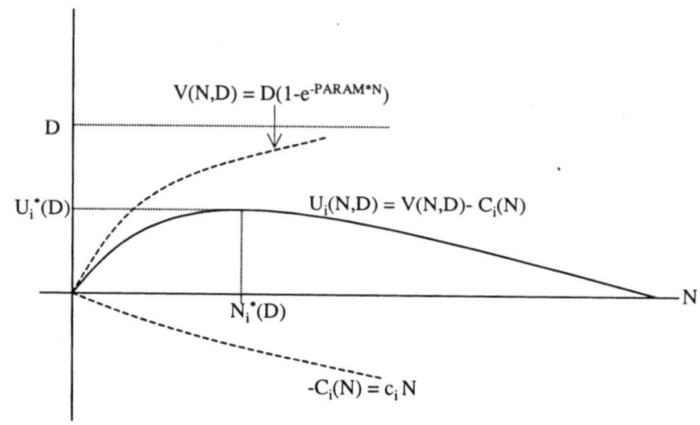

Figure 1. Intrinsic value function (V) and marginal benefit (U) function.

signed. Thus, for each search, subscribers stand to gain a maximum of \$5 to \$50, if they searched exhaustively for the ideal component. Therefore, the *value parameter* in this subscriber population can be assumed to be in the range \$5 to \$50, or D_{min} = \$5 and D_{max} = \$50. If the largest of the subscriber firms is assumed to contain 100 concurrent design teams, each developing one product per year, each of which requires one search for the particular component, then, the maximum usage volume in the subscriber population, $Q_{max} = 100$.

Subscribers are assumed to have an intrinsic value per search for information measured by V(N,D), V(0,D) = 0, for a subscriber of value parameter D searching a database of size N. The functional form of V is increasing and concave in N as illustrated in Figure 1. The rationale is that a larger database is more valuable, but because the subscriber's benefit per search caps at D, the incremental benefit reduces with database size, as in Figure 1. Subscriber D's net value per search in given a information service, can be expressed as $U_i(N,D) = V(N,D) - C_i(N)$; let $U_i(0,D) = 0$, $U_i(N,D) \geq 0$ for some N. Figure 1 assumes that $U_i(N,D)$ is concave in N for a given D with a unique maximum at $N_i^*(D)$. Simply put, a larger database is more beneficial, but after the size increases beyond $N_i^*(D)$, the search costs bring the marginal benefit down. Because of this functional form for the net value, no provider will build a database greater than $N_i^*(D)$ for subscribers with value parameter D.

In order for subscribers to switch to a third party service, that provider must exceed a subscriber's self-service benefit, which is what they would derive if they chose to collect and assemble a database of their own from source. We characterize the self-service benefit function next for a subscriber with value parameter D ($D_{min} \leq D \leq D_{max}$) that makes Q ($0 \leq Q \leq Q_{max}$) searches per year.

Given their usage volume and value parameter, a subscriber opting for self service must create a database size of size N_s, by trading the intrinsic value of a large database against the costs of building and maintaining it. The net benefit for each search for a subscriber of value parameter D is given by $U_s(N,D) = V(N,D) - C_s(N)$, for a database of size N. The optimum database size for a self-service subscriber that makes Q annual searches of value D is given by the solution to the problem:

$$W_s^\star(Q,D) = \underset{N \geq 0}{\mathrm{Max}}(V(N,D) - C_s(N))Q - K_s(N) \quad (1)$$

$$N_s^\star(Q,D) = \underset{N \geq 0}{\mathrm{argmax}}\ (V(N,D) - C_s(N))Q - K_s(N) \quad (2)$$

Next, we model the profit maximization problems for the online and package service providers. Because a package provider cannot meter usage in their service, they can only charge a fixed subscription fee, P_a. The total annual benefit of a packaged service subscriber $W_a(N_a,P_a,Q,D)$can be expressed as:

$$W_a(N_a,P_a,Q,D) = (V(N_a,D) - C_a(N_a))Q - P_a = U_a(N_a,D)Q - P_a \quad (3)$$

where N_a is the size of the database supplied by the package provider, U_a is the marginal benefit per search, and C_a is the search cost function for the package service. A subscriber (Q,D) will purchase the packaged information service if its net value W_a exceeds the value from competing alternatives. Therefore, in a duopoly, with one other provider, it should be the case that $W_a(N_a,P_a,Q,D) > \max\{W_s^\star(Q,D), W_b(N_b,P_b, p_b,Q,D)\}$, b denoting the online provider and $W_s^\star(Q,D)$ denoting the total benefit under the self-service option, for subscriber (Q,D) to pick the package service. The profit maximizing package provider would choose their database size and fixed price to solve the following problem:

$$\underset{Pa,Na}{\mathrm{Max}}\ PR_a(N_a,P_a) = P_a\ m\ MS_a(N_a,P_a) - K_a(N_a) \quad (4)$$

subject to

$$MS_a(N_a,P_a) = \int_D \int_{\{Q_{MSa}\}} g(Q|D)dQ\ f(D)dD$$
$$\{Q_{MSa}\} = \{Q: W_a(N_a,P_a,Q,D) > \max\ \{W_s^\star(Q,D), W_b(N_b,P_b, p_b,Q,D)\}\ |D\ \}$$

Unlike the package provider, the online provider can monitor a subscriber's usage, and can charge p_b for each search. The annual benefit for a subscriber buying the online service can be written as:

$$W_b(N_b,P_b,p_b,Q,D) = (V(N_b,D)\text{-}p_b\text{-}C_b(N_b))Q\text{-}P_b = (U_b(N_b,D)\text{-}p_b)Q\text{-}P_b \quad (5)$$

where N_b is the size of the database supplied by the online provider, U_b is the marginal benefit per search, C_b is their search cost function, P_b is the annual fixed charge and p_b is the marginal charge per search. As in the formulation for the package provider, the market share for the online provider consists of subscribers for whom the benefit under the online service exceeds the benefit under both the package and self-service options, expressed by the constraint, $W_b(N_b,P_b,p_b,Q,D) > \max\{W_s^*(Q,D), W_a(N_a,P_a,Q,D)\}$. The profit maximizing online provider would choose their database size, fixed price and marginal price to solve the following problem:

$$\underset{p_b,P_b,N_b}{\text{Max}}\ PR_b = mP_bMS_b(N_b,P_b,p_b) + p_b\ S(N_b,P_b,p_b) - K_b(N_b) \tag{6}$$

subject to:

$$MS_b(N_b,P_b,p_b) = \int_D\int_{\{QMSb\}} g(Q|D)dQ\ f(D)dD$$
$$S(N_b,P_b,p_b) = m\int_D\int_{\{Qmsb\}} Qg(Q|D)dQ\ f(D)dD$$
$$\{Q_{MSb}\} = \{Q\text{: } W_b(N_b,P_b,p_b,Q,D) > \max\{W_s^*(Q,D),\ W_a(N_a,P_a,Q,D)\}|D\ \}$$

The profit maximizing problem is solved as follows. For given values of N, P (and p), the subscriber with a given (Q,D) makes a choice between their service options a, b and s, based on whichever option maximizes their net benefit as characterized by equations 1, 3 and 5. This determines the market share for each provider. In a duopoly with a package and an online provider, the providers need to find optimal values of N, P, and p (for online provider). We model that game as a one-shot game, where each provider, without knowing the other provider's choices, chooses all the parameters in their control simultaneously. The equilibrium of the one-shot game (if it exists) requires a five (N_a,P_a,N_b,P_b,p_b) dimensional grid search.

4. THE SUBSCRIBER'S DECISIONS

As a first step for analyzing the profit maximizations problems in the previous section, we characterize the shapes of the marginal (U) and annual benefit (W) functions. The shape of the marginal benefit functions (U) allows us to segment the value parameter (D) dimension between users that derive more marginal benefit per search from a package (or online) provider compared to self-service. The comparison of annual benefit functions within these value parameter segments allows us to determine usage volume segments where one of the providers ensures more annual benefit for subscribers than the other alternatives. Because the general case is intractable, we analyze the problem under specific functional forms for V(N,D), K(N) and C(N) as below:

ASSUMPTION 1. We assume that V(N,D) is concave in both N and D, and $C_i(N)$ and $K_i(N)$ are linear in N; i = s (self-service), a (package), b (online), as below:

(i) search cost function $C_i(N) = c_iN$;
(ii) provider cost function $K_i(N) = K_i + k_iN$; $K_i(0) = 0$;
(iii) value function $V(N,D) = D(1\text{-}\exp(\text{-}PARAM^*N))$, PARAM is a constant

For Assumption 1, it is easy to see, as in Figure 1, that $U_i(N,D) = D(1\text{-}e^{-PARAM^*N})\text{-}c_iN$, is concave in N with a unique maximum $U_i^*(D)$ attained at $N = N_i^*(D)$. The

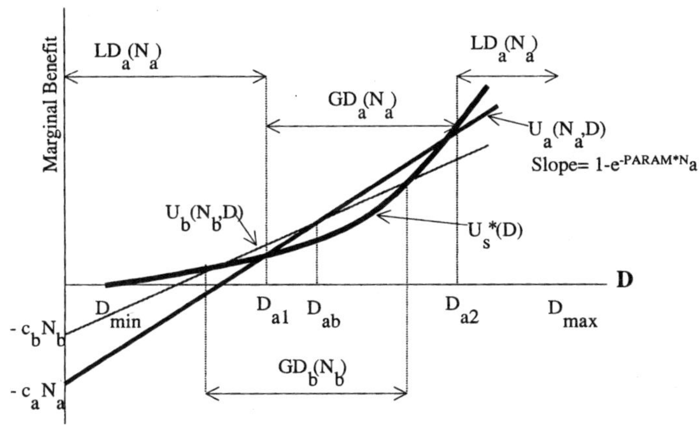

Figure 2. Segmentation in value parameter dimension.

rationale here is that a larger database is more valuable, however, as N grows large, the search costs per search become excessive and start bringing the marginal benefit $U_i(N,D)$ down.

In Proposition 1, we show that the maximum marginal benefit per search $U_i^\star(D)$ is non-decreasing in D, which is intuitive because we would expect subscribers with a higher value parameter D to derive a larger marginal benefit per search.

PROPOSITION 1. Let $U_i^\star(D) = \sup_{N \geq 0} U_i(N,D) = D(1\text{-}e^{-PARAM\star N})_{-}c_iN$; i = s (self-service), a (package), b (online). $U_i^\star(D)$ has the following properties:

(i) $U_i^\star(D)$ is convex, non-decreasing in D, and $U_i^\star(D) \geq 0$.
(ii) $N_i^\star(D) = -\ln(c_i/D\star PARAM)/PARAM$

A key implication of Proposition 1 is that a provider will never build a database of size greater than $N_i^\star(D_{max})$. Another implication, summarized in Proposition 2 below, is that we can determine sets of value parameters (D) where the self-service option provides greater marginal benefit per search than the other options. This result is used later to calculate market segments in the usage volume dimension by comparing the annual benefit functions W_a, W_b and $W_s^\star$.

PROPOSITION 2. (See Figures 2 and 3).

(a) There can be no more than two values of D that solve the equation $U_a(N_a,D) = U_s^\star(D)$.

(b) If $U_a(N_a,D_{max}) > U_s^\star(D_{max})$, then there will be at most one D that solves the equation $U_a(N_a,D) = U_s^\star(D)$.

Proposition 2 also holds for the online provider.

To use Proposition 2 for characterizing market areas, we first partition D into two sets for the package and online provider:

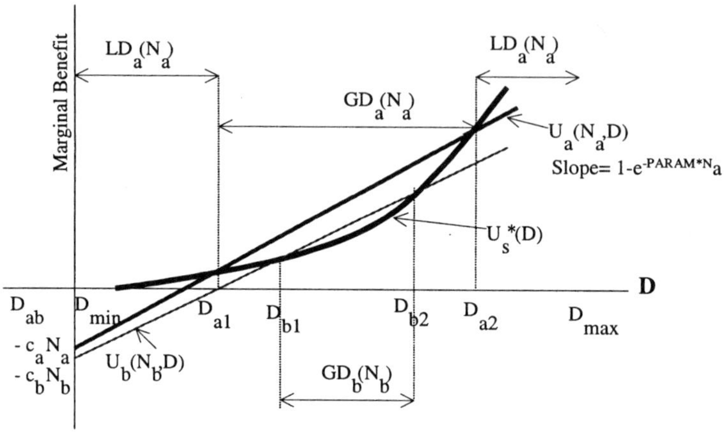

Figure 3. Second possible segmentation in the value parameter dimension.

$$LD_a(N_a) = \{D|\ \max(0,U_a(N_a,D)) < U_s^\star(D)\} \tag{7}$$
$$GD_a(N_a) = \{D|\ U_a(N_a,D) \geq U_s^\star(D)\} \text{ for the package provider}$$
$$LD_b(N_b) = \{D|\ \max(0,U_b(N_b,D)\text{-}pb) < U_s^\star(D)\} \tag{8}$$
$$GD_b(N_b) = \{D|\ U_b(N_b,D)\text{-}p_b) \geq U_s^\star(D)\} \text{ for the online provider}$$

Proposition 2 essentially states that $GD_a(N_a)$ (or $GD_b(N_b)$) is a contiguous interval of the form $[D_1,D_2]$, while the set $LD_a(N_a)$ (or $LD_b(N_b)$) may comprise at most two disjoint intervals, as in Figures 2 and 3.

Next, we show that for $D \in GD_a(N_a)$ (or $GD_b(N_b)$), the package (or online) provider will be picked by users with usage volumes Q falling inside a contiguous interval. To show this, we first determine the shape of the self-service benefit function $W_s^\star(Q,D)$ for a given D. We can show that this function is convex in Q for a given D, and has an asymptotic linear lower bound. As equations 3 and 5, and Figure 4 show, the benefit functions for the online (W_b) and package (W_a) providers are linear in Q for a given D. Overlaying these two linear functions on the convex function $W_s^\star(Q,D)$ allows us to conclude that the market segments for the online and package providers are intervals of the form $[Q_1,Q_2]$ for a given D (see Figure 4 for an example). The usage volume breakpoints for $W_s^\star$ and W_a (or W_b) can be determined using a bisection (or binary search) algorithm because we know that we have to look for at most 2 such breakpoints. This algorithm simply compares $W_s^\star$-W_a(or W_b) for a Q less than and greater than the candidate breakpoint—if the difference $W_s^\star$-W_a changes from positive to negative, we know we have a breakpoint. Lemma 1 below summarizes these observations:

LEMMA 1. Let $N_s^\star(D) = \arg\sup_{N \geq 0} U_s(N,D)$, $U_s^\star(D) = \sup_{N \geq 0} U_s(N,D)$. For a given D,$W_s^\star(Q,D)$ has the following properties:

(a) $\partial^2 W_s^\star(Q,D)/\partial Q^2 \geq 0$ and $\partial W_s^\star(Q,D)/\partial Q \geq 0$.

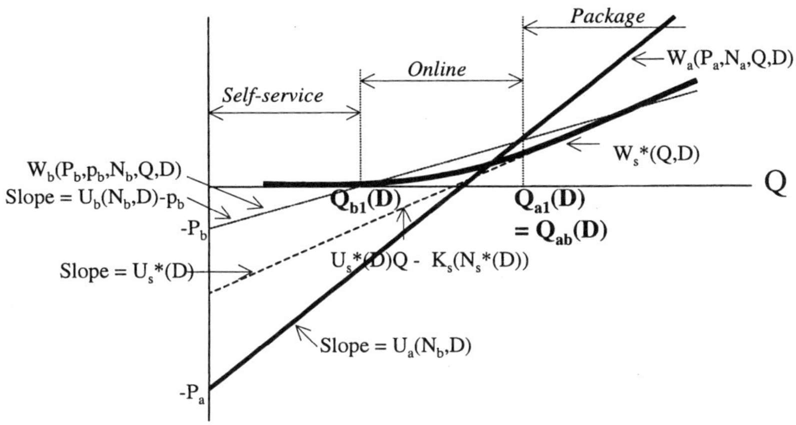

Figure 4. Segmentation in the usage volume (Q) dimension for a given value parameter D.

(b) $W_s^*(Q,D) \geq U_s^*(D)Q - K_s(N_s^*(D))$.

(c) The linear bound of Lemma 1(b) is approached asymptotically as Q becomes large. If $U_s^*(D)$ is positive, then $W_s^*(Q,D)$ is eventually increasing in Q.

For a given D, Figure 4 reveals the impact of the fixed price and database sizes on the market segment (and revenues thereof) for package and online provider. In Figure 4, as the fixed price (intercept P_a) increases, the provider loses subscribers. As the database size changes for a given D, the slope $U_a(N,D)$ changes and the provider's segment of subscribers changes. So, a provider's strategy and profits is really driven by the price and slope (which is tied to database size) of their benefit function.

Using Figure 4, Lemma 1 and Proposition 2, we can calculate usage volume segments for a given D in Proposition 3 below.

PROPOSITION 3.

(a) For every $D \in GD_a(N_a)$ (or $GD_b(N_b)$), there will be at most one Q such that $W_s^*(Q,D) = U_a(N_a,D)Q - P_a$ (or $W_s^*(Q,D) = (U_b(N_b,D) - p_b)Q - Pb$). The package provider's benefit function in Figure 4 depicts this condition.

(b) For every $D \in LD_a(N_a)$ (or $LD_b(N_b)$), there will be exactly two values of Q such that $W_s^*(Q,D) = U_a(N_a,D)Q - P_a$ (or $W_s^*(Q,D) = (U_b(N_b,D) - p_b)Q - P_b$). The online provider's benefit function in Figure 4 depicts this condition.

Based on Figure 4, it is easy to see that for a given value parameter D, the market segment for the package provider is simply an interval $[Q_{a1}(D), Q_{a2}(D)]$ and an interval $[Q_{b1}(D), Q_{b2}(D)]$ for the online provider based on where the self-service, online and package provider benefit functions intersect in the figure.

Assume that there is one online and one package provider. For any D, define $Q_{ab}(D) = (P_a - P_b)/(U_a(N_a,D) - U_b(N_b,D) + p_b)$ (see Figure 4). $Q_{ab}(D)$ represents the break-even

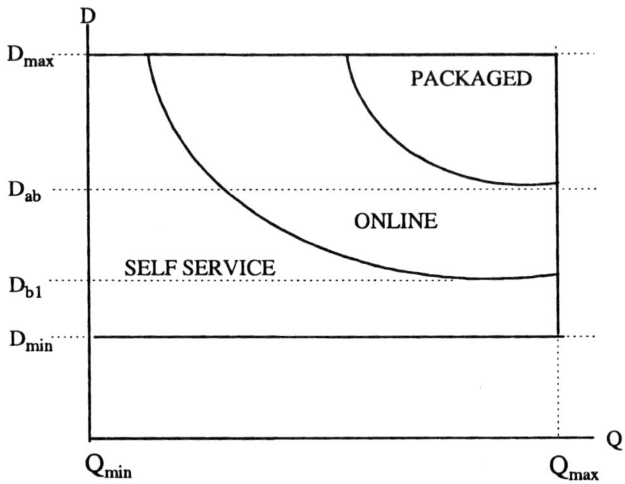

Figure 5. One possible shape of market areas for package and online providers.

volume between the online and package service. When $P_a > P_b$ and $N_a > N_b$, the market segment for the online provider is simply $(Q_{b1}(D),\min(Q_{ab}(D),Q_{b2}(D)))$ for every D. That is, the online provider serves low volume users of value parameter D. Of course, if $Q_{b1}(D) \geq Q_{ab}$ for any D, then the online provider does not capture any subscribers of that value parameter, irrespective of their volume of searches. Likewise, for the package provider, the market segment is given by $(\max(Q_{ab}(D),Q_{a1}(D)),Q_{a2}(D))$ for every D, when $P_a > P_b$ and $N_a > N_b$.

When $P_a > P_b$ and $N_a > N_b$, we can show that the market areas for the package and online provider may look like either Figure 5 or Figure 6 (see Appendix A, Proposition 8 for details).

The results of this section enable efficient algorithms for computing the market segments ($\{Q_{MSa}\}$ and $\{Q_{MSb}\}$). Specifically, for a given D, Figures 2, 3, 4 and Propositions 2,3 are used to rapidly calculate the breakpoints $Q_{a1}(D),Q_{a2}(D),Q_{b1}(D),Q_{b2}(D)$ and $Q_{ab}(D)$ for a given choice of database sizes (N_a,N_b) and prices (P_a,P_b,p_b), compare the annual benefit functions at each breakpoint and determine market segments where the package, online or self-service options win. These segments are then used in equations (4) and (6) to determine profits for the package and online provider in a duopoly. In the following sections, we analyze the equilibrium in a duopoly based on the results of this section.

5. THE DUOPOLY MODEL

In this section, we consider a market with two service providers. Potential subscribers choose between self-service and the third party providers. Consider the case where one firm provides a packaged database, and the other an online one. Define **A** and

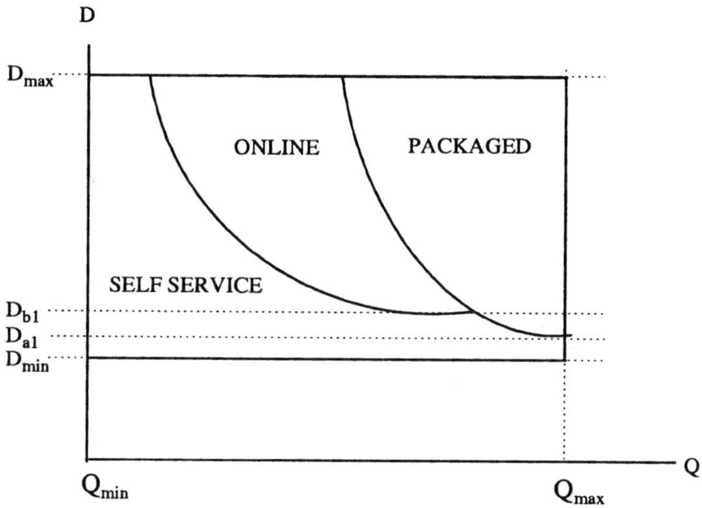

Figure 6. Another possible market area shape in duopoly.

B as the strategy spaces for the package and online service providers respectively. We characterize them below:

$$\mathbf{A} = \{(N_a, P_a) | \ 0 \leq N_a \leq N_a^\star(D_{max}), \ 0 \leq P_a \leq P_a^{max}\} \subseteq \mathbf{R}_+^2$$
$$\mathbf{B} = \{(N_b, P_b, p_b) | \ 0 \leq N_b \leq N_b^\star(D_{max}), \ 0 \leq P_b \leq P_b^{max}, \ 0 \leq p_b \leq U_b^\star(D_{max})\} \subseteq \mathbf{R}_+^3$$

where $N_i^\star(D) = \arg\sup_{N \geq 0} U_i(N,D)$, $i = s,a,b$

The upper bound on N_a and N_b follow from Proposition 1. For any D, $N_i^\star(D_{max})$ is the largest database size a third party provider would choose since $N_i^\star(D)$ is increasing in D, from Proposition 1(ii). Likewise, since $U_b^\star(D)$ is increasing in D, the maximum value for the variable charge for the online provider is $U_b^\star(D_{max})$. Notice that the provider with smaller search costs has a larger set of feasible database sizes compared to their competitor.

Firms play a one shot game in which they choose all the size and price parameters in their control simultaneously. Let **a** and **b** denote strategies of the two players, with $PR_i(\mathbf{a},\mathbf{b})$ denoting the profits for firm i. We define a Nash equilibrium for that game as follows:

DEFINITION 1. A Nash equilibrium is a combination $(\mathbf{a}^\star,\mathbf{b}^\star) \in \mathbf{A} \times \mathbf{B}$ that satisfies $PR_a(\mathbf{a}^\star,\mathbf{b}^\star) \geq PR_a(\mathbf{a},\mathbf{b}^\star)$ for $\mathbf{a} \in \mathbf{A}$ and $PR_b(\mathbf{a}^\star,\mathbf{b}^\star) \geq PR_b(\mathbf{a}^\star,\mathbf{b})$ for $\mathbf{b} \in \mathbf{B}$. PR_a and PR_b are defined in equations (4) and (6).

Though the strategy spaces **A** and **B** are compact, it is not possible to establish the concavity of $PR_i(N_a,P_a,N_b,P_b,p_b)$, $i = a,b$, and the necessary and sufficient conditions

Table 1. Test matrix for market models

Parameters	Value
Q_{max}	100
Q_{min}	0
Mode of G(Q)	10
D_{max}	50
D_{min}	4.5
Mode of F(D)	10
m	700
b	0.2
$K(0):(K_s(0),K_a(0),K_b(0))$	(0,30000,3000)
$k: (k_s,k_a,k_b)$	(0.2,3,1)
$c: (c_s,c_a,c_b)$	(0.9,0.1,0.2)

for the existence of an equilibrium in the general case (Friedman 1977). Nonetheless, we establish context-specific sufficient conditions for an equilibrium in the following result. In Section 6, we solve the simultaneous moves one shot game numerically.

PROPOSITION 4. If both providers survive in equilibrium, then:

(a) they will choose different database sizes and prices. Specifically, if $P_a > P_b$, then $U_a(N_a,D) > U_b(N_b,D)-p_b$ for all D and vice versa.

(b) the lower search cost provider will pick the larger database size and provide greater marginal benefit (U(N,D)). As result, they will serve high volume subscribers of any given value parameter. Specifically, if $(C_a(N) < C_b(N))$, then the package provider will serve high usage volume subscribers of any value parameter D with a database size such that $U_a(N_a,D) > U_b(N_b,D)-p_b$ and $P_a > P_b$.

6. NUMERICAL RESULTS

In this section, we simulate the duopoly model developed in the previous sections. The test parameters are displayed in Table 1. The value parameter D has a triangular distribution on $[D_{min},D_{max}]$ with the given mode. The value function for a search is given as $V(N) = D(1 - \exp(-N/5))$, that is, PARAM = 0.2. This implies that the first 5 alternatives generated by a search deliver 63% of the maximum value D, which is realistic. Usage volume Q has a triangular distribution on $[0,Q_{max}]$ with the given mode, and is independent of the value parameter distribution.

We perform a 5-dimensional grid search where we cycle through the online provider‘s strategy space (P,N,p) first for a given choice (P,N) by the package provider and test the conditions in Definition 1 to see if an equilibrium is reached. Propositions 1–4 are used heavily at each candidate equilibrium (or grid point) to calculate market shares (from breakeven Q‘s for a given D) and profits there of.

To restrict the search space for the enumeration, we derive bounds on the fixed charge $p_i$max, i = a,b from the profit maximizing conditions for a the provider assuming that the usage volume distribution is independent of D, and both f(D) and g(Q) are triangular in Proposition 5 below. The lower bound on all the price parameters is, of course, zero.

Table 2. Database size, pricing, market and profits for table 1 parameters

Market structure	Duopoly		Monopoly		Monopolist with Both service designs	
Parameter	Packaged	Online	Packaged	Online	Packaged	Online
N	18.62	14.88	18.09	12.06	14.66	
P	125.86	18.04	283.18	0.00	409.04	0.00
Ph	—	1.91	—	5.59	—	6.31
PR	17,884.09	12,490.96	66,948.71	112,171.33	92,116.39	
MS	0.54	0.43	0.49	0.80	0.22	0.52

PROPOSITION 5. For any N_i, the maximum price that a monopolist provider would charge is $U_i(N_i,E(D))\max\{Q_{max}/3,\sqrt{m_g Q_{max}/3}\}$, where m_g is the mode of g(Q) and E(D) is the expected value of D. Thus, the upper bound for the fixed price is given by $P_i^{max} = U_i^\star(E(D))\max\{Q_{max}/3\sqrt{m_g Q_{max}/3}\}$. The maximum marginal price that an online provider may charge is $U_b^\star(D_{max})$. i = a (package), b(online) and s(self-service).

The pricing, profit and market penetration under different database sizes and market structures are shown in Table 2. For reference we show results for markets where the online or package provider is a monopolist. For the parameters in Table 1 and 2, $N_s^\star(D_{max}) = 12.04$, $N_a^\star(D_{max}) = 23.03$ and $N_b^\star(D_{max}) = 19.56$. For each of the scenarios in Table 2, we see that $\min(N_a,N_b) \geq N_s^\star(D_{max})$, $P_a > P_b$ and $N_a > N_b$.

From Table 2, we see that the lower search cost package service is able to command a higher fixed price in a duopoly, and serves high volume users of any given value parameter. Conversely, the online provider covers the low volume subscribers of a given value parameter, consistent with Proposition 2. Since the package provider's has higher fixed costs of building their database, its database size does not differ significantly when the market changes from monopoly to duopoly. Since the package provider's search cost is only half that of the online provider, price competition in a duopoly market causes its fixed charge to drop by more than half the monopoly level, accounting for the sharp drop in profits even when the change in market share is only 5%. The online provider's profits drop sharply when the market structure changes from monopoly to duopoly, because of a corresponding decline in market share.

The online provider's market coverage in the duopoly is 97%, whereas it is 80% when that provider is a monopolist and 74% when the packaged service provider offers online access. Thus, subscribers with low value parameters benefit in the duopoly.

In the duopoly, $N_b = 14.22$. Choosing a lower size would limit the fixed price that provider could charge (see Proposition 5), making it impossible to recover fixed costs. As a monopolist, the provider could afford a zero fixed price since revenues from marginal usage of the database more than recover his fixed costs. However, in the duopoly the online provider is constrained to serve the low volume subscribers of each value parameter, whose marginal usage is too low to make a zero fixed charge feasible. Likewise, when the package provider provides online access in a monopoly, high volume users of all value parameters pay a high premium over their fixed charge in the duopoly because they subsidize a zero fixed charge for online users. Finally, note

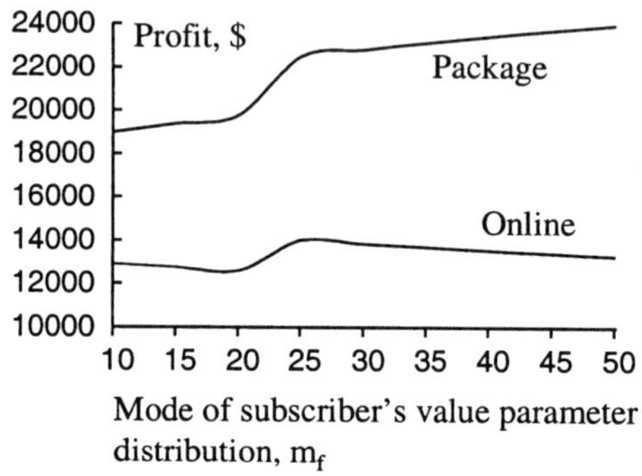

Figure 7. Effect of the mode of subscribers' value parameter distribution on profits for package and online providers in a duopoly.

that as a monopolist, the online provider makes larger profits than a monopolist package provider allowing online access. This is simply because the package provider incurs ten times higher fixed costs for building and maintaining his service in Table 1.

EFFECT OF VALUE PARAMETER DISTRIBUTION

As the mode m_f of the subscriber's value parameter distribution increases, both the package and online service providers increase their database sizes. This is because the fixed charge the market is willing to support indicated in Proposition 5 increases with m_f since $E(D) = (D_{max} + D_{min} + m_f)/3$ increases. Since the fixed costs for the service providers dominate the marginal costs in Table 1, the small increase in marginal costs for higher database sizes are made up by larger profits (Figure 7) from an increasing market share for the package provider. The online provider loses both profits and market share corresponding to the gains by the packaged provider because for any value parameter, the provider is constrained to serve the low volume subscribers.

EFFECT OF SUBSCRIBER VOLUME DISTRIBUTION

As the mode m_g of the subscriber size distribution increases, the fixed charge of either provider in the duopoly first drops and then increases. From Proposition 5, the maximum price a provider may charge is seen to be proportional to $\max\{Q_{max}/3, \sqrt{m_g Q_{max}/3}\}$. When $m_g \leq Q_{max}/3$, the maximum fixed charge is independent of m_g. For a given decrease in fixed charge, the package provider gains a larger market compared to the online provider because their larger database size implies a greater marginal utility for subscribers (see Proposition 1). Thus, when $mg \leq Q_{max}/3$, the packaged provider reduces fixed charge, gains a larger market and makes larger profits (Figure 8). Correspondingly, the online provider loses both profits and market shares. When $m_g > Q_{max}/3$, the maximum fixed charge service providers may charge is proportional to m_g. Thus, we see the fixed charge for both providers going up, although only the package provider posts a significant increase in market shares or profits,

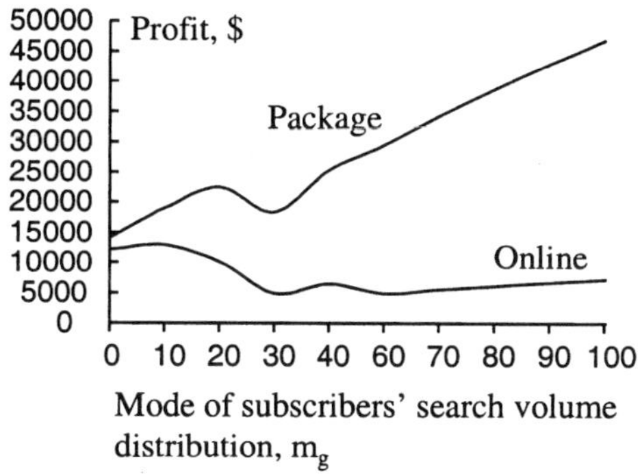

Figure 8. Effect of the mode of subscribers' usage volume distribution on profits for package and online providers in a duopoly.

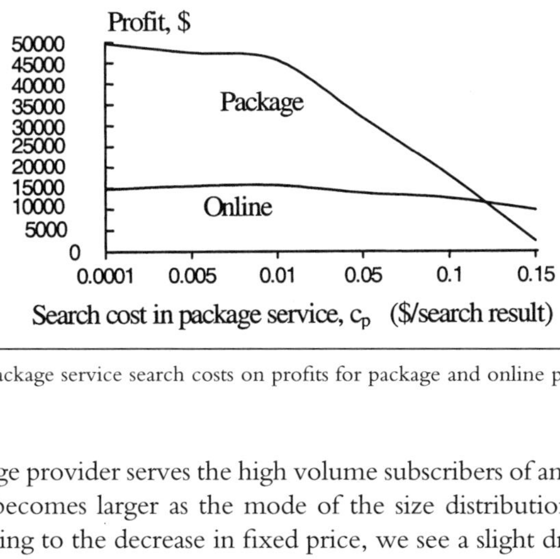

Figure 9. Effect of package service search costs on profits for package and online providers in a duopoly.

because the package provider serves the high volume subscribers of any value parameter, whose segments becomes larger as the mode of the size distribution is moved to the right. Corresponding to the decrease in fixed price, we see a slight drop in the database sizes.

EFFECT OF SEARCH COSTS

Figures 9 displays the effect of the package service provider's search costs on the duopoly equilibrium. Profits for the package provider drop steadily as search costs increase because while fixed costs remain the same, the price that the market is willing to support drops. The online provider also loses profits because price competition intensifies as the package provider's search costs increase to approach that of the online provider. We also see in Figure 9 that the package provider would be the first to drop out of the market when the search costs for their service approach that of the online provider. This is because there is little to distinguish the services in the database size dimension (see Proposition 1(b): the maximum size that a service provider can choose is inversely related to the search costs of the service), whereby intense price competition erodes profits. In the limit, when the search costs are equal, the packaged service provider stays out of the market because of higher fixed costs, and the online provider becomes a monopolist. Thus, improvements in online technology have serious consequences for the packaged service provider, given their large fixed costs for creating their database. This is likely to happen in the near future, as more subscriber firms connect to high bandwidth online access. Providing online access or drastically reducing fixed costs may be the only strategic options for the packaged service provider in the future. By going online, such a firm would capture high volume users of a given value parameter through the packaged design and low volume users through online access, as seen from Table 2.

7. CONCLUSIONS AND FUTURE RESEARCH

We investigated information services markets where potential subscribers are indexed by both their annual number and value of searches. We show that in a duopoly, the higher search cost provider serves low volume subscribers of any value parameter, while the lower search cost provider covers high volume subscribers. We also showed that reducing the search costs for online access would seriously jeopardize the viability of a package service provider in a model where the package provider starts out with the lower search cost.

Several extensions of the present work are also possible. First, information latency (or tardiness of updates) was ignored in this paper, but may quite a significant driver of benefits for subscribers that rely on current information, such as those in Financial Services. In a future paper, we will extend the current benefit models to account for information latency. We anticipate that the results of this paper will remain largely unchanged. Second, we could relax the assumption that providers charge a single price and use one database size for all subscribers regardless of their value parameter D. By charging a separate price and providing a different database size, we hypothesize that providers will capture more subscribers and make larger profits than in the current paper. Such a strategy is being used by Aspect Development in the semiconductor parts business. Aspect has many subscribers that are in different industry verticals such military, networking and medical products. They provide a separate database tailored for each of these subscriber types, there by maximizing market coverage and revenues. Third, we assumed that both service providers choose all the design and pricing variables in their control simultaneously in the duopoly game. Relaxing this assumption to allow service providers to first commit to a service design, and then choose prices (in the spirit of Moorthy 1988) in a sequential game may also provide valuable insights on the equilibrium model.

APPENDIX A: PROOFS Of PROPOSITIONS 1-8

Proof of Proposition 1. By definition, $U_i^\star(D) = \sup_{N \geq 0} U_i(N,D) = D(1\text{-}e^{-PARAM\star N})\text{-}c_iN$. So, $N_i^\star$ satisfies: $D\star PARAM\star\, e^{-PARAM\star N_i^\star} - c_i = 0$, which is the first order differential of $U_i(N,D)$ w.r.t. N_i. Thus, $N_i^\star(D) = -\ln(c_i/D\star PARAM)/PARAM$, leading to $U_i^\star(D) = U_i(N_i^\star,D) = D(1 - c_i/D\star PARAM) + c_i^\star \ln(c_i/D\star PARAM)/PARAM$. It is easy to verify that $d^2 U_i^\star/dD^2 = 1 > 0$.

Proof of Proposition 2: The proof follows from the convexity of $U_s^\star(D)$ and linearity of $U_a(N_a,D)$. See Figure 2 and 3.

Proof of Lemma 1.

(a) For a given D, observe that $W_s^\star(Q,D)$ is the maximum over a set of functions linear in Q, and is therefore convex (Rockafellar 1970).

(b) For a given D, $W_s^\star(Q,D) = U_s(N_s^\star(Q,D))Q - K_s(N_s^\star(Q,D)) \geq U_s^\star(D)Q - K_s(N_s^\star(D))$, from the definition of $N_s^\star(Q,D)$.

(c) It is sufficient to show that $N_s^\star(Q,D) \to N_s^\star(D)$ for large Q. The first order condition for the self-service problem may be written as $\partial U_s(N,D)/\partial N = (dK_s(N)/dN)/Q$. As long as $dK_s(N)/dN$ is bounded above, the above first order condition approaches the solution to $\partial U_s(N,D)/\partial N = 0$. $dK_s(N)/dN$ would be bounded from above if, for example, $K_s(N) = K_s + k_sN$; $K_s(0) = 0$ (ie a function consisting of a fixed cost and a marginal cost per record) which is indeed the case in practice, and as stated in Assumption 1.

Proof of Proposition 3: Segmenting D into $D \in GD_a(N_a)$ and $D \in LD_a(N_a)$ allows us to perform a more efficient grid search for calculating market shares and equilibrium there of. For $D \in GD_a(N_a)$, we can show that the segment of subscribers opting for the package service will be of the form $[Q_{a1},Q_{max}]$ for a given D (see Figure 4 for an example). For $D \in LD_a(N_a)$, we can show that the segment of subscribers opting for the package service will be of the form $[Q_{a1},Q_{a2}]$, $Q_{a1} < Q_{a2} < Q_{max}$. Given this result, the usage volume breakpoints for $W_s^\star$ and W_a (or W_b) can be determined using a bisection (or binary search) algorithm because we know that we have to look for at most 2 such breakpoints. This algorithm simply compares $W_s^\star - W_a$(or W_b) for a Q less than and greater than the candidate breakpoint—if the difference $W_s^\star - W_a$ changes from positive to negative, we know we have a breakpoint. The proof of the proposition follows:

(a) For every $D \in GD_a(N_a)$ (or $GD_b(N_b)$), the slope of the package (online) provider's benefit function in the Q dimension (which is linear in Q) exceeds $U_s^\star(D)$, which is the asymptotic slope for the subscriber's self-service benefit function, as seen in Figure 4. Thus, there will be at most one Q such that $W_s^\star(Q,D) = U_a(N_a,D)Q - P_a$ (or $W_s^\star(Q,D) = (U_b(N_b,D) - pb)Q - P_b$). The package provider's benefit function in Figure 4 depicts this condition.

(b) For every $D \in LD_a(N_a)$ (or $LD_b(N_b)$), the slope of the package (online) provider's benefit function in the Q dimension (which is linear in Q) is less than $U_s^\star(D)$, which is the asymptotic slope for the subscriber's self-service benefit function, as seen in Figure 4. Thus, there will be exactly two values of Q such that $W_s^\star(Q,D) = U_a(N_a,D)Q - P_a$ (or $W_s^\star(Q,D) = (U_b(N_b,D) - p_b)Q - P_b$). The online provider's benefit function in Figure 4 depicts this condition.

The proof follows from the convexity of $W_s^\star(Q,D)$ and linearity of W_a. See Figure 4.

Proof of Proposition 4.

(a) Assume that if both suppliers use identical product design and pricing strategies, they split the market equally. Without loss of generality, index the providers as 1 and

2. Suppose both suppliers survive in the market, but $N_1 = N_2 = N$, $P_1 = P_2 = P$ in equilibrium. Keeping a database size of N, Provider 2 can then price at a level infinitesimally smaller than provider 1, and capture the entire market, since provider 1 is dominated for every value parameter D (see Figure 4). Provider 1 reacts likewise, and the two providers undercut each other till one provider charges the smallest price $P_0(N)$ such that $mP_0(N)MS(N,P_0(N)) - K_i(N) = 0$. The other provider then stays out of the market, which reverts to a monopoly. A similar argument applies when both attempt to change their product designs while charging identical prices in equilibrium. Thus, if both providers survive, they utilize different pricing and design strategies. Clearly, the higher priced provider will then use a larger database, otherwise, they will be dominated for all value parameters. Visualize this using Figure 4.

(b) For a given N, the lower search cost provider ensures greater marginal benefit based on the definition of U(N,D). In the volume dimension (see Figure 4), this translates into a greater slope for the benefit line for the lower cost provider. Based on Proposition 1(b), the lower search cost provider has the larger range of possible database sizes because N_i^* is decreasing in c_i. Two cases are possible. First, as in Figure 3, the lower search cost provider dominates the marginal benefit provided by the higher search cost provider. In this case, if the lower search cost provider charges the lower fixed price, then their benefit function will dominate the competitor's benefit line. Visualize this by manipulating the benefit lines in Figure 4. So, if both providers survive, it must be the case that the lower search cost provider chooses the larger price and serves high volume users of a given D. In the second case, depicted in Figure 2, the lower search cost provider picks their database size such that $D_{min} < D_{ab} < D_{max}$. For $D \in [D_{ab},D_{max}]$ the lower search cost provider provides greater marginal benefit. So, regardless of the price they charge, they will serve high volume users—if they charge the lower price, then the higher search cost provider is eliminated from serving any customers in $D \in [D_{ab},D_{max}]$. For $D \in [D_{min},D_{ab}]$ the lower search cost provider provides lesser marginal benefit than the higher search cost provider. If they charge the higher price they will not serve any low volume users with $D \in [D_{ab},D_{max}]$ because the higher search cost provider dominates.

Proof of Proposition 5: For ease of exposition, we prove the result for the package provider (the package provider is a special case of the online provider with $p = 0$). It is sufficient to consider $D \in GD_a(N_a)$ to derive a upper bound. Also, assume that $LD_a(N_a) = \phi$.

The cumulative probability for a triangular distribution is given by (DeGroot 1989):

$$G(Q|D) = \begin{cases} \dfrac{Q^2}{Q_{max} m_g}, & 0 \leq Q \leq m_g \\ 1 - \dfrac{(Q_{max} - Q)^2}{Q_{max}(Q_{max} - m_g)}, & m_g < Q \leq Q_{max} \end{cases}$$

For any given D, notice in Figure 4 that the lower break-point $Q_{a1}(D) >$

$P_a/U_a(N_a,D)$. First consider the case when $P_a/U_a(N_a,D_{min}) < m_g$. For this condition, $\{Q_{MSa}\} = [Q_{a1}(D), Q_{max}]$ for a given D. Thus,

$PR_a(N_a,P_a) = P_a\, m \int_D (G(Q_{max}) - G(Q_{a1}))\, f(D)dD - K_a(N_a)$. Substituting the approximation for $Q_{a1}(D) \sim P_a/U_a(N_a,D)$ in the profit function, along with the cumulative triangular distribution results in:

$$\begin{aligned} PR_a(N_a,P_a) &= P_a\, m \int_D (1 - Q_{a1}^2/m_g Q_{max})\, f(D)dD - K_a(N_a) \\ &= P_a\, m \int_D (1 - (P_a/U_a(N_a,D))^2/_{m}gQ_{max})\, f(D)dD - K_a(N_a) \end{aligned}$$

Differentiate the above with respect to P_a to obtain the following first order condition $P_a^{max}(N_a)$:

$$\int_D (1 - 3(P_a/U_a(N_a,D))^2/(m_g Q_{max}) f(D)dD = 0$$

Since $U_a(N_a,D)$ is linear in D, $1/U_a(N_a,D)$ and $1/U_a(N_a,D)^2$ are convex functions of D. Therefore, we can use Jensen's inequality to simplify the solution to the above first order condition, which therefore gives us a candidate $P_a^{max}(N_a) = U_a(N_a,E(D))\sqrt{(m_g Q_{max}/3)}$.

Next consider the case when $P_a/U_a(N_a,D_{min}) \geq m_g$. (That this covers all cases follows from Proposition 6(a) where it is shown that $\partial Q_{a1}(D)/\partial D \leq 0$ when $D \in GD_a(N_a)$). In a similar fashion, the first order condition for P_a is obtained:

$$\int_D (Q_{max} - P_a/U_a)(Q_{max} - 3P_a/U_a) f(D)dD = 0$$

Using Jensen's inequality to simplify the solution to the above first order condition, gives us another candidate $P_a^{max}(N_a) = U_a(N_a,E(D))Q_{max}/3$. Thus, $P_a^{max}(N_a) = U_a(N_a,E(D))\max\{Q_{max}/3, \sqrt{(m_g Q_{max}/3)}\}$. Substituting $U_a(N_a) = U_a^*(D_{max})$ gives the bound in the proposition.

PROPOSITION 6. Let $N_s^*(D)$ be defined as in Proposition 1. Then,

(a) If $N_a \geq N_s^*(D_{max})$, then $\partial Q_{a1}(D)/\partial D \leq 0$ and $\partial^2 Q_{a1}(D)/\partial D^2 > 0$, while $Q_{a2}(D) \equiv Q_{max}$.

(b) If $N_a < N_s^*(D_{max})$, then there are two cases:

 (i) $\partial Q_{a1}(D)/\partial D \leq 0$ and $\partial Q_{a2}(D)/\partial D \geq 0$ for D such that $N_s^*(D) \geq N_a < N_s^*(D_{max})$,

 (ii) For D such that $N_s^*(D) > N_a$, $\partial Q_{a1}(D)/\partial D \geq 0$ and $\partial^2 Q_{a1}(D)/\partial D^2 > 0$, while $\partial Q_{a2}(D)/\partial D \leq 0$.

Proof of Proposition 6: Here, we show that $Q_{a1}(D)$ is either monotonically decreasing or first decreasing and then increasing in D, while $Q_{a2}(D)$ is either equal to Q_{max} or, monotonically increasing, then decreasing in D. Similar results hold for the online provider.

Let the superscripted variables denote partial derivatives with respect to the indicated variable(s). For a given D, the lower critical volume solves: $U_a(N_a,D)Q - P_a = W_s^*(Q,D)$

Differentiate the above with respect to obtain:

$$(U_a - \partial W_i^\star(Q,D)/\partial D)dQ/dD = Q(\partial V/\partial D - U_a^d) \quad (1A)$$
$$= Q(e^{-PARAM\star N_a} - e^{-PARAM\star N_s});\ N_s = N_s(Q,D)$$

Differentiate (1A) again with respect to D, and substitute the required terms from Proposition 2 to get:

$$(U_a - \partial W_i^\star(Q,D)/\partial D)d^2Q/dD^2 = Q(\partial^2 V/\partial D^2 - U_a^{dd})$$
$$+dQ/dD(\partial V/\partial D - 2U_a^{dd} + \partial^2 W_i^\star/\partial D^2) \quad (2A)$$
$$= Qbe^{-bN_s} + dQ/dD((e^{-bN_a} - e_s^{-bN^\star})$$
$$+ (e^{-bN_a} - e^{-bN_s^\star}) - ^{(D)}))$$

Properties of lower critical volume (Q_{a1})*:*
The lower critical volume satisfies $U_a - \partial W_s^\star(Q,D)/\partial D > 0$. If $N_a \geq N_s$, then $e^{-bN_a} - e^{-bN_s} \leq 0$ and vice-versa. If $N_a \geq N_s^\star(D_{max}) > N_s$, then $dQ/dD < 0$ from (1A) and $d^2Q/dD^2 > 0$ from (2A). If $N_a < N_s^\star(D_{max})$, there are two cases:
Case 1. $N_s < N_a < N_s^\star(D_{max})$
Here, $dQ/dD < 0$ from (1). The sign of d^2Q/dD^2 is not determinate from (2A).
Case 2. $N_a < N_s < N_s^\star(D_{max})$
Here, $dQ/dD > 0$from (1) and $d^2Q/dD^2 > 0$ from (2A).

Properties of upper critical volume (Q_{a2}):
The lower critical volume satisfies $U_a - \partial W_s^\star(Q,D)/\partial D < 0$. If $N_a \geq N_s$, then $e^{-bN_a} - e^{-bN_s \leq 0}$ and vice-versa. If $N_a \geq N_s^\star(D_{max}) > N_s$, then $Q_{a2}(D) = Q_{max}$ from Proposition 3. If $N_a < N_s^\star(D_{max})$, there are two cases:
Case 1. $N_s < N_a < N_s^\star(D_{max})$
Here, $dQ/dD > 0$ from (1). The sign of d^2Q/dD^2 is not determinate from (2A).
Case 2. $N_a < N_s < N_s^\star(D_{max})$
Here, $dQ/dD < 0$ from (1). The sign of d^2Q/dD^2 is not determinate from (2A).

PROPOSITION 7. Define $Q_{ab}(D) = (P_a - P_b)/(U_a(N_a,D) - U_b(N_b,D) + p_b)$. If $N_a > N_b$ and $P_a > P_b$, $Q_{ab}(D)$ is convex, decreasing in D.

Proof of Proposition 7: If $N_a > N_b$, then $\partial U_a(N_a,D)/\partial D > \partial U_b(N_b,D)/\partial D$. Now, $Q_{ab},(D) = -(P_a - P_b)(\partial U_a(N_a,D)/\partial D - \partial U_b(Nb,D)/\partial D)/(U_a - U_b + p_b) < 0$ under the conditions of the Proposition. Also,

$$Q_{ab}''(D) = 2(P_a - P_b)(U_a - U_b + p_b)(\partial U_a(N_a,D)/\partial D - \partial U_b$$
$$\times (N_b,D)/\partial D)^2/(U_a - U_b + p_b)^2 > 0$$

if $(U_a - U_b + p_b) > 0$. If $(U_a - U_b + p_b) \leq 0$ for any D, then the packaged provider is dominated by the online provider and does not capture any subscribers of that value parameter under the conditions of the Proposition.

PROPOSITION 8. If the providers choose their database sizes such that $\min(N_a,N_b) \geq N_s^*(D_{max})$, then the market area in the duopoly looks either like in Figure 5 or Figure 6.

Proof of Proposition 8: Let $D_{ab} = (a_aN_a - (a_bN_b+p_b))/(e^{-bN_b}-e^{bN_a})$ (see Figure 2). D_{ab} is the break-even value parameter between the online and packaged service, such that for $D > D_{ab}$, subscribers derive greater marginal utility $(U_a(N_a,D) > (U_b(N_b,D)-p_b))$ from the package service, while the converse holds for $D < D_{ab}$. When $N_a > N_b$ and $P_a > P_b$, two situations may arise in the value dimension. In the first case, $D_{min}<D_{ab} <D_{max}$, depicted in Figure 2. The market area for the duopolists when they choose their designs such that $\min(N_a,N_b) \geq N_s^*(Dmax)$, $LD_a(N_a) = \phi$ and $LD_b(N_b,p_b) = \phi$ is depicted in Figure 5. With such a choice of service designs, we know that $Q_{b1}(D)$ is convex decreasing from Proposition 6(a), and $Q_{b2}(D) = Q_{max}$, $Q_{a2}(D) = Q_{max}$, from Proposition 3. As noted previously, the market for the online provider is simply $(Q_{b1}(D),\min(Q_{ab}(D),Q_{max})$, and that for the packaged provider is $(\max(Q_{ab}(D),Q_{a1}(D)), Q_{max})$ for every D, resulting in the market areas of Figure 5. We designate $Q_{b1}(D)$ as the *lower critical volume* and $Q_{ab}(D)$ as the *upper critical volume* in the duopoly. Notice for every **D**, subscribers with volumes above the upper critical volume opt for the packaged design, those with search volumes in between the lower and upper critical volumes prefer the online design, and firms with search volumes less than the lower critical volume use self-self service. It is interesting to note that the market area for the package provider is convex because $\max(Q_{ab}(D),Q_{a1}(D))$ is convex. Note that for $D < D_{ab}$, the package provider is dominated by the online service, and does not capture any subscribers. Of course, if $D_{ab} \geq D_{max}$, the package provider stays out of the market which reverts to a monopoly for the online provider. The second case, $D_{ab} < D_{min}$, is depicted in Figure 3. The market area for the duopolists when they choose their designs such that $\min(N_a,N_b) \geq N_s^*(D_{max})$, $LD_a(N_a) = \phi$ and $LD_b(N_b,p_b) = \phi$ is depicted in Figure 6, using an analysis similar to the above.

REFERENCES

Bashyam, T.C.A. 1996. Competitive Capacity Expansion Under Demand Uncertainty. *European Journal of Operational Research* **95,** 89–114.

Bashyam, T.C.A and U.S. Karmarkar (2000), "Aspect Development Inc (A)", *Managing the Global Corporation: Case Studies in Strategy and Management,* J. De La Torre, Y.Doz and T. Devinney (editors), 2nd Edition, pp 335–348, McGraw Hill, New York.

Boeri, R. and M. Hensel (1995), "What Good is SGML?", *CD-ROM Professional* v8, n4, 108–110.

Davis, S. M. and B. Davidson. 1991. *2020 Vision.* New York : Simon & Schuster.

Electronic Engineering Times, Two Launch Component Information Services, November 1, 1993, 16–18.

DeGroot, M.H (1989), *Probability and Statistics,* Addison-Wesley Publishing Company, Reading, Massachusetts.

Friedman, J. (1977), *Oligopoly and the Theory of Games,* North Holland, Amsterdam.

Moorthy, K.S. (1988), "Product and Price Competition in a Duopoly", *Marketing Science* 7(2), 141–168.

Oren, S.S. and S.A. Smith (1981), "Critical Mass and Tariff Structure in Communications Markets", *Bell Journal of Economics* 12, 467–487.

Oren, S.S., S.A. Smith and R.B. Wilson (1983), "Competitive Nonlinear Tariffs", *Journal of Economic Theory* 29, 49–71.

Rayport, J.F. and J.J. Sviokla. 1995. Managing in the Marketspace. *Harvard Business Review* November-December, 141–150.

Reisman, R.R. 1995. The Missing Link. *CD-ROM Professional* **8**(4), 66–75.

Rockafellar, R.T. (1970), *Convex Analysis*, Princeton University Press, Princeton, New Jersey.

Takayama, A. (1974), *Mathematical Economics*, The Dryden Press, Hinsdale, Illinois.

8. COLLABORATION IN E-BUSINESS: TECHNOLOGY AND STRATEGY

AMIYA K. CHAKRAVARTY

ABSTRACT

With constantly improving information technology, it has become feasible, and in many cases a competitive necessity, to outsource not only component and systems production, but also portions of design. This has led companies to narrow their focus on fewer business processes in a value-chain. To effectively manage the resulting globally distributed system, collaboration among supply chain partners is critical. Collaboration enables sharing of data, processes, and resources, enhancing compatibility between organizations. Some of the areas where companies may collaborate with *suppliers* are, information visibility, logistics management, production planning, product development, and resource sharing. Collaboration with *customers*, on the other hand, is largely aimed at revenue enhancement, demand management, order fulfillment, and customer acquisition. Technologies that enable different aspects of collaboration include, access, data sharing, application sharing, process sharing, and free-form interaction. In this chapter we discuss (1) how collaboration creates value in different business contexts with different strategic implications, (2) how online technologies facilitate collaboration, (3) how an optimal technology portfolio can be to chosen for investment, and (4) how a technology portfolio can be deployed. Our objective is to develop an approach for *need based selective collaboration* with partners.

1. INTRODUCTION

To remain competitive, organizations must constantly review and reorganize its portfolio of activities: those they undertake in-house, and those they contract out to suppliers. This implies that companies must manage a large number of relationships with business partners, simultaneously (Callahan and Pasternack 1999). The sheer complexities in doing so are compelling companies to narrow their focus to only a few activities; they collaborate with partners as part of a larger value chain (Chakravarty 2001a). Collaboration reduces cost and enhances quality of decisions by enabling sharing of data, processes, and resources, which require that the organizations be compatible with one another (Baldwin and Clark 2000). Sources of incompatibility include, data formats, production schedules, proprietary information, delays in fund-flow, and partner accountability. Collaboration may limit the degrees of freedom a company may possess in performing its operations, however (Ticol, Lowy, and Kalakota 1998).

In a recent survey carried out by Deloitte Research (2000) both suppliers and customers listed collaboration attributes (in decreasing order of importance): reduced logistics costs (58%), gaining competitive edge (47%), increased customer loyalty (42%), increased sales (35%), increased speed to market (34%), reduced inventory (32%), reduced material cost (29%), and increased shareholder value (18%). Thus while the largest gain was seen to accrue to suppliers through logistics management, gains to customers were spread widely over many attributes.

Baldwin and Clark (1997) have defined the foundations of a business enterprise in terms of architecture, interfaces with partners, and the standards of information exchange. The architecture and interface together determine partner-capabilities and degree of inter-process coordination through rules of sharing information and managing resources and tasks. Standards, on the other hand, facilitate inter-process communication and performance evaluation. Thus, architecture, interface, and standards (together) define a business model. Such models, as we shall see later, can be conceptualized in terms of the suppliers, customers, and other service providers.

Collaboration with suppliers is primarily aimed at minimizing unused inventory, excess capacity, and delivery lead time. The major functions where collaboration has been shown to be beneficial include, logistics management, production planning, order fulfillment, product development, and resource sharing. Sharing of data from the databases of different companies, and sharing of business applications such as ERP and accounting systems have now become common. Clearly, the supply chain information should be integrated with front-end customer information to enhance customer satisfaction and fulfillment. For coordinating events in real time (production, logistics, and customer orders), technologies such as Workflow (Basu and Kumar 2002) become useful. Good examples of supply chain collaboration come from Dell Computers, and Li & Fung of Honking. In its Business-to-business (B2B) model Dell computers collaborate with logistics companies such as UPS who pick up computers from its facilities in Texas and an appropriate number of monitors from Sony's factories in Mexico. UPS is responsible for matching computers with monitors before delivering them to the customers. Clearly, Dell must make a large portion of its customer orders accessible to UPS. In the apparel industry Li & Fung manage a complex value chain by dissecting it, and allocating different slices to many partners, dispersed globally (Magretta 1998).

Collaboration with customers, on the other hand, is aimed at revenue maximization, demand management, fulfillment, and customer acquisition. Collaborative forecasting may permit a real time exchange of information between retailers and suppliers to smooth out forecasts and orders (Aviv 2001). Online trading where customers bid for the supplier's excess inventories and provide demand visibility to the supplier can help increase sales (Chakravarty and Zhang 2002). Mass customization allows customers to collaborate in designing the products or services they buy. This may take several forms such as customers designing their own products by choosing appropriate modules, and soliciting customer preferences online. Companies may increase revenues and acquire new customers through collaborative advertising in each other's territory (Huang and Li 2001), and through affiliate marketing that routes customers to its partner (Hoffman and Novak 2000).

Technology for supporting a variety of collaboration modes is now available. The major categories are access, data sharing, application sharing, process sharing, and free-form interaction (Linthicum 2001). While the access technology is of the most basic kind, process sharing using workflow technology and messaging techniques can be quite complex. Companies can *access* their partners through the Web browser or through electronic data interchange (EDI). In this mode it is not possible to interactively update data or pull information from the company's internal applications such as payroll and accounting. In the *data level* collaboration the input of one entity is matched with the output of another, without affecting the internal processes of participants. *Application sharing* enables the partners to run each other's applications to extract and exchange data between their disparate systems. *Process collaboration* allows networked companies to share processes through workflow systems. Shared processes are managed in a project management environment where different parts of the project could be assigned to different partners in real time. Finally the *free form interaction* allows companies to interact over the Internet to create, discover, and communicate concepts, questions, ideas, and needs. It is clearly the most flexible format and can be used for online discovery, response, and negotiation.

Our objective is to develop an approach for need based selective collaboration with partners. We first discuss value of collaboration in different business contexts (manufacturing, marketing, and product design) and Internet technologies that facilitate collaboration. We then present a framework for technology portfolio choice, and its deployment for partner collaboration. Specifically, in section 2 we describe e-business models and position them relative to one another to identify scenarios where collaboration would be meaningful. In sections 3 to 5, we describe collaboration needs in three broad business contexts: supplier relations, product development, and customer and market relations. In section 6 we discuss the major information exchange modes and technologies. In section 7 we outline strategy and describe a decision model for selecting a technology portfolio. Conclusions appear in section 8. Technical aspects of information exchange are discussed in the appendix.

2. BUSINESS MODELS

We discuss business models to emphasize that collaboration may not be equally valuable in all business scenarios. By business model we imply an architecture that positions information-flow, product-movements, and capital-transfers, between business partners in a logical framework. The model also describes the sources of revenues and costs, and the potential benefits to the participants. Importance of business models has increased in recent years, as value chains and business interfaces can now be restructured rapidly using Internet technology (Applegate 2000, Raisch 2001). Timmers (1998) has identified several business models for e-business: e-sell, e-procurement, e-auction, e-marketplace, value-chain service provider, value-chain integrator, collaboration platform, and informediary.

There are major differences between *e-sell* and *e-procurement* models in terms of emphasis placed on business processes. The *e-sell* emphasizes fulfillment services that check real-time capacity commitments against customer demand, and tracking of customer orders until their fulfillment. It manages customer orders with an ERP: tracking fulfillment of the order internally, and coordinating the customer desired changes externally. It also provides for vendor-managed inventory (VMI) that requires customers to share their inventory status, production plans, and forecasts with suppliers (Clark and McKenney 1994). The *e-procurement* model, on the other hand, addresses a company's need for direct materials, and components. It automates the request for quotation (RFQ), and makes part-drawings and specifications visible to suppliers over the Web, and enables choice of suppliers based on their performance. It provides for sharing of forecasts with suppliers for timely receipt of goods and services, and aggregates all product demands from internal sources (through the ERP) to obtain the best price quotation.

In contrast, electronic auctions (*e-auctions*) are used in B2B, primarily to dispose of excess inventory and to purchase unused capacity. They support multiple bidding schemes and, in some cases (ChemConnect), bilateral negotiations between the parties. Buyers gain from reduced purchase costs and availability of a broad product range. Suppliers gain from reduced surplus stock, and a better utilization of production capacity (Chakravarty and Zhang 2002).

An *e-marketplace* is an aggregation of *e-sell, e-procurement,* and *e-auctions* sites; augmented by common services such as a consolidated product catalog, order transaction, logistics, and payment systems (Kaplan and Sawhney 2000). They can take three different forms: independent e-marketplaces, private exchanges, and consortium e-marketplaces. An independent e-marketplace is owned and run by a third party (not a buyer or seller). These have primarily evolved into exchanges for commodity products. A private exchange, in contrast, is owned by a dominant buyer or supplier, and it sets the rules of business transactions for all. In a consortium e-marketplace ownership is shared by a subset of buyers and sellers and they make the rules. Clearly, collaboration would be more meaningful in a consortium or private exchange than an independent e-marketplace.

A *value-chain service provider* possesses competitive advantage on a specific narrow segment of the value chain. Examples are payment systems and logistics service providers. The service-provider model also includes brokers such as search agents, and matchmakers. Application service providers, and content providers are other examples. Note that it is the electronic version of the widely used outsourcing model that permits a finer segmentation of the value chain. *Value-chain integrators,* on the other hand, focus on reconstructing the value chain by integrating multiple steps of the chain

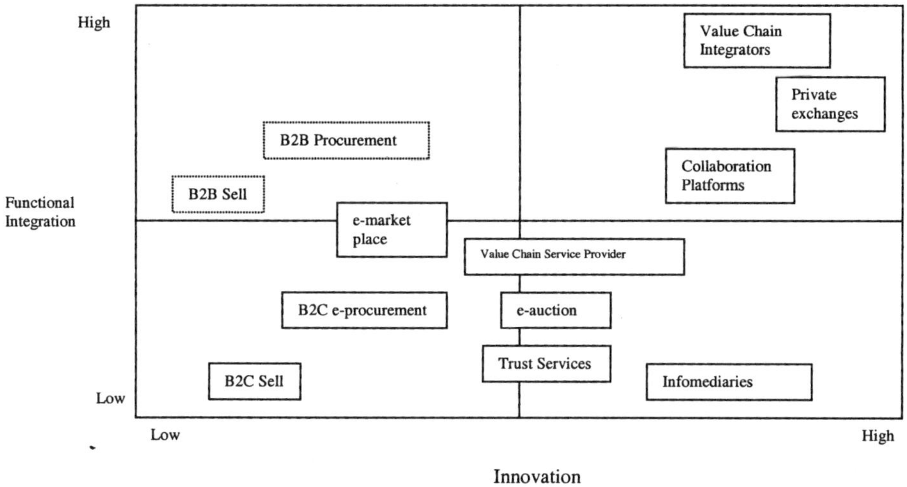

Figure 1. Business models.

(Magretta 1998). As suggested in Wise and Morrison (2000) they can be configured as specialist-originators, transaction-hubs, e-speculators, solution providers, and sell-side asset exchanges. Revenues come from consulting fees as in the originator that matches buyers with suppliers, transaction fees as in a transaction hub, spreads and margins as in an e-speculator, and asset-swap prices as in asset exchanges.

Collaboration platforms provide a set of web services that may be used directly by any two parties wishing to collaborate in their business operations. These tools help in managing application sharing, and communications. Collaboration platforms can be valuable in buy-sell operations, fulfillment, vendor-managed inventory (VMI), asset sharing, and product design. The .net initiative of Microsoft is a good example of a collaboration platform.

Infomediaries collect and sell information to other businesses. Data about consumers and their buying habits can be extremely valuable when information is carefully analyzed and used to target markets and products (may use datamining techniques). Special categories of informediaries are trust services (Raisch 2001). They act as electronic notaries, and recommender systems that allow users to exchange information about the quality of products and services they might have bought and/or sold.

Timmers (1998) classifies these business models in two dimensions: degree of innovation, and functional integration as shown in Figure 1. Clearly, the collaboration needs would be highest when the functional integration is critical.

Thus, the greatest impact of collaboration would be in scenarios where the business model involves a large number of relationships but business processes are integral. In Figure 1, these would be the models in the upper right quadrant - supply-chain integration, product design collaborations, and collaborative platforms such as consortium and private exchanges. It is also clear that business models such as B2B in the upper left quadrant would benefit from collaboration more than a B2C model in the lower left quadrant.

Clearly, the business context is very important in understanding the collaboration needs and opportunities associated with a business model. A cursory examination reveals that the business models in Figure 1 can be classified into three major business domains: supplier relationship management (SRM), collaborative product commerce (CPC), and customer and market relationship management (CMRM). We next discuss major business processes within these three domains to highlight information exchange and conflict resolution.

3. SUPPLIER RELATIONSHIP MANAGEMENT

Supplier relationship management (SRM) is aimed at supplier related functions that include, sourcing, procurement, manufacturing, replenishment, logistics, technology transfer, and knowledge sharing. Common denominators would be information sharing, coordinated planning, and project management. Information visibility creates value by minimizing the impact of "bull-whip" effect, which is the amplification of demand variance through the multiple tiers in a supply chain (Lee, Padmanabhan, and Whang 1997). Coordinated planning permits partners to synchronize their production and order schedules, and project management facilitates real time coordination.

Information exchange involves data sharing, and application sharing. The simplest data sharing architecture is an information hub shown in Figure 2. To enable data exchange, data such as product catalog, terms of transactions, customer preferences, demand forecasts, order acknowledgement, and promotions must be converted to a format that is understood by its partner. Ideally, all data should first be converted to a standard format, and then to partner's data format, as and when needed. However, in the extreme case one would need different conversion software for each pair of partners. Security protection is provided through protocols, which can also be utilized to restrict data sharing to designated clusters of partners. If some data categories are not available directly, they can be derived from partner's applications such as ERP, human resource management, and accounting. For example, order quantities can be obtained by first accessing the customer's ERP, and then extracting the requirements for specific items. However, for this to be possible the ERP systems of suppliers and customers must be able to communicate seamlessly.

Clearly information access and exchange, by themselves, do not ensure synchronization of operations. For this to happen the partners need to agree how and for what purpose the shared information is to be used. For example, if the customer's orders are not in sync with the supplier's production plans, they must establish rules for taking corrective actions so as to smooth the differences. Collaborative Planning Forecasting and Replenishment (CPFR) is based on just such an idea of setting up rules of conflict resolution (Aviv 2001). CPFR facilitates collaboration through the use of electronic bulletin boards on the Internet (VICS 2001). Cisco Systems, for example, intends to coordinate supply and demand planning across its supply chain using an intelligent planning software (Lee and Whang 2001, Cisco Systems 1998). This soft-

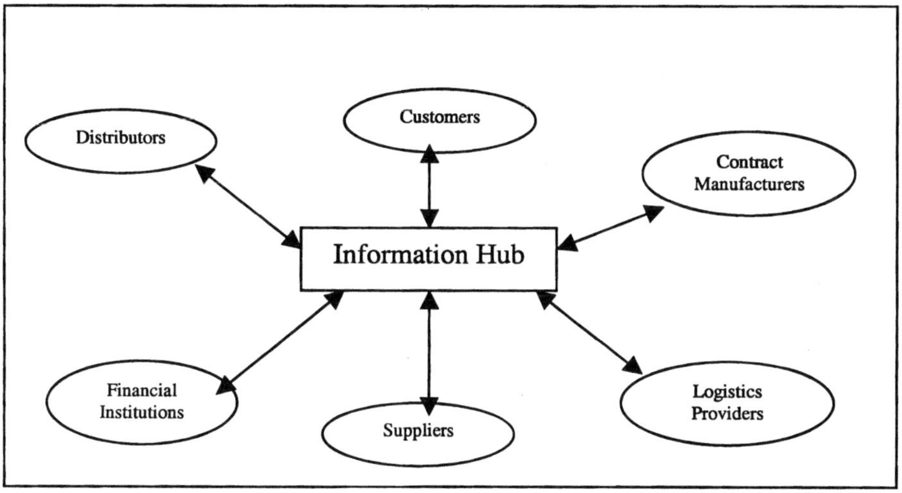

Figure 2. Information hub.

ware would help in early identification of supply chain conflicts, and suggest resolutions. McAfee (2001) describes a different coordination problem faced by Adaptec (based in California) with its major supplier TSMC in Taiwan. They used *alliance* software (supplied by Extricity) to make six-month production forecasts, which are also visible to TSMC. TSMC, in turn, transmits detailed work-in-progress data to Adaptec in a format that can be accessed and understood by Adaptec seamlessly, their disparate systems notwithstanding. TSMC also alerts Adaptec about products that have completed production and shipped to ASAT, which is a contract assembler for Adaptec.

Coordination in time is an additional dimension in collaboration that requires a project management orientation. For example, at the time of placing an order (an event) a customer would like to know whether the supplier could supply it all. The customer can make this determination by running an electronic workflow (Basu and Kumar 2002), which is linked to the supplier's production plans. The workflow uses the customer's ordering event to trigger the relevant processes of the supplier, simultaneously updating all documents. This ensures that at any time (event) the customer would have the most recent information. Several Web based procedures have been developed that automate the customer's procurement process from requisition to order, and the supplier's processes from order to payment. In a project management environment, several suppliers may collaborate to collectively satisfy a customer's total needs. That is, they could be delivering different components that, though interrelated, could differ in order quantities, delivery schedules, and specifications. Even in scenarios where suppliers perform different operations on the same product of a customer, collaboration can bring dividends. It would require event-driven production and sharing of component specifications among the suppliers, as and when new events occur. It may also require transfer of resources between partners (event driven) so that work may begin at different locations at planned times. Accounting of resource sharing, control of task executions, progress monitoring, and sequencing of tasks that share critical resources are the prerequisites of such a collaboration. The workflow tool can be most useful in such collaboration scenarios.

SRM requires well-defined channels of communication with clearly established roles for participants. There must also exist well-understood metrics for performance evaluation, accountability, and corrective actions (VICS 2001). Last, but not the least, the incentives of partners must be carefully aligned with the overall objectives of the collaboration. This requires an equitable sharing of both gains and risks among the members (Chakravarty 2001b). Finally, the standards for data and process collaboration, which are growing rapidly, need to be addressed (UDDI 2000, Linthicum 2001).

4. COLLABORATIVE PRODUCT COMMERCE (CPC)

Unlike SRM product design is a non-repetitive activity. Performance of a complex product is dependent upon many intricate components from a diverse set of specialist suppliers. Since a single designer is unable to completely design such a product, collaboration with partners becomes necessary. In addition to working out product specifications the designer must determine whether the components can be made (given the constraints of technology), identify suppliers that can make them, and test the product's integrity by configuring it online with chosen components. Product integrity may require several revisions in product design. Aberdeen (1999) organizes the design process into several functions: *sourcing* to determine suppliers' capabilities and availability, *virtual configuration* to ensure the product's viability, *product data management* (PDM) for coordinating design data, and *design portals* for sharing information in real time.

The design process is complex because segments of design tasks must be assigned to different teams of designers. These teams are responsible for designing specific features of the product, and they may well be located at different geographical locations. For example, the Ford Motor Company's "world" car design teams were located in Germany, France, U.S., Japan, and several other countries. Each team was responsible for one aspect of the car, which could be the engine, the chassis, the brake system, or the body. In such a scenario, design changes and revisions involved multiple exchanges of information among the teams. All changes had to be carefully recorded, and the design documents had to be moved systematically from one team to another through modifications, reviews, and approval.

To manage CPC functions (sourcing, virtual configuration, PDM, design portal) there needs to be an overall project management platform that acts as a moderator and/or coordinator. Note that the four CPC functions have different intellectual needs that span from the routine to the innovative, and they become active in different phases of a product's life cycle - from concept generation to product retirement. In Figure 3, observe that while the sourcing applications are routine, the project manage-

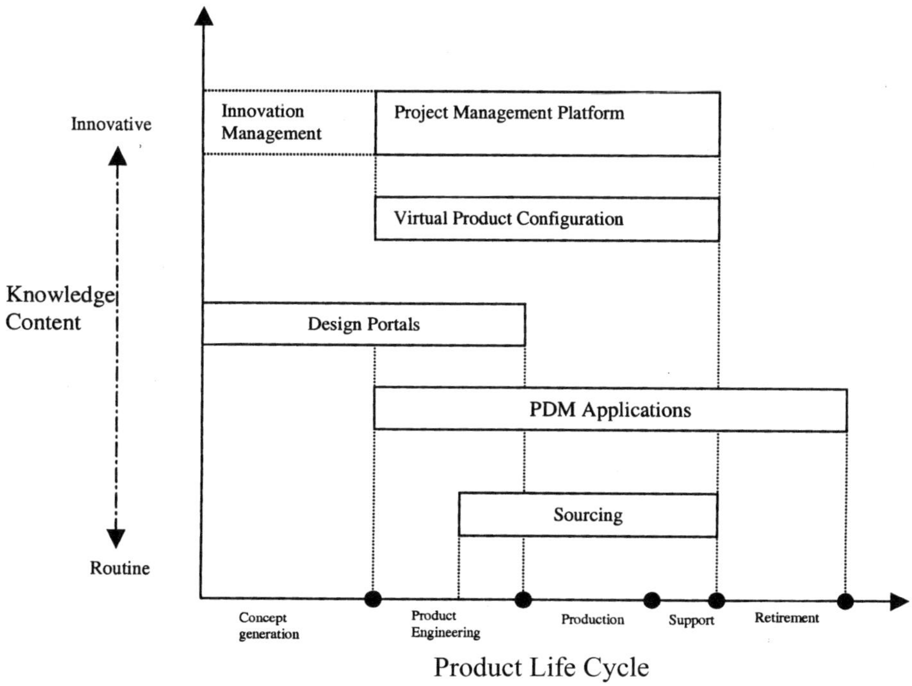

Figure 3. Managing the CPC functions.

ment platform must be innovative. A design portal plays an active role only up to the production phase, but the PDM continues to be relevant up to the product retirement phase. Clearly, there is nothing to be managed in a project-sense until product engineering begins. However, there needs to be a management structure during the concept generation phase. This is shown as innovation management in the diagram.

PDM is perhaps the most important function of CPC. It manages the relationships among product attributes through a relational database. The three important features of PDM are classification, product-structure, and data-access. It classifies the data into categories such as product features and components. It organizes the components and documents in a hierarchical network structure, so that the relationships between a component and its associated documents are always retained. It uses a structure like a bill-of-material (BOM) to record relationships, such as that between an assembly and its constituent parts. Different data structures—classification tree, product structure, and key words—can be used for extracting information relevant to a component or assembly.

The project management platform uses the PDM to control the progress of the design project, using transitions through multiple stages of the design process. These transitions are facilitated by a routing list and a "triggering" mechanism. Individuals or groups may represent stages in design tasks. Information *cannot* be moved from one group to another without changing the stage marker of the document. Workflow technology is used to trigger completion of design tasks such as initiate, submit, approve, and release. This ensures that all team members can share documents, and it allows tracking of all modifications. Audit trails are used to keep track of events and movements, and document backtracking is permitted for problem resolution.

The *sourcing* function enables component-based design to minimize the number of used parts. It identifies part-suppliers based on performance parameters, and enables negotiations with them. It can carry out optimization with respect to purchase cost, delays, quality control, and inventory reduction. It also ensures that all partners can simultaneously access the most recent (updated) information.

Virtual configuration management ensures design integrity by providing real time information on components as and when they are produced by partners (specifications, material used, performance, inputs/outputs etc.). The project manager uses this information to verify that components of different suppliers would work together as a whole, and thus avoids unnecessary design iterations. Whenever a supplier makes changes to its components it can accumulate the status of all data associated with that component, and make them visible to its partners simultaneously. Specialized visualization software permit viewing of configured products for any choice of suppliers and components. Virtual configuration can also provide decision support based on simulation, time-to-market (Chakravarty 2003) and/or product-content analysis. Although virtual configuration can be used for any product, it would be most beneficial for integral products such as the wing of the Boeing 777 jet plane, and the frame of a modern motorcycle (Fine 1998, Anderson and Parker 2002).

5. CUSTOMER AND MARKET RELATIONSHIP MANAGEMENT (CMRM)

Unlike SRM that emphasizes sourcing, procurement, logistics, and manufacturing, the emphasis in CMRM is on revenue generation through demand management and sales, customer satisfaction through fulfillment and other customer services, and acquisition of new customers. Although the information exchange needs are similar to that in SRM (or CPC), business processes can be very different, providing different collaboration opportunities.

Demand Management

Retailers frequently modify their order quantities and schedules in response to changes in demand forecasts. If the suppliers could foresee these changes, they would plan better and reduce their costs. In a collaborative environment, the supplier would have access to the retailer's point-of-sale (POS) data, through a direct link to the retailer's checkout counters. Access to this information alerts the supplier about forthcoming order modifications. Collaborative initiatives such as vendor-managed-inventory (VMI) and continuous-replenishment (Clark and McKenney 1994), where the supplier replenishes retailers' warehouse without requiring the retailer to place orders, depend on similar exchange of information. In addition to the POS data, now the supplier must also have access to the retailer's inventory status.

In a trading network, collaborative arrangements with different partners may be required for different business processes such as, order placement, order fulfillment, logistics, and payment. Additionally, a large number of requests involving orders for new products, returns, and inventory status may need routing to appropriate partners. Clearly, the system would degenerate into chaos if information were not synchronized well. Synchronization enables distributed entities to work off the same information, and the workflow coordinates the access and modification of the shared information.

Demand brokering is a concept that business partners use to cope with unusual changes in demand. Large orders may require splitting and assigning to suppliers at different locations. How to split the order, which suppliers to assign them to, and how to consolidate suppliers' output are some of the decisions involved in demand brokering. Online exchange of information helps in multiple ways: selection of partners (suppliers) based on their capability and availability, negotiation of commitment and remunerations, coordination of service delivery, and resolution of execution problems.

For complex products such as machine tools that require intensive information for their usage, a supplier may wish to collaborate with a consortium of consultants, who can walk the customer through the intricacies of using the product. Similarly, if the product requires a complex set of manufacturing steps, a collaboration of suppliers with diverse manufacturing expertise would be needed to manufacture it. In such scenarios, techniques such as quality function deployment (QFD) (Hauser and Clausing 1988) can be used to match the features desired by the customer with the required manufacturing expertise. Suppliers possessing that expertise can then be identified. Note that this is similar to the sourcing function in CPC, but now the attribute matching is being performed at the expertise level. It should be possible to adapt the QFD to e-business so that suppliers making different components may collaborate through the QFD to fulfill customer orders that could not be done otherwise. UDDI (UDDI 2000) attempts to match customers with business service providers through a directory that lists various business services and links them to suppliers who can provide those services.

Collaborative Advertising

Vertical collaborative advertising is an interactive relationship between a manufacturer and a retailer (Huang and Li 2001). The manufacturer pays a part of the cost of advertising in the retailer's local markets. The manufacturer's sales (to the retailer) increase, as the total sales increase. Clearly, the level of local advertising by the retailer would depend upon the manufacturer's share of the advertising cost. Typically the manufacturer, as the channel leader, would declare its advertising subsidy, and the retailer would react by choosing the quantity to order from the manufacturer. The authors show that an equilibrium solution, which produces gains for both parties, does exist. In the E-business context, the manufacturer instead of advertising in local markets would advertise the retailer's site in different forums. The retailer would get to decide the quantity to order from the manufacturer based on this advertising budget, as before.

Hoffman and Novak (2000) describe a different form of collaborative advertising involving two retailers in an e-marketplace. First, retailer A's customers purchase retailer B's product due to a positive consumption externality in the e-marketplace. B also advertises on A's website to draw additional customers without impacting A's sales. Of course, B would only be interested if the cost of advertising on A's site does not exceed the surplus generated from these additional customers. Clearly, it would be in both A and B's interest to maximize the number of customers clicking through A's site, and so A may want to invest in contextual services to increase the number of customers attracted to it. There would be a trade-off between A's investment in contextual services and B's investment in banner advertisement, both on A's web site.

The affiliate model, in contrast to the generalized portal that only seeks to maximize the volume of traffic, provides purchase opportunities to buyers wherever they may be. It typically offers a financial incentive to partners, who provide a click-through link from its site to the supplier. Hoffman and Novak (2000) describe a model involving CDNow and Geffen Records. Geffen established links on its site to direct its customers to CDNow site. CDNow sold more records as a retailer, and Geffen sold more records to the CDNow as a wholesaler. Affiliate marketing has evolved rapidly in sophistication—it can now use datamining approaches to link content to the products. Fizzylab (www.fizzylab.com) has developed automated procedures that can link a storyline that appears on a web site with a product. Clearly, the number of customers drawn from partner sites would increase with the quality of such software, but the software can be expensive.

Product Promotion

Retailers use product promotions to capture short-term demand from competitors, and/or to dispose off inventories. Promotions usually take the form of targeted advertising, coupons, mail-in rebates, and other such short-term drives that result in savings for customers. Promotions may cause sudden (short term) surges in demand, and if not coordinated well with production, they can lead to shortages and/or excessive production related costs. Clearly, there would be a trade-off between production cost and the additional revenues accrued from increased sales. Note that if a promotion is planned only in reaction to accumulated inventories, it will not constitute a profit maximizing strategy—inventory accumulation should be an endogenous part of a promotion plan. The proactive strategy would be to produce enough in prior periods to cover demands generated from a planned promotion. Therefore the two crucial decisions are, when to promote the product, and when and how much to produce. For complex scenarios involving multiple products and multiple periods, it is not easy to optimize these decisions (Chakravarty 2001a, Sogomonian and Tang 1993). What is required is to create information visibility between retailers and a supplier that would not only provide inventory status, but also information about cost of promotion, size of market, and cost of setting up for production. These information items can be made available, in an e-business context, to support online negotiations between the two parties.

CRM

In many companies, customer-related information is scattered throughout multiple departments, across numerous divisions, and in various formats. Rather than acting as a unified team, businesses find themselves operating as divided enterprises when interacting with customers, and many opportunities for customer acquisition are lost. Customer relationship management (CRM) is a comprehensive approach to managing the needs of a customer-centric business (Rigby, Reichheld, and Schefter 2002). It connects seamlessly with supply-chain management, financial management, and business intelligence, to make the same information available to all concerned. CRM typically includes functionalities such as sales force automation, e-marketing campaigns, customer service, reporting, and analysis. It allows a company to conduct interactive dialogs with customers, provides secured access of information resources (to the customer), facilitates collaboration between the selling and service teams, captures transaction data, delivers timely and critical information to employees, business partners, and customers, and provides a single source of accountability to improve operational efficiency.

Call centers are often responsible for working with their customers on sales, billing, and service. It has the ability to consolidate customer data from different applications to provide a total view of the customer, and to more effectively meet each customer's needs. This enables the customer service representative to access a wide range of business applications in resolving customer issues. A call center portal application is shown in Figure 4 (McAfee, Christ, Dhillon, and Reary 2001). The portal provides information needed to assess the nature of the call, and tracks all transactions with that customer. This includes information on service contracts, installed products, quotes and product repairs. It enables browsing product specifications, and customer profile: preferences, contacts, and addresses. It uses flexible business rules to route customer questions for quick and accurate responses, and to check inventory and order status. Personalization of catalogs and shopping lists are also possible.

6. INFORMATION EXCHANGE MODES

The nature of collaboration desired by partners is clearly a function of the frequency and value of interactions between them. We have shown five modes of collaborative information exchanges that have different capabilities and technological needs in Table 1. It is clear that partners motivated by an arms-length relationship would not be interested in process level collaboration. The mode of collaboration would therefore depend on business processes, cost of technology, and strategic partners. As we shall discuss later, a company may choose to collaborate in all five modes—each mode applied to a specific segment of partners, or in a single mode "focused" at a subset of partners. E-marketplaces such as WebMD and eBay emphasize the access mode and structured data transfers. ChemConnect also supports chat-rooms in a free-form interactive mode for buyers and sellers to negotiate price and other parameters of interest. Other e-marketplaces such as Oracle and SAP emphasize linking customer orders to a supplier's back-office ERP through application sharing. Yet other e-marketplaces—e2open, Converge, and Covisint—are consortium e-marketplaces that emphasize process collaboration.

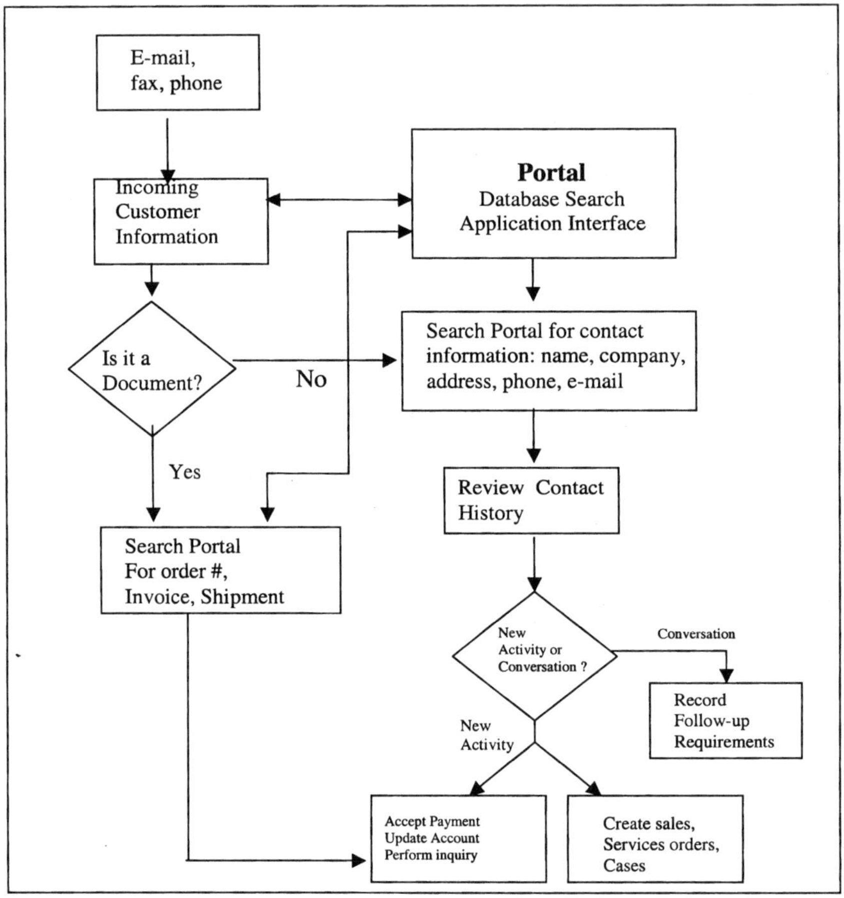

Figure 4. Call center operations (Adapted from McAfee et al. 2001).

The technology required for each of the five modes of information exchange is described briefly next. The *access* mode is often used for instant search and discovery with the client. It is a simple way of collaboration involving portals on the World wide-web. Access can be provided to a predetermined set of data items on a host's web page. As shown in Figure 5, a client can request the web pages of companies A, B, and C through a Web server, which acts like a file-server.

The web server delivers the host's web page to the client's web browser. The client can select information of interest, on the host's web pages, and have them delivered to him/her by the web server. It is a noninvasive approach that can provide instant access. However, information may not flow in real time and so response to business events cannot be automated.

Table 1. Modes of information exchange

Mode	Number of partners	Synchronization	Technology	Information format	Function	Application
Free-form interaction	Unbounded	None	Interactive	Message	Linked access	Chat, Create, Negotiate
Access	Many	Little	Web browser	HTML	Instant access	Discover, Search
Data Exchange	Moderate	Low	Database interface, File Transfer	Structured Data Transfer	File update	Transactions
Application Sharing	Moderate	Medium	ERP, EAI, Link to back-office	3-tierinforma-tion link	Data extrac-tion, Data transforma-tion	Real time updates
Process collabora-tion	Few	High	Workflow, Public Processes	Event Based, Mixed format	Autonomous data trans-fer between processes	Production Planning, Product Design

In *data exchange,* structured transaction data is moved from a client's database to that of the host (and vice versa). To complete this exchange, at least three pieces of information are necessary. These are: the data categories that require moving, the frequency of movement, and the method for moving the data. Such higher-level information items are known as metadata. As an example consider exchanging sales data whenever a sales event occurs. If the metadata is known technologies for moving data, called middleware (discussed in the appendix), can be triggered to extract data every time a sales event occurs and update the destination database through a database server. The system can be structured to transfer data at specific time increments as well.

The problem in data exchange is in binding the logic to data. That is data type,

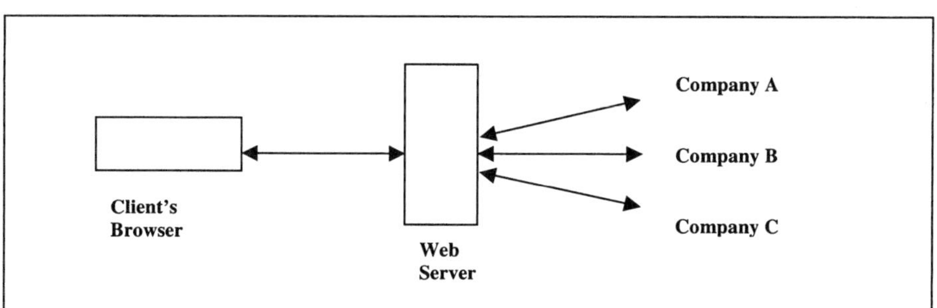

Figure 5. Information access in worldwide web.

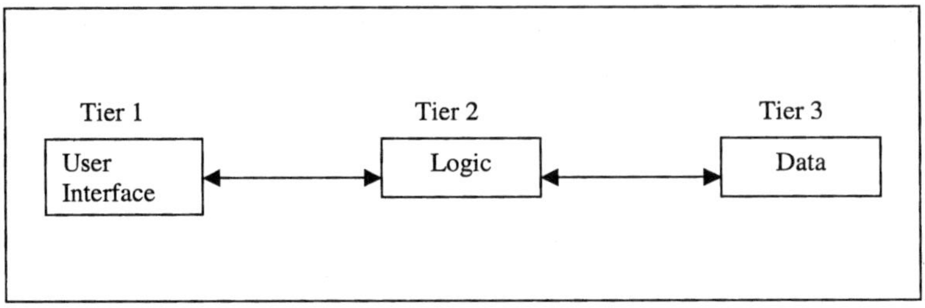

Figure 6. Three-tier application architecture.

its location and its transfer schedule must be determined by metadata, which is exogenous to the database. For data transfers involving multiple parties and a large number of databases, this becomes a major problem. In applications such as ERP, data to be exchanged is created by running the application. Thus the problem of identifying the data and its transfer schedule is avoided by permitting real time information to be extracted through an application server.

Application Sharing

Sharing information at the application level provides several advantages in addition to overcoming the problem of binding logic to data. A typical business application (payroll or inventory control) can be segmented into three layers (pieces): user interface, business logic, and data. These segments are separated and retained on three distinct tiers. A three-tier architecture is shown in Figure 6. Note that the client only need interact with the user interface. The databases act as the repositories of data, while the application server determines the appropriate business logic and databases for processing a client request. The user interface can capture a client's request for information (such as inventory status) through menus, windows, and graphical representations, and the client-application's interactions with the user interface are automated. The logic layer is organized into a well-defined set of services, called business services. The application server has access to a directory of the business services, specifying the required data inputs and information outputs from each service. The host application server can invoke these services to extract all relevant information from appropriate databases for processing queries such as "Get Quantity Available" or "Get Account Balance". The output information is then placed within a middleware for transmission to the client's system.

Process Sharing

Business processes such as placing an order, shipping a product, and assembling a product may each require several business tasks for completion. Consider an assembly process. On completion, it should trigger a shipping (or warehousing) process for the assembled product, and an order placing process for the components that would be assembled next. Order placing triggers the supplier's shipping operation, as shown in

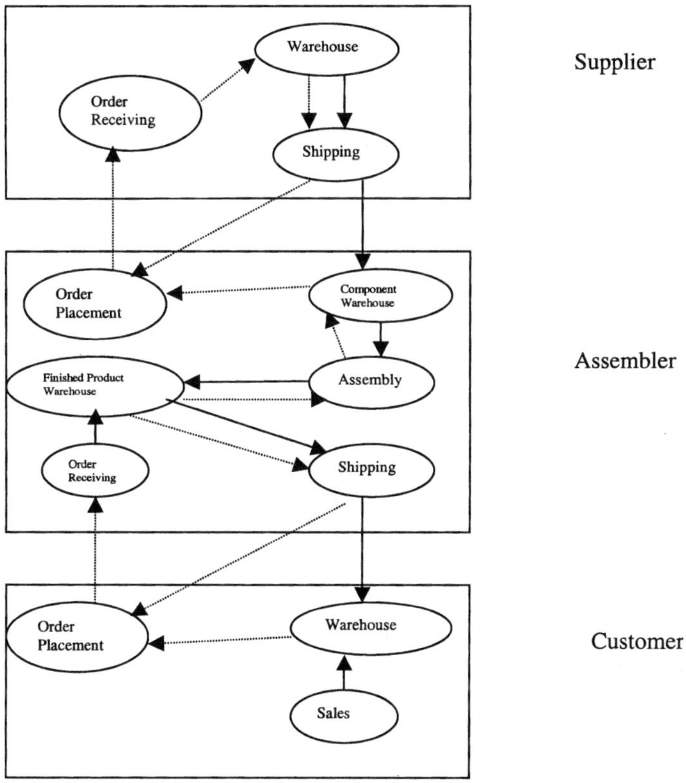

Figure 7. Inter-company operations.

Figure 7. The arrows with solid and dotted lines show flows of products and information, respectively.

When an order is received from a customer, the assembler triggers its finished goods warehouse to fulfill the customer's order from its stock. The warehouse in turn triggers two other processes: ship the product to the customer's warehouse, and start assembly of the next unit of the product. The assembly process triggers the assembler's component-warehouse for component delivery, which in turn places an order with a supplier for a replacement part. The supplier, in a similar way, must trigger its warehousing and shipping after receiving an order from the assembler.

From this example, it is clear how business processes between supplier, assembler and customer are inter-related. To understand collaboration, first consider a traditional business where assembler places an order for Q units of a component. When supplier receives an order, it checks its warehouse inventory. If it does not have enough to supply Q units assembler is informed, who may then look for another supplier to make up the shortfall. In a collaboration, on the other hand, assembler's order placement process will have visibility of supplier's warehouse. Thus assembler can adjust the order-quantity online by amount short (=q), and place an online order for Q-q units. Assembler could also track supplier's shipping process to ensure on time delivery. Clearly, supplier's warehousing and shipping processes are being made visible to assembler's order placement process. Supplier, in turn, would like to have the visibility of assembler's production plan or component inventory status. Some manufacturers prefer to keep their production planning processes private from its suppliers, as information seepage to competitors can occur since some of the suppliers may not be sole-suppliers. Sharing inventory status, on the other hand, is easier to accomplish and is a common practice in vendor-managed inventory (VMI). As a consequence, many collaboration service providers incorporate "public" processes that are visible to all, and "private" processes that are hidden from the partners. In the above example, warehousing and shipping will constitute public processes, whereas, production planning will be a private process.

The IBM Technology Group (ITG) recently considered such an approach for collaboration with a PC manufacturer, who was its customer (McAfee 2000). ITG and its customer defined the public processes that specified the format of both the electronic order document and the information that accompanied the ITG's disk drives (such as the performance data). The customer's private processes included the decision rules for allocating order to ITG. The ITG's private processes included customer qualification for its credit worthiness. Three types of capabilities facilitated process integration: process definition, electronic document creation, and application integration. Process definition used a graphical user interface (GUI) that included decision rules embedded in the process. The electronic documents transferred between the partners were created to conform to the EDI and other standards. The system also created several interfaces to integrate with back-office applications. For example, ERP could trigger action for a purchase order when certain inventory conditions were met. An example of public and private processes (McAfee 2000) is shown in Figure 8.

As mentioned earlier in section 3, McAfee (2001) discusses another application of collaboration between Adaptec that made products for data transfer between computers and TSMC that supplied fabricated microchips, using the Extricity software. The companies collaborated on engineering design, demand forecast, purchase order, work in process inventory, and shipping.

To implement process collaboration the partners (together) must possess capabilities of modeling and execution of business processes. Process modeling entails a graphical representation of processes, flows of information between them, the logic controlling the flows, and the resources required for completing the processes. A simple process model (Sparks 2000) is shown in Figure 9. Note that the book catalog is a resource for the process "sell books on-line", and user inquiry is an event. The outputs are orders, ship orders, and delivered orders. A process-model conforms to the rules of the business, ignoring technical details of the system. It can be executed using adapters and middleware that retrieve the required information as inputs to the process. The process engine performs the system updates triggered by transactions.

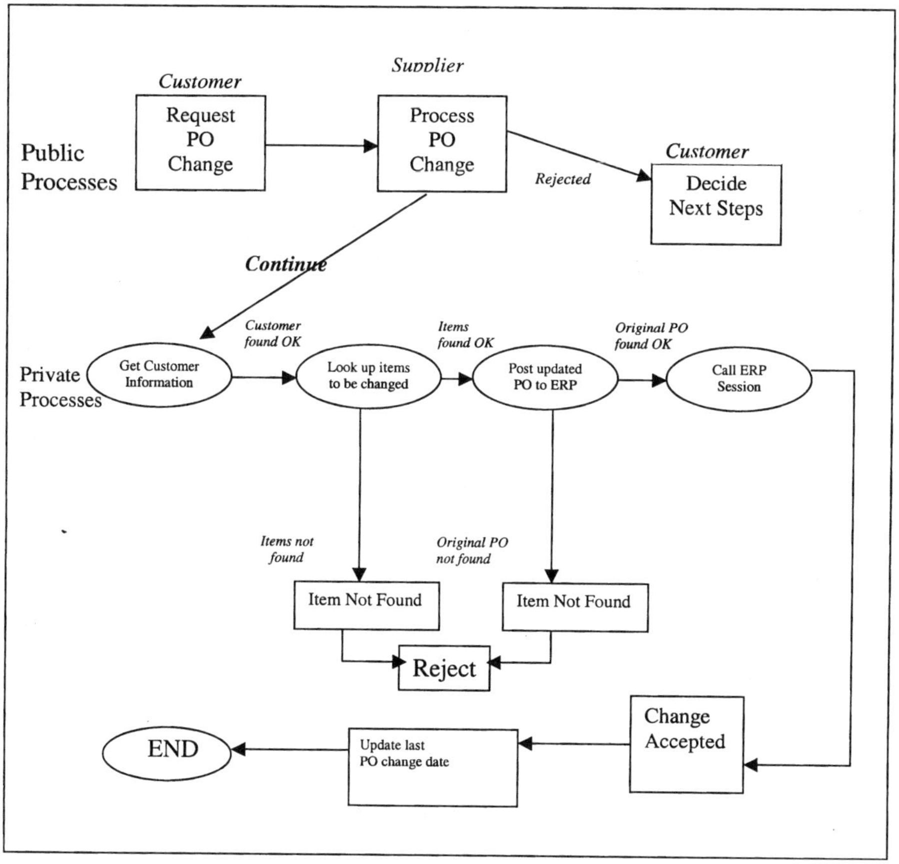

Figure 8. Process sharing.

Another way of implementing process collaboration is to use the workflow tool (Basu and Kumar 2002). The workflow, like a process model, follows the business rules to pass tasks information or documents between two parties. Additionally, it can automate the routing of work (or transactions) to different partners, identify required resources to accomplish the work, update data and status, and activate external systems and applications. Workflow engines can trigger events, and can automatically assign adequate resources for execution of an activity. A workflow facilitates viewing of the set of outstanding tasks and priorities. Users select specific tasks for execution from this set, and workflow activates appropriate programs for the selected tasks. If required resources are not available, it places the tasks in a queue. It captures execution parameters including wait and performance times, and determines task-status in terms of these parameters. Finally, it can add new processes or modify the existing ones to improve performance, if there are resource bottlenecks and other inefficiencies.

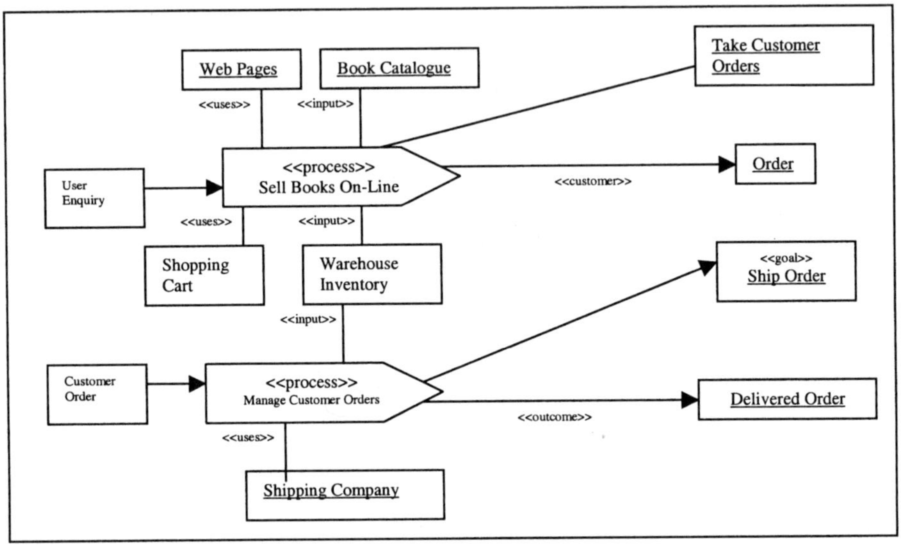

Figure 9. Business process model.

Clearly, the process level collaboration can be invasive, and may require partners to reengineer many of their existing business processes (Davenport 1993). It can also be risky, as partners must now share the ownership of their processes with others. Individual partners stand to lose certain degrees of freedom in their business operations (Ticol, Lowy and Kalakota 1998). In addition, adding new partners may be painstaking, as shared processes would invariably need redefining and modifications.

7. COLLABORATION STRATEGY

Clearly, the strategy for collaboration is shaped by market opportunities a company wishes to exploit, and collaboration capabilities it possesses. Market opportunities, in this context, are product commerce, supply chain management, and customer relationships. Collaboration creates opportunities in product commerce in terms of customization, time to market (Chakravarty 2003), and cost reduction. In a supply chain collaboration creates opportunities for vendor-managed-inventory, demand-variance reduction, and CPFR. Opportunities in customer relationship would be product promotion, customer solutions, and demand brokering (Callahan and Pasternack 1999).

The two major capabilities necessary for collaboration are information exchange, and conflict resolution. Information exchange may require different degrees of technology sophistication depending on whether or not it is invasive. As discussed earlier

Capabilities	Opportunities		
	Product Commerce	Supply Chain	Customer Relation
Information Exchange	Design hub	VMI	Customer solution
Conflict Resolution	Customized Configuration	Resource Allocation	Demand Brokering

Figure 10. Collaboration strategies.

access technology is non-invasive, whereas, process sharing can be invasive. The conflict resolution capability includes corrective action and resource allocation capabilities, and is invasive.

Interactions between opportunities and capabilities produce several models of collaboration, as shown in Figure 10. For example, a company that only possesses the non-invasive information exchange capability (technologically unsophisticated) can create a design hub for product development, explore VMI opportunities in its supply chain, and win new customers by offering solutions to customer problems in concert with other suppliers. A company that is technology savvy, on the other hand, may explore virtual configuration and visualization for product customization, and workflow models of process sharing that can split work and allocate resources in a supply chain. It can also pursue demand-brokering arrangements with its suppliers and customers.

Strategically, the company needs to decide where in the two-dimensional space (Figure 10) it would like to belong, with the possibility that it may belong to more than one cell. Clearly, the company needs to know which opportunities it should target and what capabilities it should possess. However, since its partners may possess varied degrees of technology sophistication and since collaboration with all of them may not be equally valuable to the company, the optimal strategy for the company would be to think in terms of a portfolio of relationships with its partners.

Such relationships can be conceptualized as exploratory, occasional, significant, and intensive. The explorative format is actually a pre-relationship exchange of ideas that benefits from free-form interaction. An occasional relationship between partners signifies a low frequency of interaction, and it would be adequate if the partners were only interested in access and information visibility. A relationship is called significant when the partners start exchanging structured data using databases and application programs. This would require interfacing of the front-office customer queries-with the back-office production systems. Finally an intensive relationship indicates joint decisions in a project management environment. This would enable the partners to coordinate their activities and make real time resource adjustments. The syn-

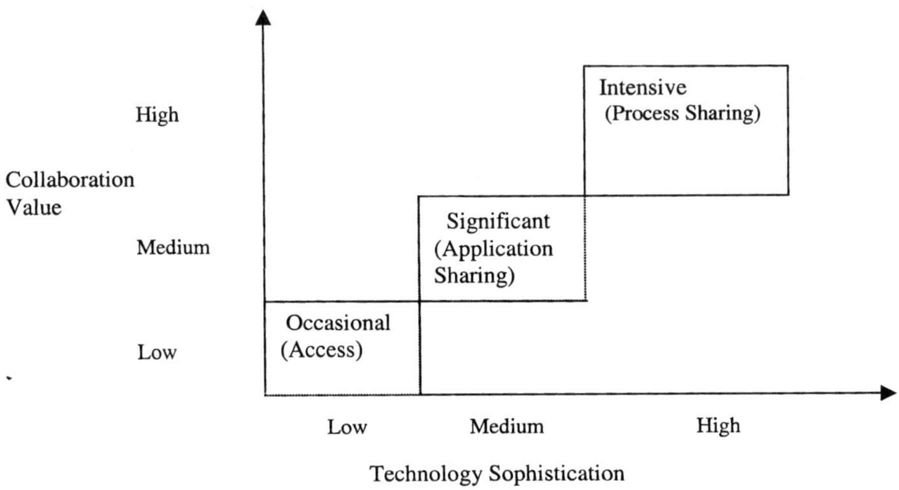

Figure 11. Collaboration strategy.

ergy between value of collaboration and technological sophistication is shown in Figure 11.

Decision Making

Depending on the importance of suppliers and customers, the company should be able to decide in which collaboration technologies it wishes to invest. As discussed earlier the problem may be quite complex—a company with a variety of customers (value and technology sophistication), for example, may not be able to justify VMI for all. Similarly it may have a variety of suppliers—contract manufacturers that can justify process sharing collaboration, and banks (supplying service products) that may only require occasional access. The decision-making problem facing such a company with respect to its partners is shown in Figure 12. Clearly there are two major decisions facing the company: which of the collaboration technologies (from access to process-sharing) it should invest in, and which collaboration technologies it should use with each of its partners. The example in Figure 12 tells us that the company has invested in two forms of collaboration technology: access, and application sharing. It shares applications with partners B and C, whereas, it collaborates with A in the access mode.

To answer such questions we explore the decision-making problem in more detail. In Figure 13, a typical manufacturer is shown with its transaction partners. Assume that the manufacturer conducts business transactions (directly or indirectly) with suppliers, resellers, retailers, banks, warehouses, and logistics service providers. It would like to explore whether collaboration with its partners is feasible and profitable. First the manufacturer would have to determine the appropriate mode of collaboration for each of its partners—what is appropriate for collaboration with a bank or a third-

Partners	Collaboration Mode				
	Access	Free-form	Data Exchange	Application Sharing	Process Sharing
A	x	--	--	--	--
B	--	--	--	x	--
C	--	--	--	x	--

Figure 12. Collaboration decision.

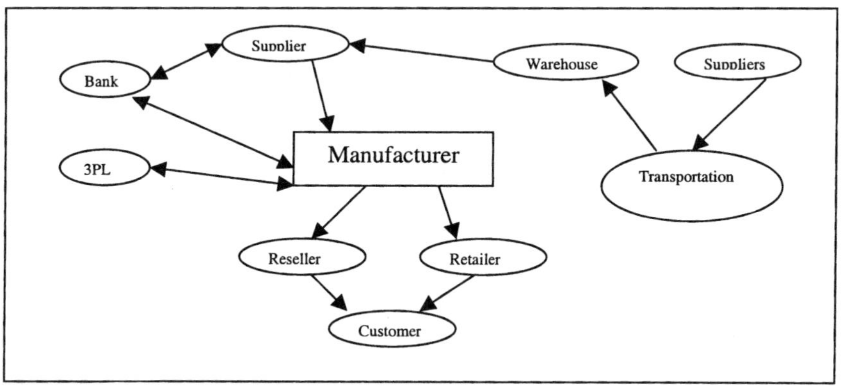

Figure 13. Manufacturer's value web.

party logistics provider (3PL) may not be adequate for a supplier or a reseller. The access mode of collaboration may be more than sufficient for financial transactions with a bank. For joint product development and coordinated supply chain decisions a process sharing collaboration (with a supplier) would be more appropriate.

The two major determinants in such decisions are the value and cost of collaboration. Factors that impact value creation and costs in any collaboration include: frequency of interactions with a partner, revenue and cost per interaction, and cost of setting up collaboration. The setup cost would have two components: partner specific, and collaboration-mode specific. For example, if a partner is technologically unsophisticated but can create significant values, the manufacturer may wish to subsidize its partner for it to become technology savvy (partner specific setup cost). Mode specific setup cost, on the other hand, would be high or low depending on the collaboration mode chosen. For a complex information exchange mode such as process sharing, mode specific setup cost would be high. The value (and cost) of collaboration would obviously depend on the frequency of interactions between the parties—higher the frequency, higher the value and the cost. Clearly, to a manufacturer collaboration

Table 2. Collaboration cost and value

	Operating Surplus (R − c)		Cost of setting up private processes (s)		Interactions frequency (v)
	Access	Process-Sharing	Access	Process-Sharing	
Supplier	1000	10,000	5000	100,000	100
Bank	4000	5,000	5000	200,000	15
3PL	2000	8,000	5000	100,000	40

with suppliers and retailers would be of more value than that with banks and freight-agents.

Consider a technologically unsophisticated supplier. To setup a process sharing mode of collaboration with this supplier, the manufacturer not only has to invest in creating connectivity and the public processes but it also has to subsidize the supplier in creating its own private processes. Denoting these costs as F and s, respectively, the total cost of process-creation would be F + s. However, F would be "split" with other process-sharing partners, and hence it should be allocated as an overhead. Next, assuming that the number of interactions per period is v, revenue per interaction is R and cost per interaction is c, the benefit of collaboration per period can clearly be expressed as v (R−c). Value of R would obviously depend on the importance of partner and the nature of information being shared. Since the share of F charged to a particular partner will depend on the number of other partners with whom the manufacturer shares processes, the appropriate modeling approach would be to consider the system wide costs and benefits, in contrast to considering just the partner specific costs.

As an example consider the case of three partners and two collaboration modes (access, and process-sharing). Assume that the costs, benefits, and the frequency of interactions are as shown in Table 2.

Assume that the costs of setting up connectivity and public processes (F) are 50,000 for the access mode, and 500,000 for the process-sharing mode.

The value of collaboration (= v (R−c)−s) can be verified to be as shown in Table 3.

Since the net values of process sharing with the bank and with the 3PL are negative, manufacturer will not consider process sharing with them. Thus if an investment of $500,000 is made in process sharing, it will be beneficial only with respect to the supplier, with a *net* value of $400,000 (= 900,000 − 500,000) to the manufacturer. With the access mode, the net profit will be $80,000 (= 55,000 + 75,000 − 50,000) if only the bank and the 3PL use it. If the supplier also uses the access mode, it would add another $95,000, with a total value of $175,000. This would be so if all three partners collaborate in the access mode. However, since process sharing with the supplier generates $400,000 by itself, it is clear that the manufacturer will not use the access mode with the supplier. Therefore the optimal collaboration solution will be:

Table 3. Value of collaboration

Partners	Access	Process-Sharing
Supplier	95,000	900,000
Bank	55,000	−125,000
3PL	75,000	−68,000

access mode for both the bank and 3PL, and process-sharing mode for the supplier. The optimal net profit will be $480,000. Note that the access mode will not be used with the supplier so long as F value for process sharing is below $805,000.

If the number of partners is not large and the types of collaboration modes being considered are also not large, the optimal solution can be found as discussed in the example above. For a large problem, however, the number of options to be considered would be huge, and the mathematical model discussed below will be required to find the optimal solution.

For the general case we denote a partner by the subscript i, and a collaboration mode by subscript j. We also define a binary variable $x_j(x_j = 1 \text{ or } 0)$. The value of x_j equals 1 if an investment is made for mode j; equals zero otherwise. Similarly, the binary variable Y_{ij} ($Y_{ij} = 1$ or 0) determines whether partner i is chosen for the j th mode of collaboration.

We express the total net benefit of collaboration as,

$$\Pi = \sum_{j \in J}\left[\left\{\sum_{i \in I} v_I\left(R_{ij} - c_{ij}\right) - s_{ij}\right\} y_{ij} - F_j x_j\right] \tag{1}$$

where

F_j is investment in common process j (such as public processes) that can be used by any partner.

s_{ij} is the investment in processes specific to the ith customer and the jth process.

v_i is the number of desired interactions per period with partner i.

R_{ij} is the value of collaborating with the i th partner (per interaction) using the j th collaboration mode.

c_{ij} is the cost of transaction (per interaction) with partner i using mode j of collaboration.

x_j = 1, if the j th collaboration mode is chosen for investment (0, otherwise)

y_{ij} = 1, if the i th partner is chosen for the collaboration mode j.

Clearly the optimization problem will be expressed as

$$\text{Maximize}\,\Pi = \sum_{j \in J}\left[\left\{\sum_{i \in I} v_i\left(R_{ij} - c_{ij}\right) - s_{ij}\right\} y_{ij} - F_j x_j\right]$$

Subject to

$$y_{ij} \leq x_j, \text{ for } \forall_{i,j} \tag{2}$$

and

$$\sum_{j \in J} y_{ij} \leq 1 \tag{3}$$

The manufacturer obviously wishes to maximize Π, the total benefit of collaboration. Constraint (2) requires y_{ij} to be zero if $x_j = 0$, and constraint (3) stipulates that partner i may participate in at most one mode of collaboration.

Clearly the above decision model needs to be embedded in the strategy framework shown in Figure 10. That is, if the company wishes to pursue opportunities in a particular domain such as supply chain management, it would determine the optimal investment in technology for a given set of suppliers and customers. If the company wishes to enhance its capability by including collaboration *services* in other domains such as product commerce, it must evaluate the benefits against possible new investments. A joint decision model for investment in technology for collaboration with its supply chain partners, and for providing collaboration services in CPC and/or CRM would help determine the strategic positioning of the company.

8. CONCLUSIONS

We have framed the collaboration decisions in terms of choice of a portfolio of collaboration options defined by technologies and partners: where to invest (technology), and who to include in collaboration (partners). Clearly, such decisions must be tailored to a company's long term strategy—the interplay between market opportunities and the company's technological capabilities. We have shown how business processes, cost of technology, and strategic partners, may impact such choices. Given that business processes may vary widely (supply, design, or customer related processes), technologies may possess unique characteristics in terms of applicability, and company objectives may be heterogeneous, a large number of possibilities exist in portfolio selection. We have shown how an optimal portfolio comprising technologies and collaboration partners can be conceptualized.

We have provided qualitative discussion of salient factors to underscore how value can be created, how costs can be discerned, and how a framework for analytical decisions can be structured for collaboration. However, there is a long way to go before we can build decision models that would integrate all aspects of collaboration, discussed in this chapter. One way to proceed would be to first build such models for specific domains (SRM, CPC, and CRM) to better understand the underlying tradeoffs (as we have done), and then to integrate them for a single investment decision for the company.

Clearly, our decision framework is structured for a market-dominant company (monopoly). In an oligopoly partners get to decide whether or not to enter into collaboration. This would lead to a multi-party gaming situation. The most likely

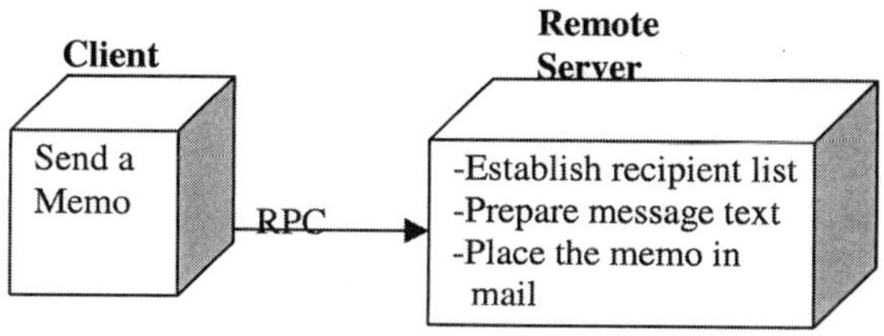

Figure A1. RPC link.

scenario would be a Stackelberg game where the partners react to one another's technology decisions. In future research we intend to address this issue.

We have also assumed that the value of collaboration, R, is known or that it can be determined easily. Even in simple cases where collaboration emphasizes only information visibility, value of collaboration would depend upon market dynamics of price, demand elasticity (Li 2002), and forecast revisions (Aviv 2001). With multiple collaboration objectives value would also be a function of the competitive behavior among partners. Customer demand may also be subject to a random variation, and the manufacturer may decide to hold some inventory in response. Research is in progress to model these issues (Chakravarty 2002).

APPENDIX A: MIDDLEWARE TECHNOLOGY

A middleware is software that enables the communication and integration of business processes between two or more companies across diverse platforms and networks. It insulates the application user from the details of operating system and the network interfaces (API). Available middleware for collaboration can be categorized as remote process calls (RPC), database interface (DBI), message oriented middleware (MOM), and transaction processing monitors (TPM) (See for details, Lintchicum 2001). A middleware can be synchronous or asynchronous. A synchronous middleware, (also called the request/respond model) tightly couples the applications, so that the calling application must pause and wait for the remote application to respond. An asynchronous middleware, on the other hand, can decouple itself from the application. The source application places its message in a queue and the target application can retrieve the message from the queue at its convenience. Such middleware can be designed for one-to-one or many-to-many connectivity.

Remote Process Call (RPC)

RPC is a synchronous middleware. The calling application can invoke a function residing on a remote machine, as shown in Figure A1.

RPC's are simple to use, but they require a large number of instructions (exceeding 10,000) to process a request (Linthicum 2001). This is so because additional calls are required for access-control, and translation services. This adds to the inefficiencies from having to wait for response from the remote application.

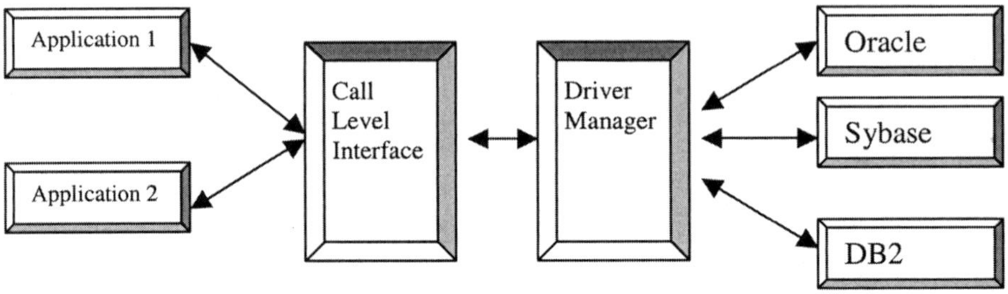

Figure A2. Database interface middleware.

Database Interface (DBI)

The DBI middleware first converts a query from an application so that the remote database can understand it. It then sends the converted query over a network, and processes it at the other end (target). The data conversion (source application to remote application) is done by the interface software. The interfacing architecture is simple in a one-to-one connection, but when multiple databases must be connected (many-to-many), the interface must be flexible and operate through a driver manager for linking with multiple databases. The interface is called Call Level Interface (CLI), as shown in Figure A2.

Message Oriented Middleware (MOM)

There are two types of messaging software: message queuing, and publish/subscribe. Both are asynchronous models (Whetton 2001). Message queuing is used when messages must be delivered to a single destination, as in confidential financial transactions. It is reliable, but the message may not be received in real-time at the destination. The MSMQ (Microsoft) and the MQSeries (IBM) allow multiple messages to be retrieved at a time in the order they arrive, and provide a router that determines the least cost solution to send the message. A message queuing architecture is shown in Figure A3.

The publish/subscribe software, on the other hand, delivers to more than one receiver at a time. Producers (source) publish information on multiple communication channels, and information consumers (destination) subscribe to channels of interest. It is asynchronous, anonymous and immediate. Producers do not need to wait for consumers, or know their identities. For example, to make the accounts receivable information available the application would post it on the publish-subscribe software, and all subscribers to the account receivable topic would start receiving this information immediately. This requires the information to be intelligently routed so as to reach only those who subscribe to that topic. Therefore the router is endowed with the capability of identifying the message source, and invoking the rules that control the processing and distribution of messages. The rule engine can be made sophisticated by including advanced capabilities such as dynamic routing, information sharing logic, and message processing frequency (batch or real-time).

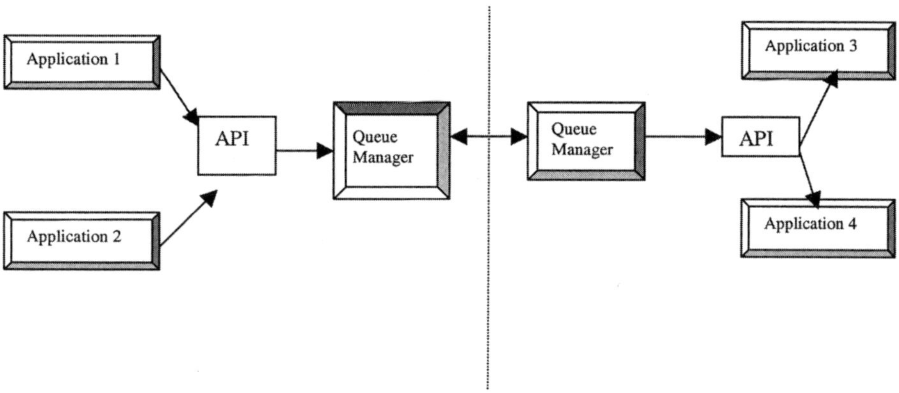

Figure A3. Message queuing architecture.

Transaction Processing Monitor (TPM)

TPM controls transaction applications and performs business logic computations. It provides the ability to divide complex applications into modular units called transactions. For any client/server link, it builds a "list" of necessary transactions using the modular units, and processes them from beginning to end. TPM would be a natural choice in the process-sharing mode of collaboration. A client can invoke the transaction services that reside on the TPM, and those services can share the same databases. The load-balancing feature enables the TPM to prioritize tasks, and process them according to their priority labels.

REFERENCES

Aberdeen (1999), "Collaborative Product Commerce: Delivering Product Innovations at Internet Speed", Aberdeen Group Inc., One Boston Place, Boston, MA, USA.

Applegate, L. (2000), "Overview of E-Business Models", Case Study #9-81-172, Harvard Business School, Boston, MA.

Aviv, Y. (2001), "The Effect of Collaborative Forecasting on Supply Chain Performance", *Management Science,* Vol. 47, No. 10, pp. 1326–1343.

Baldwin, Y. and K. Clark (1997), "Managing in an Age of Modularity", *Harvard Business Review*, September-October, pp. 84–93.

Baldwin, Y. and K. Clark (2000), Design Rules: The Power of Modularity, MIT Press, Cambridge, 2000.

Basu, A. and A. Kumar (2002), "Research Commentary: Workflow Management Issues in e-Business", *Information Systems Research*, Vol. 13, No. 1, pp. 1–14.

Callahan, C. and B. Pasternack (1999), "Corporate Strategy in the Digital Age", *Strategy and Business*, Quarter II, Issue 15, pp. 10–18.

Chakravarty, A. (2003), "Overlapping Evolving Product Design with Prototyping" Forthcoming, *IIE Transaction.*

Chakravarty, A. (2001a), Market Driven Enterprise: Product Development, Supply Chains, and Manufacturing, Chapters 9 and 10, John Wiley and Sons, New York, NY, USA.

Chakravarty, A. (2001b), "Designing a Digital Marketplace for Supplier Aggregation and Integration", Working Paper, A.B. Freeman School of Business, Tulane University, New Orleans, LA, USA.

Chakravarty, A., and J. Zhang (2002), "Collaborative Trading in a Digital Marketplace", Working Paper, A.B. Freeman School of Business, Tulane University, New Orleans, LA, USA.

Clark, T. and J. McKenney (1994), Campbell Soup Company: A Leader in Continuous Replenishment Innovations, Case Study, Harvard Business School, Boston, MA.

Cisco Systems (1998), "Networked Manufacturing for the 21st Century", Kaiser Associates, Vienna, VA.

Davenport, T. (1993), Process Innovation: Reengineering Through Information Technology, Harvard Business School Press, Boston.

Deloitte Research (2000), "Collaborative Commerce: Going Private to Get Results", B2B Insider, www.dc.com/obx/pages.php?Name=B2Binsiders.

Fine, C. (2000), Clock Speed: Winning Industry Control in the Age of Temporary Advantage, Perseus Books, Reading, MA.

Hauser, J. and D. Clausing (1988), "The House of Quality", *Harvard Business Review*, May-June, pp. 63–73.

Hoffman, D. and T. Novak (2000), "How to Acquire Customers on the Web", *Harvard Business Review*, May-June, pp. 97–103.

Huang, Z. and S. Li (2001), "Co-op Advertising Models in Manufacturer-Retailer Supply Chains: A Game Theory Approach", *European Journal of Operational Research*, Vol. 135, pp. 527–544.

Kaplan, S., and M. Sawhney (2000), "E-Hubs: The New B2B Marketplaces", *Harvard Business Review*, May-June, pp. 97–103.

Lee, H., and P. Padmanabhan and W. Whang (1997), "Information Distortion in a Supply Chain: The Bullwhip Effect", *Management Science*, Vol. 43, No. 4, pp. 546–558.

Lee, H., and S. Whang (2001), "E-Business and Supply Chain Integration" SBSCMB-W2-2001, Stanford Global Supply Chain Management Forum, Stanford University, Palo Alto, California, USA.

Li, L. (2002), "Information Sharing in a Supply Chain with Horizontal Competition", *Management Science*, Vol. 48, No. 9, pp 1196–1212.

Linthicum C. (2001), B2B Application Integration, Addison-Wesley, Reading, MA.

Magretta, J. (1998), "Fast, Global, and Entrepreneurial Supply Chain Management: Hong Kong Style", *Harvard Business Review*, September-October, pp. 103-114.

McAfee, A (2000), "IBM Technology Group", HBS Case-Study, 9-600-010, Harvard Business School, Boston, MA, USA.

McAfee, A. (2001), "Extricity Inc.", HBS Case Study, N9-601-113, Harvard Business School, Boston, MA, USA.

McAfee, R., K. Christ, K. Dhillon, and B. Reary (2001), "Collaborative Supply Chain Management", PeopleSoft Inc., www.peoplesoft.com/corp/en/products/line/scm/white_papers.asp.

Parker, G. and E. Anderson, (2002), From Buyer to Integrator: The Transformation of the Supply Chain Manager in the Vertically Disintegrating Firm, *Production and Operations Management,* Vol. 11, No 1, pp 75–91.

Raisch, W. (2001), E-Marketplaces: Strategies for Success in B2B Commerce, McGraw Hill, New York.

Rigby, D., F. Reichheld, and P. Schefter (2002), "Avoid the Four Perils of CRM", *Harvard Business Review,* February, pp. 101–109.

Sogomomian, A. and C. Tang (1993), "A Modeling Framework for Coordinating Promotion and Product Decision Within A Firm", *Management Science,* Vol. 39, No. 2, pp. 191–203.

Sparks, G. (2000), "The Business Process Model", Enterprise Architect, www.sparksystems.com.au.

Ticol, D., A. Lowy, and R. Kalakota (1998), "Joined at the Bit", Working Paper, Georgia State University.

Timmers, P. (1998), "Business Models for Electronic Markets" www.electronicmarkets.org/netacademy/publications.nsf/all_pk/949.

UDDI (2000), "Executive White Paper", www.uddi.com.

VICS (2001), CPFR: Collaboration Planning, Forecasting, and Replenishment, www.cpfr.org.

Whetten, B. (2001), "Message-Based Computing: The Fourth Wave of Integration" ebizQ, www.ebiz-Q.net.

Wise, R., and D. Morrison (2001), "Beyond the Exchange: The Future of B2B", *Harvard Business Review,* November-December, pp. 86–96.

IV. FIELD STUDIES AND REAL-WORLD EVIDENCE

9. LEVERS FOR IMPROVING NPD TIME AND FINANCIAL PERFORMANCE

AKHILESH BAJAJ, SUNDER KEKRE, and KANNAN SRINIVASAN

ABSTRACT

In this chapter, we discuss research conducted at a leading avionics and missile guidance systems manufacturer to identify the key managerial drivers in the design phase of new product development (NPD) projects. Three factors were found to significantly influence the time and financial performance metrics in the design and manufacturing phases. Our model recommends higher levels of managerial oversight, lower levels of specialization and selective interaction with customers in environments where manufacturing budgets dominate. The model also captures the interdependencies between the design and manufacturing phases of NPD. It reinforces concepts such as design for manufacturing and systems integration by project managers. The factors we examine integrate several concepts from marketing, operations and engineering management for high technology environments.

1. INTRODUCTION

The aerospace industry provides several opportunities to investigate the drivers of new product development (NPD) efforts for complex products requiring intensive outlays to meet the ever-changing needs of the defense sector. Examples of such products include missile guidance systems, advanced avionic systems and flight simulators. Manufacturers such as Lockheed Martin, United Technologies and Boeing constantly face intense pressures to satisfy the diverse needs of their customer base, and at the same time have tight schedule commitments and budget constraints. Typical products in this industry require significant design outlays, with unique challenges to manage the NPD projects and coordinate across the large number of specialists and generalists involved in the various stages of project execution. Budget overruns and schedule slippages are key metrics deployed to track the performance of projects. This chapter reports findings from a recent field study in this industry to identify strategies to improve NPD performance.

This issue of NPD performance has been extensively researched in the academic literature Several concepts such as concurrent engineering, design for manufacturing (DFM) and the use of cross functional teams have been proposed to improve the time and financial performance (Cohen, Eliashberg, & Ho, 2000; Eppinger, Fine, & Ulrich, 1990; Krishnan & Ulrich, 2001; Brown & Eisenhardt, 1995; Chakravarty, 2001; Novak & Eppinger, 2001). From a project management standpoint, numerous factors such as design team composition, intensity of customer interaction, degree of oversight by the project manager and communication between team members have been identified to significantly influence project performance (Ulrich & Ellison, 1999; Pahl & Beitz, 1988).

We position our study in this context and define NPD to be the process of designing, manufacturing, assembling and testing a new product. We combine the conceptualization and design development into one phase that we call *design*; similarly we aggregate the manufacturing, assembly and testing phases into a single phase termed *manufacturing*. Given the nature of the NPD environment described above, we model three choices faced by project managers. First, what is the appropriate balance between deploying specialist designers versus generalist designers when developing advanced defense systems? Second, what is the appropriate level of oversight by the project manager to ensure time and budget constraints? Finally, what is the appropriate intensity of customer interaction, which the project manager should establish? Each of these choices made by the project manager has a cost and a potential benefit associated with it. In the next section, we elaborate on the motivation to examine these issues and develop the basis for a detailed model of NPD performance.

2. UNRAVELING NPD PERFORMANCE

Eisenhardt and Tabrizi (1995) provide a comprehensive survey of the most influential empirical studies in the area of NPD, and cite some of the key factors that have been identified to influence NPD performance. Prior studies underscore factors such as organizational structure (matrix v/s heavyweight), location (co-location v/s dispersion) and the degree of supplier content v/s in-house content. In this study, organizational structure and location were similar across all projects and are hence controlled. Unlike earlier studies, we probe deeper into the nature of in-house NPD efforts, and its interaction with factors such as the degree of customer involvement. The explanatory variables that we use to study this are defined next.

2.1 Specialization

Specialization can be operationalized along multiple dimensions: disciplinary, component and process. Disciplinary specialization refers to narrow area expertise such as heat transfer, finite element analysis, and electronic hardware design. Generalist designers can span disciplines. Component specialization refers to designers who specialize in sub-systems such as in avionics, power systems or guidance systems. Specializa-

Table 1. Illustrative example of specialization

	Heat transfer	FEA	Vibration analysis	Generalist designers
Avionics Sub System	0	0	6	3
Power Sub System	10	2	1	2
Propulsion Sub System	18	14	1	2
Systems Level (span sub systems)	2	4	2	33
	30	**20**	**10**	**40**

tion by process refers to expertise in the different stages of design such as system design, detailed design, production and testing. It should be noted that in this study, generalist designers span components, and handle system level tradeoffs across disciplines. Specialist designers tend to focus on single components, since each component typically demands expertise in a subset of disciplines. An illustrative profile of engineering hours on these dimensions of specializations is shown in table 1.

In table 1, the partitioning by columns is by disciplines, and rows by components. The numbers in each box represent design hours[1]. Given the confidential nature of the projects, disaggregate data such as shown in table 1 is not available. Instead aggregate hours for specialist versus generalist designers were available by project (analogous to the 60/40 split in table 1).

The research site for the field study was Missile Systems Inc.[2], a leading designer and manufacturer of advanced missile guidance systems. Guidance systems are highly customized applications and involve the simulation of flight-guidance of missiles, with precise monitoring of parameters such as vibration, temperature, and rotational and linear speeds. In this environment, there are usually many aspects of design, such as kinematics design, electrical design and thermal design. Project managers typically enlist designers that have generalist engineering design skills, as well as some with very specialized skills in areas such as strength of materials, materials fracture, thermal effects, electronic hardware and control circuits. Since several of the design tasks can be potentially accomplished by either generalist designers or by specialist designers, project managers have discretion in the choice of the intensity of specialization of the design team. According to project managers at Missile Systems Inc., generalist designers command lower salaries but take longer time because of more iterations from trial and error approaches to complex design tasks. Specialist designers, on the other hand, can leverage their prior experience and expertise to make faster and better designs. Thus, the intensity of specialization represents an important managerial decision in

1. In the example in table 1, the total engineering hours from specialist designers (the totals of columns 1, 2 and 3) is 60. These 60 hours represent resource deployment aggregated across various disciplines. The generalist hours are 40, and represent hours spent by designers who span disciplines and work on multiple components.
2. The identity of the company is masked for confidentiality.

projects at Missile Systems Inc., and there is a wide variation in this regard across projects of similar complexity.

We elaborate on the ramifications of this choice. There are two schools of thought regarding the costs versus benefits of specialization. While having a higher specialization is potentially useful for complex design processes, negative impacts can also occur. For instance, specialist designers are bottleneck resources and are assigned to multiple projects concurrently, leading to issues of congestion (Adler, Mandelbaum et al. 1995). Second, as mentioned above, specialist designers are more expensive, leading to potentially greater costs and budget overruns in the design phase. Third, specialists tend to have a deep and narrow sub-system focus. Project managers thus have to provide greater co-ordination across specialists to bring in the system perspective to ensure integration across sub-system interfaces.

These negatives have to be weighed against potential benefits. First, specialists can optimize the design of their assigned sub-systems for manufacturing, thereby enhancing downstream manufacturing performance. Second, specialists are better able to understand detailed customer requirements at the sub-system level, and can tailor the design accordingly. Third, specialists can also leverage their prior project experiences leading to lower design hours (Nobeoka and Cusamano 1997).

Given these tradeoffs, we wanted to ascertain the correlation of specialization with performance metrics at our research site, *i.e.,* did higher levels of specialization in some projects pay off at Missile Systems Inc.? Figures 1 and 2 show a scatter plot of the time and financial performance respectively of the 42 NPD projects used in our study. On the Y-axis are performance metrics plotted against the degree of specialization, on the X-axis, for each project[3]. A cursory inspection of the scatter plot for project schedule performance in figure 1 seems to suggest little, or at best slightly negative impact of higher specialization on time performance. The correlation between project schedule performance and specialization is −0.51. On the other hand, figure 2 suggests the opposite trend; where higher specialization leads to little, or at best, slightly better financial performance of projects. The correlation between the project financial performance and specialization is 0.46. Clearly, both the plots exhibit a wide scatter, possibly arising because of the confounding effects of other variables that have not been explicitly controlled for. We conducted further interviews to ascertain the likely variables that could influence these metrics. These detailed interviews highlighted two additional factors that varied significantly across projects: oversight and customer interaction.

2.2 Oversight

The effects of oversight in NPD projects are not well understood in the academic literature. Eisenhardt and Tabrizi (1995) assert that more frequent milestones (implying greater oversight) are more suited for the design environments of immature

3. Details of the fields study, research site, data collection and analysis are in (Bajaj, Kekre, & Srinivasan, 2002). Intensity of specialization is defined later in the chapter.

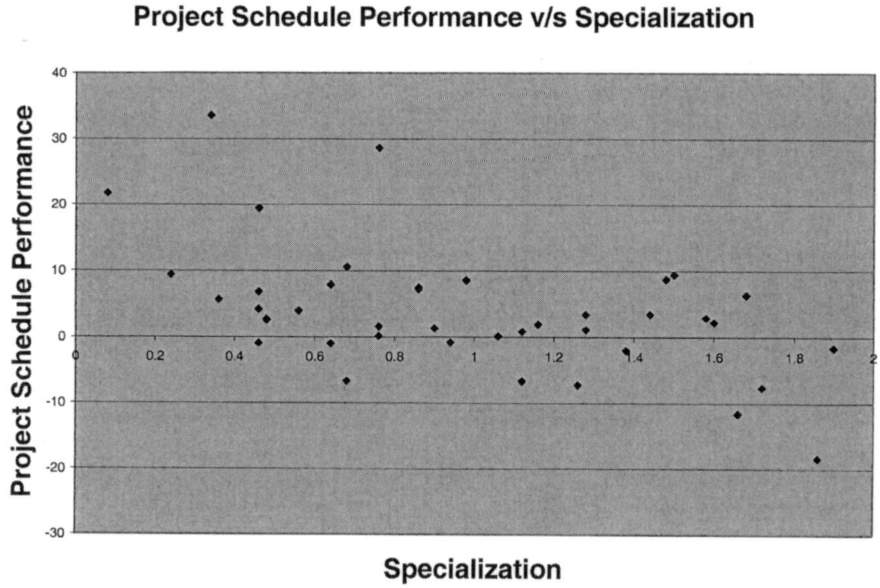

Figure 1. Project schedule performance with varying specification.

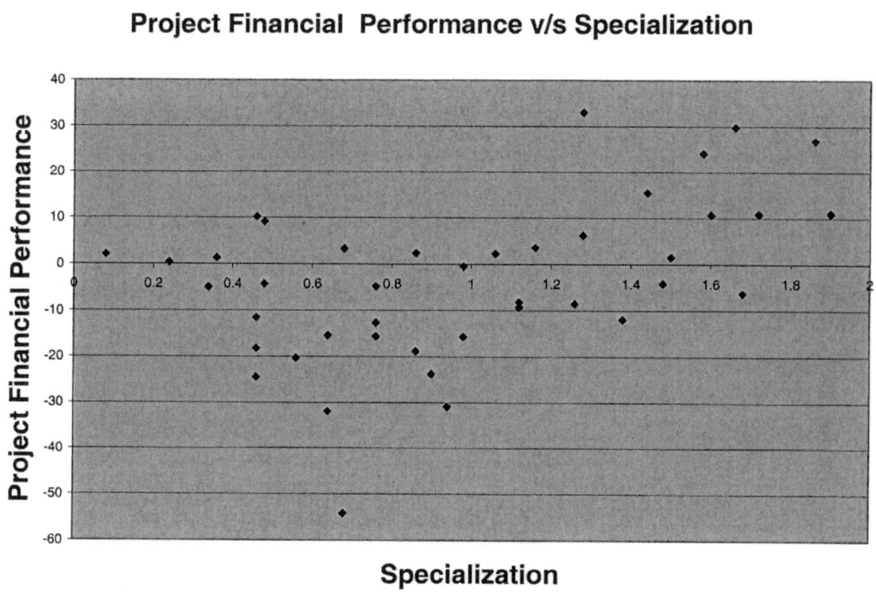

Figure 2. Project financial performance with varying specification.

products. Ha and Porteus (1995) caution that excessive as well as inadequate oversight are both harmful. Excessive oversight can lead to poor time performance because of micro management by project managers, whereas inadequate oversight causes communication gaps between design team members, leading to poor quality and increased rework. Krishnan and Ulrich (2001) echo this sentiment: ". . . practitioners seem to struggle to strike the right balance between excessive intervention and inadequate oversight."

In the "process management" literature, there has been extensive work on controls by project managers. Process controls occur when managers specify the means used to achieve goals, and monitor the activities pursued. Here again, there seem to be two diverse opinions on the effects of process control on NPD outcomes. One body of empirical work argues in favor of increased oversight during NPD (Cooper 1993; Wheelwright and Clark 1992). This has led to several specific process management methodologies such as the phase-gate process (Cooper 1993), the quality function deployment method (QFD) (Hauser and Clausing 1988) and the design structure matrix (DSM) (Eppinger, Fine et al. 1990; Eppinger 2001). On the other hand, the counterproductive impact of oversight has been emphasized by (Ouchi and Maguire 1975; Ouchi 1979; Jaworski, Stathakopoulos et al. 1993). Thus, oversight has been shown to be both harmful as well as beneficial. We were interested in studying the nature and magnitude of the impact of oversight in complex NPD environments such as Missile Systems Inc.

2.3 Customer interaction

The third factor we identified from interviews and project records was the intensity of customer interaction. Like the previous two factors, customer interaction varied across projects and could also impact project performance either positively or negatively. On the plus side, increased customer interaction can lead to shorter lead times (Cooper 1993; Gupta and Souder 1998). According to Thomke and Nimgade (2000), "Designers often seek perfection which could potentially lead to cost and time overruns—also known as *creeping elegance*". Increased customer interaction potentially mitigates this creep. On the negative side, customers can place ever-increasing demands and change their requirements leading to excessive engineering changes and design rework. Higher levels of customer interaction can thus impede or enhance project schedules and lead to excessive consumption of resources.

2.4 Interaction effects

Some other secondary effects also had to be accounted for. Project managers at Missile Systems Inc. articulated that the effects of their oversight were more pronounced in projects where the customer interacted more frequently. Likewise, specialists also revealed that projects with higher customer involvement tended to conform better to both schedules and budget. Both these insights seemed to suggest an interaction between the intensity of customer interaction and the managerial levers deployed by project managers during design (oversight and specialization). In our model, we quantify these interactions and assess their separate impact from both financial and time performance viewpoints.

2.5 Linking design and manufacturing phases

The aggregate project performance metrics shown earlier in figures 1 and 2 did not reveal the direct impact of increased specialization. As mentioned earlier, this could be attributed to confounding effects from other variables, such as the degree of oversight and the intensity of customer interaction. Another possibility could have been that aggregation of the two phases (design and manufacturing) might have masked the differential effect of the levers on each individual phase. To investigate this further, we collected data on the performance in *each* phase separately and modeled the effect of the drivers.

We also had to capture the interdependency between the two phases. An interesting aspect of our analysis was to isolate the "linked" impact of design and manufacturing. The issue was: to what extent does performance in the design phase impact the subsequent manufacturing phase? We estimate the differential impact of project management levers in each phase to address this question. To summarize, our model, described next, explored the effects of the managerial levers: degree of design oversight, intensity of specialization in design and the intensity of interaction with the customer. We explored the relative impacts of these drivers on the time and financial performance in *both* the design and manufacturing phases separately. Next, we describe briefly the operationalization of the variables and the statistical estimation procedure of the model parameters.

3. NPD PERFORMANCE MODEL

The unit of analysis for our study is the individual NPD project. Our data set represents 42 distinct NPD projects, completed during 1993-2001 at geographically dispersed locations of Missile Systems Inc. All these projects were similar and involved the development of customized applications in the aerospace sector, with similar levels of technical complexity. There were differences however, in terms of the size of the project, the extent of customer involvement and the nature of internal resources (specialization and oversight) deployed, resulting in differential performance. The dependent variables in our study are the financial and time performance for the design and manufacturing phases of an NPD project. The independent variables are specialization, oversight and customer interaction. These are operationalized next.

3.1 Independent management levers

Specialization in design represents the intensity of deployment of input from specialists during design. This intensity can be captured along two different dimensions: a) the percentage[4] of total design hours allocated to specialist designers, and b) the number

4. We use percentage rather than absolute dollars to account for scaling effects of different project sizes.

of different specialties in a project[5]. The first dimension reflects the overall "intensity" of specialization and the second captures the "variety" of specializations deployed. At our research site, we found a high correlation[6] between a) and b) and detailed data on b) were restricted due to the confidential nature of the defense projects. Hence, we decided to use a) to jointly capture both the overall intensity and the diversity of specialization. Thus, we operationalize the intensity of specialization in the design phase to be:

$$SP = \frac{\textit{(Budget For Specialist Designers)}}{\textit{(Total Design Budget)}} \times 100$$

The second independent variable in our study is *oversight*. To compare the intensity of monitoring by project managers across projects of different design budgets, we normalized the number of tasks monitored in the design phase by the size of the overall design budget. Hence oversight is defined as:

$$OV = \frac{\textit{(Number of tasks monitored in the Design Phase)}}{\textit{(Total Design budget)}}$$

This operationalization captures the involvement of project management and the intensity of project reviews in the design phase.

The third driver of performance is the intensity of *customer interaction* during the design phase. Here also, we controlled for the size of design budget and captured the intensity of customer interaction during design as:

$$CUSTINT = \frac{\textit{(Number of customer signoffs in design phase)}}{\textit{(Total Design Budget)}}$$

The above operationalizations also permitted us to assess the interaction between intensity of customer interaction and the other two independent variables (specialization and oversight). These interaction variables were defined as the following product pairs:

SP-CUSTINT = SP* CUSTINT
OV-CUSTINT = OV* CUSTINT

5. This equals the number of specialists since the size of projects at our research site are not large enough to warrant more than one specialist for a given specialty in any project.
6. This high correlation implies that whenever specialists are brought in, irrespective of their specialty, they are engaged for similar percentages of the design budget.

3.2 Dependent performance metrics in each phase

The organization uses standard procedure to establish both the financial and the time budgets, that recognizes the size and nature of customization for each project. Given the confidential nature of the defense projects, the details of the budgeting process are not available. The budgeting process however has been stable over the observation period, allowing a meaningful comparison of deviations from both the financial and the time budgets. For each phase, we gathered primary data on the time and financial performance of each phase of the project, relative to the budgeted amounts. Positive values of the metrics indicate performing ahead of schedule or below budgeted cost. To facilitate inter project comparison, we normalized the deviation from budget by the budgeted resource in each phase. We therefore defined the following time and financial performance metrics for the design and manufacturing phases, respectively:

$$\text{DES-SCHED} = \frac{(\text{Time ahead of schedule for the design phase})}{(\text{Time budgeted for the design phase})} \times 100$$

$$\text{MFG-SCHED} = \frac{(\text{Time ahead of schedule for the Manufacturing phase})}{(\text{Time budgeted for the manufacturing phase})} \times 100$$

$$\text{DES-SVNGS} = \frac{(\text{Cost Savings in the Design Phase})}{(\text{Total Design budget})} \times 100$$

$$\text{MFG-SVNGS} = \frac{(\text{Cost Savings in the Manufacturing Phase})}{(\text{Total Manufacturing budget})} \times 100$$

3.3 Structural model for each phase

For brevity, we do not discuss the hypotheses leading to the following structural system of four equations. The first two equations represent time and financial performance for the design phase. Equations 3 and 4 represent performance in the downstream manufacturing phase.

$$\begin{aligned}\text{DES-SCHED} = {} & \alpha_1 + \beta_{11}\text{OV} + \beta_{12}\text{SP} + \beta_{13}\text{CUSTINT} \\ & + \beta_{14}\text{OV-CUSTINT} + \beta_{15}\text{SP-CUSTINT} + \varepsilon_1\end{aligned} \tag{1}$$

$$\begin{aligned}\text{DES-SVNGS} = {} & \alpha_2 + \beta_{21}\text{OV} + \beta_{22}\text{SP} + \beta_{23}\text{CUSTINT} \\ & + \beta_{24}\text{OV-CUSTINT} + \beta_{25}\text{SP-CUSTINT} + \varepsilon_2\end{aligned} \tag{2}$$

$$\text{MFG-SCHED} = \alpha_3 + \beta_{31}\text{OV} + \beta_{32}\text{SP} + \beta_{33}\text{DES-SCHED} + \varepsilon_3 \tag{3}$$

$$\text{MFG-SVNGS} = \alpha_4 + \beta_{41}\text{OV} + \beta_{42}\text{SP} + \beta_{43}\text{DES-SVNGS} + \varepsilon_4 \tag{4}$$

Table 2. Estimated coefficients of the model

Variables	DES-SCHED (Equation 1)		DES-SVNGS (Equation 2)		MFG-SCHED (Equation 3)		MFG-SVNGS (Equation 4)	
INTERCEPT	13.88	(2.66)*	52.79	(0.29)+	112.58	(2.28)	39.65	(3.75)
OV	−9.38	(−5.48)	−160.99	(−2.66)	−2.2	(−0.14)+	13.53	(1.69)
SP	6.62	(1.85)	342.2	(2.72)	74.4	(1.38)+	−26.45	(−2.00)
CUSTINT	−15.25	(−4.46)	−5.19	(−0.04)+	NA		NA	
OV-CUSTINT	5.38	(5.32)	−37.17	(−1.04)+	NA		NA	
SP-CUSTINT	6.26	(2.29)	−63.30	(−.65)+	NA		NA	
DES-SCHED (instrumental)	NA		NA		−16.7	(−5.24)	NA	
DES-SVNGS (instrumental)	NA		NA		NA		−7.84	(−2.45)

**Values in parentheses are t-statistics.*
+ Insignificant at 10% level.

Note that our structural model consists of two sets of recursive equations (1, 3 and 2, 4), with one set modeling the linkages between the time performance of the design and manufacturing phases, and the second set modeling the linkages of the financial metrics. We estimate equation (1) by the Ordinary Least Squares (OLS) method and use the predicted values of the dependent variable DES-SCHED in equation (3) and estimate the resulting equation[7]. We can also allow for the residuals of the equations in each set to be correlated and estimate the parameters allowing for such a correlation. In this case, we checked the rank and order conditions (Theil 1971).

Similarly, we estimated the second set of equations linking DES-SVNGS to MFG-SVNGS. The results of our estimation are shown in Table 2, with *t*-statistics in parentheses.

Figures 3 and 4 summarize the effects of the key variables on each phase and reveal the dependency between the two phases. The bold values of the βs represent significant effects. With respect to time performance[8] in the *design* phase, SP, OV and CUSTINT in figure 3 are all seen to have direct impact on design time performance. Increased OV and CUSTINT affect the time performance adversely, as exhibited by the negative values of the coefficients. SP however is beneficial, supporting the view that higher intensity of specialization speeds up the design phase. The interaction variables (OV-CUSTINT and SP-CUSTINT) are also significant and positive, implying that increased oversight and specialization interact with customer interaction favorably. The *manufacturing* time performance comprises of two effects: a direct effect by SP and OV (shown as dotted lines) and a ripple effect by DES-SCHED (shown as a bold line).

7. See (Theil 1971) and (Judge et al. 1985) for a discussion of the 2SLS analysis performed here.
8. In this work, we use the terms *schedule performance* and *time performance* interchangeably.

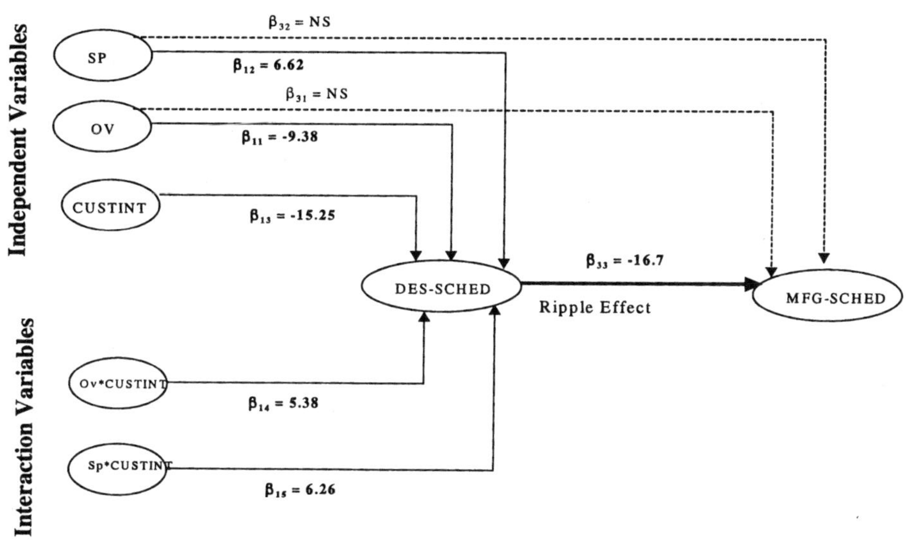

Figure 3. Drivers of time performance in each phase of NPD.

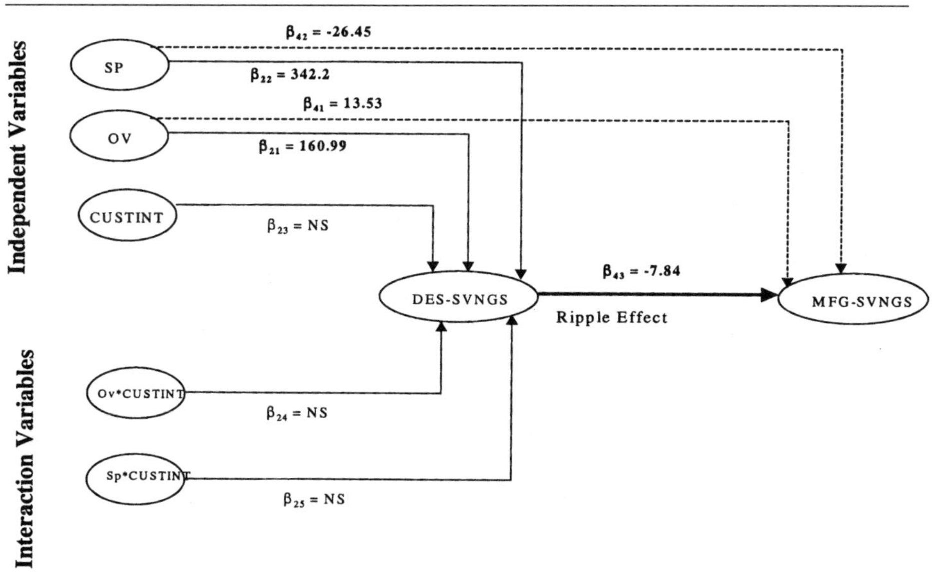

Figure 4. Drivers of financial performance in each phase of NPD.

Likewise, figure 4 depicts the financial performance of the two phases. Here again, the ripple effect of DES-SVNGS on MFG-SVNGS is significant and pronounced. Unlike the time performance model, the interaction variables (OV-CUSTINT and SP-CUSTINT) are not found to be significant.

Next we describe the managerial implications of our findings.

4. MANAGERIAL IMPLICATIONS

The phase dependent model enabled us to resolve the dilemma of the combined effects of specialization and oversight. The marginal impact of specialization (figures 5 and 6) and oversight (figures 7 and 8) at our site are discussed next.

4.1 Effect of increased specialization

Figure 5 shows the marginal impact of higher levels of specialization on time performance separately in design and manufacturing. The top left hand graph shows that higher specialization is beneficial in the design phase. This arises from the positive value of β_{12} in figure 3, which represents the direct impact of specialization on design time performance. This confirms the greater efficiency of specialist designers at our site. An additional positive impact is obtained from the interaction effect between specialization and customer interaction, as shown by the positive value of β_{15} in figure 3. Thus, we conclude that specialists can leverage the input from customers and execute the design of their sub-systems faster.

The effect of specialization on manufacturing time performance comprises of both the direct effect (found to be insignificant at our research site) and a ripple effect (represented by a negative β_{33} in figure 3). The net impact of a unit change in specialization is $(\beta_{33} \star (\beta_{12} + \beta_{15} \star \text{CUSTINT}))$. This net effect is negative at our site, implying that increasing specialization adversely affects manufacturing schedule performance. This is confirmed by the top right hand graph in figure 5 where the line is downward sloping.

Taken together, the implication is that while higher specialization benefits time performance in the design phase, the manufacturing schedules tend to slip. We can explain that this phenomenon occurs due to subsystem focus on the part of designers, leading to delays in resolving assembly problems in manufacturing. The effect of specialization on the overall project performance is shown in the lower graph of figure 5. Note that the overall effect is negative because of the dominance of manufacturing in the project timeline. Since the overall schedule is a weighted sum of the individual phases, if the design phase were to dominate the time line, then the effect of increasing specialization on the overall project time performance could become positive. For this to occur, the critical ratio of design to manufacturing time from our model is:

$$\frac{|(\beta_{33} \star (\beta_{12} + \beta_{15} \star \mathit{CUSTINT})|}{\beta_{12} + \beta_{15} \star \mathit{CUSTINT}}$$

$$= |\beta_{33}| = 16.7$$

Based on the estimates in our field study, the design timeline needs to be more

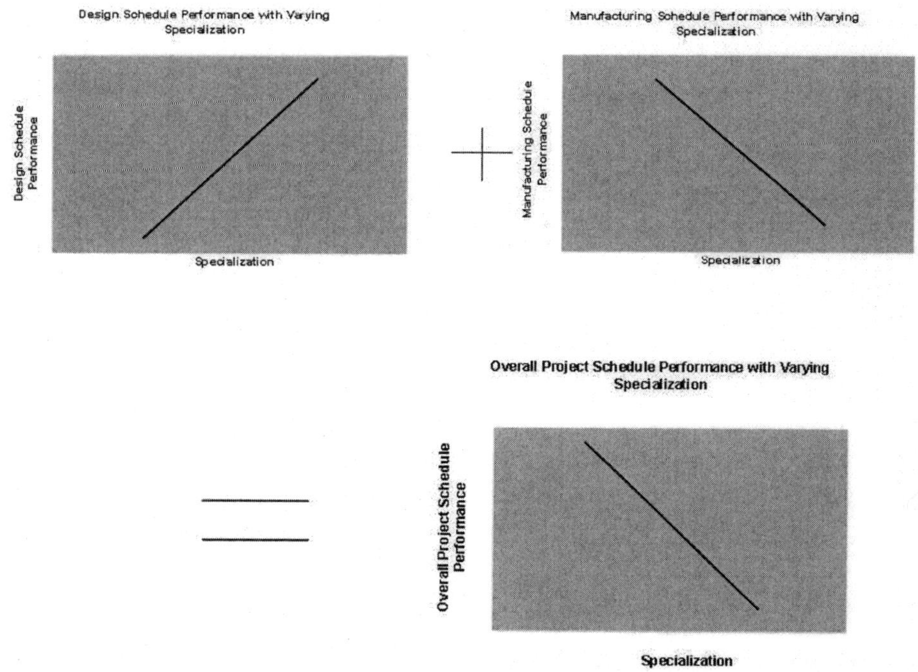

Figure 5. Overall project schedule performance with varying specialization as aggregation of performance in design and manufacturing phases.

than 16.7 times the manufacturing timeline for specialization to affect the overall project time performance positively. This high value suggests that in most practical environments, design specialization should be reduced for overall project time performance.

Likewise, the effect of specialization on financial performance is shown in figure 6. Here the directional effects are similar to design schedule. Design costs are lower because of higher efficiencies of the specialists, as well as their ability to leverage earlier experiences in their own areas of specialization. However, this sub-system focus causes downstream problems in manufacturing. Assembly costs increase to resolve inter-subsystem conflicts. The overall financial performance of the project also deteriorates, since manufacturing budgets dominate design budgets. Here too, the ripple effect from β_{43} is strong. For a unit change in intensity of specialization, the marginal change in design savings and manufacturing savings is (β_{22}) and ($\beta_{22}{*}\beta_{43} + \beta_{42}$). Note that the impact on manufacturing savings comprises of the ripple effect of specialization via design savings ($\beta_{22}{*}\beta_{43}$), as well as the direct effect of specialization on manufacturing savings (β_{42}). As in the case of time performance, the marginal impact of specialization on the total project financial performance is adverse, as shown in the bottom graph of figure 6. This occurs because manufacturing budgets dominate. Similar to the dis-

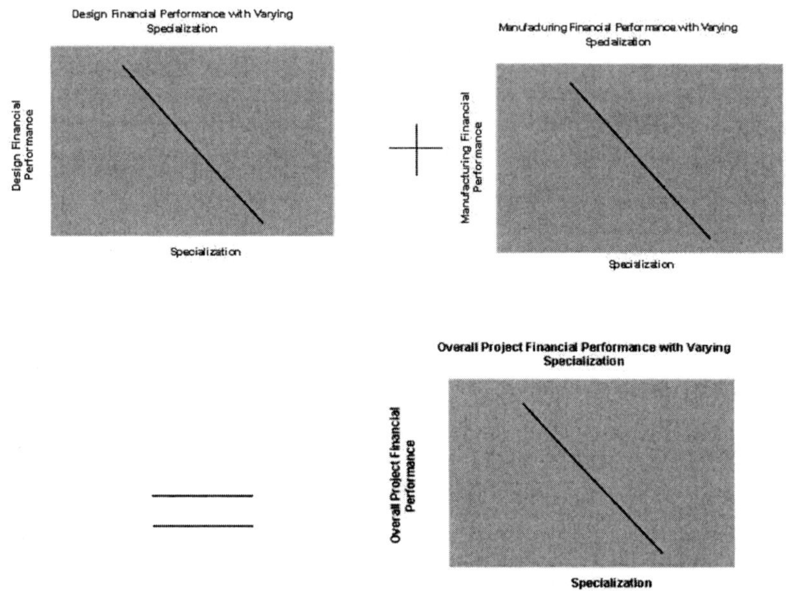

Figure 6. Overall project financial performance with varying specialization as aggregation of performance in design and manufacturing phases.

cussion for the time performance, the marginal impact of specialization would have been positive if design budgets had dominated manufacturing by a factor of

$$\frac{|(\beta_{22} \star \beta_{43} + \beta_{42})|}{\beta_{22}}$$

$= 7.91.$

4.2 Effect of increased oversight

Figure 7 shows the marginal effect of oversight on time performance in each phase as well as for the overall project. First consider the impact of higher oversight on time performance in the design phase. As depicted in figure 3, the marginal impact of oversight comprises of two effects: a direct effect (β_{11}) and an interaction effect with CUSTINT ($\beta_{14}\star$CUSTINT). The direct effect is negative, while the interaction effect is positive. The positive interaction term suggests that project managers can leverage the customer interaction to provide a systems approach and speed up the design phase. Iansiti (1998) characterizes this as the "T" role of project managers, who have both the depth and the breadth to bridge customer requirements to internal design capabilities. At the mean level of CUSTINT in our data set, the net effect is still negative, as shown by the top left hand graph in figure 7. This suggests that higher oversight

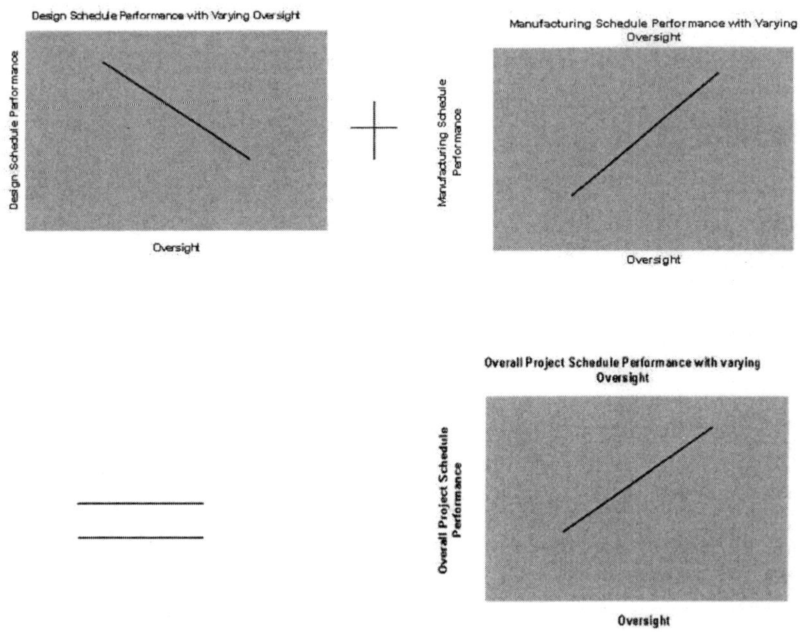

Figure 7. Overall project schedule performance with varying oversight as aggregation of performance in design and manufacturing phases.

leads to greater time slippage in design. This slippage in the design phase could be either due to extraneous activities such a presentations and meetings or due to activities that would be useful to downstream manufacturing. This ambiguity is resolved by examining the net impact of oversight on time performance in the subsequent manufacturing phase.

The top right hand graph of figure 7 confirms that higher oversight improves manufacturing time performance. This occurs because of the ripple effect of time performance in the design phase ($\beta_{33} < 0$). This shows that time slippages in the design phase are in reality an investment of time that accrue significant benefits later in manufacturing with the overall project schedule improving. This is shown in the bottom right hand graph of figure 7. Our findings strongly support the concepts behind DFM and reinforce the role of design to enhance project performance.

The effect of oversight on the financial performance in the design and manufacturing phase (see figure 8) portray trends similar to those shown in the time performance discussed earlier. We leave it to the reader to map these trends to the structural model shown in figure 4.

4.3 Combining the two levers

Our model leads to the formulation of the appropriate strategy for levers that have been identified in the design phase. This is depicted in figure 9. It shows that project

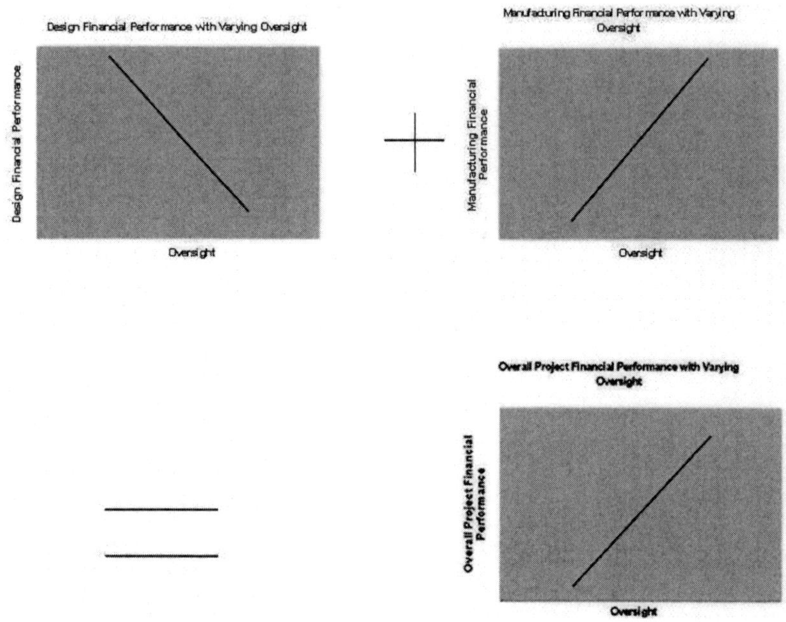

Figure 8. Overall project financial performance with varying oversight as aggregation of performance in design and manufacturing phases.

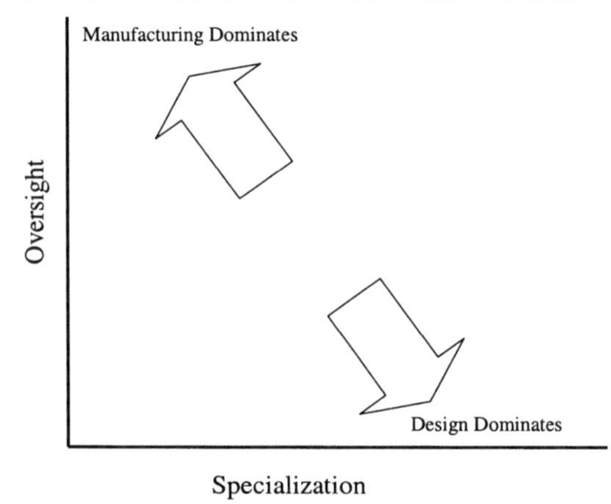

Figure 9. Strategies for deployment of managerial levers in NPD.

managers who are in an environment where manufacturing budgets dominate, should position their strategy in the top left hand region. They should select higher levels of oversight and lower levels of specialization in the design phase. Vice versa, if design were to dominate, higher specialization pays off together with lower levels of oversight. Increasing interaction with the customer is helpful, as seen from the interaction with specialization and oversight discussed earlier. Note however, that one has to weigh the benefits against the costs incurred. This suggests that the nature of interactions should be selective, so as to minimize low content exchanges with the customer, and strive to maximize the payoff from these interactions.

5. CONCLUSION

In this chapter, we have provided strong evidence for the levers that project managers can deploy in the design phase of NPD to improve time and financial performance. The three key levers we identified are the intensity of specialization in the design phase, the level of oversight by project managers in the design phase, and finally the degree of interaction with the customer. The choice of the setting of each lever depends on the relative resource budgets in design and manufacturing. In cases where manufacturing dominates, such as at our research site, our model indicates a strategy of higher levels of managerial oversight, lower levels of specialization and selective customer interaction. Key features of our model include: a) separation of the performance in the design and manufacturing phases, and b) three types of effects (direct, interaction and ripple). We also find strong support for the concepts underlying design for manufacturing and the systems integration role of project managers.

We believe that this stream of research holds a lot of promise to apply our methodology to other NPD environments. This will provide further evidence and generalization to better manage and co-ordinate NPD projects and this work represents an important initial step in this direction.

Acknowledgement

The authors acknowledge the support provided by the TIDE Manufacturing Program at Carnegie Mellon University, and the executives at Boeing and Missile Systems Inc. for their input during the research study. We also express our thanks to the editors and reviewers for their helpful comments and feedback.

REFERENCES

Adler, P. S., A. Mandelbaum, et al. (1995). "From Project to Process Management: An Empirically-based Framework for Analyzing Product Development Time." *Management Science* 41(3): 458–484.

Bajaj, A., Kekre, S., & Srinivasan, K. (2002). Managing NPD: Cost and Schedule Performance in Design and Manufacturing. *Working Paper: Carnegie Mellon University.*

Brown, S. L., & Eisenhardt, K. M. (1995). Product Development: Past Research, Present Findings, New Directions. *Academy of Management Review,* 20(2), 343–378.

Chakravarty, A. K. (2001). Overlapping design and build cycles in product development. *European Journal of Operations Research,* 134(2), 392–424.

Cohen, M. A., Eliashberg, J., & Ho, T. H. (2000). An Analysis of Several New Product Performance Metrics. *Manufacturing & Service Operations Management,* 2(4), 337–349.

Cooper, R. G. (1993). *Winning at New Products*. Reading, MA, USA, Addison-Wesley Publishing Co.
Eisenhardt, K. M. and B. Tabrizi (1995). "Accelerating Adaptive Processes-Product Innovation in the Global Computer Industry." *Administrative Science Quarterly* 40(March): 84–110.
Eppinger, S., Fine, C., & Ulrich, K. (1990). Interdisciplinary Product Design Education. *IEEE Transactions on Engineering Management* (November).
Eppinger, S. D. (2001). "Innovation at The Speed of Information" *Harvard Business Review* 79(1): 149–158.
Gupta, A. K. and W.E. Souder (1998). "Key Drivers of Reduced Cycle Time." *Research Technology Management* 41(4): 38–43.
Ha, A. Y. and E. L. Porteus (1995). "Optimal Timing of Reviews in Concurrent Design for Manufacturability" *Management Science,* 41(9): 1431–1440.
Hauser, J. R. and D. P. Clausing (1998). "The House of Quality." *Harvard Business Review* 66(3): 63–73.
Iansiti, M. (1998). *Technology Integration*. Cambridge, MA: Harvard Business Press.
Jaworski, B. J., V. Stathakopoulos, et al. (1993). "Control Combinations in Marketing: Conceptual Framework and Empirical Evidence." *Journal of Marketing* 57(1): 57–69.
Krishnan, V., & Ulrich, K. T. (2001). Product Development Decisions: A Review of the Literature. *Management Science,* 47(1), 1–21.
Nobeoka, K. and M. A. Cusamano (1997). "Multiproject Strategy and Sales Growth: The Benefits of Rapid Design Transfer in New Product Development." *Strategic Management Journal* 18(March): 169–186.
Novak, S., & Eppinger, S. D. (2001). Sourcing by Design: Product Complexity and the Supply Chain. *Management Science,* 47(1), 189–204.
Ouchi, W. G. and M. A. Maguire (1975). "Organizational Control: Two Functions." *Administrative Science Quarterly* 20 (December): 559–569.
Ouchi, W. G. (1979). "A Conceptual Framework for the Design of Organizational Control Mechanisms." *Management Science* 25(9): 833–847.
Pahl, G., & Beitz, W. (1988). *Engineering Design: A Systematic Approach*. New York: Springer.
Thomke, S. and A. Nimgade (2000). IDEO Product Development Case. Cambridge, MA, Harvard Business School.
Ulrich, K. T., & Ellison, D. E. (1999). Holistic Customer Requirements and the Design Select Decision. *Management Science,* 45(5), 641–658.
Wheelwright, S. and K. Clark (1992). *Revolutionizing Product Development*. New York, NY, The Free Press.

10. PLATFORM PRODUCTS: THEORY AND EVIDENCE FROM INTERNATIONAL PRACTICE LINKING MANUFACTURING, MARKETING, AND PRODUCT DESIGN

K. RAVI KUMAR and GEORGE C. HADJINICOLA

ABSTRACT

In this chapter, we model a firm with operations in two countries (termed as an international enterprise). The aim of the models is to capture the principles of platform products and their impact on profitability. Manufacturing and product design factors incorporated in the model include factory location, inventory, economies of scale, level of product attributes, and postponement. Marketing factors include product positioning and pricing. The model also includes international factors such as the exchange rate and transportation cost. We model and compare eight operational options that the firm may adopt in this context. The first two options are based on the principles of platform products and are termed core product options. In these options, the international enterprise manufactures a "core product" in a single facility located in one of the countries. Under the first core product option, this facility also performs additional operations on the core product to develop custom-tailored products. Under the second core product option, customization of the core product takes place in facilities located in both countries. The other six options are characterized by standardization/customization of product and pricing policies as well as centralization/decentralization of the production function. The main result of the analysis is that under certain production-related conditions, the core product options dominate the other six manufacturing-marketing options. This result has significant managerial implications since it suggests that platform products implemented by the core product options exploit the best of both worlds, customized pricing and product policies as well as savings derived from economies of scale from the centralized production of the core product.

1. INTRODUCTION

Designing and producing products that meet the needs of consumers in various markets or countries can be characterized as ***the*** activity that ensures the future survival and success of a firm. However, the process of designing and producing new products has become a major challenge to firms due to a number of complicating trends and factors, both external and internal to the firm. External factors and trends, among others, include the appearance of smaller market segments creating the need for mass customization (Pine, 1993; Pine et al., 1993), time-based competition stemming from shorter product life cycles, competitive pressures to introduce products in a shorter period of time (Stalk and Hout, 1990), and the cultural diversity of the workforce. Internal factors include the effort to overcome functional barriers and improve the communication mechanisms during the new product development process, reduction of the new product development costs, and reduction of the final cost of the product.

Managers developing new products for global markets are confronted with even more challenges since they have to consider other economic factors and make strategic choices. Economic factors include exchange rate fluctuations and the presence or absence of tariffs and trade barriers. A manufacturing-related strategic choice is the location of the production function, also referred to as the configuration and coordination of the production function (Porter, 1990; Ferdows, 1989; Huchzermeier and Cohen, 1996; Kogut, 1985; Kogut and Kulatikala, 1994; Shi and Gregory, 1998). This will ultimately impact the cost of the final product. A marketing-related strategic choice is the type of product policy that a firm should adopt, a problem that in international marketing is referred to as the classic debate of standardization versus customization.

The firm, is others words, must decide whether to offer a standardized or a customized product world-wide. Levitt (1983) states that due to the emergence of global consumers and the homogenization of preferences, a global corporation "sells the same things in the same way everywhere." Other researchers though, challenge this view by arguing that substantial heterogeneity across countries prevents the adoption of standardized marketing programs (Boddewyn et al., 1986; Douglas and Wind, 1987). In addition, the marketing function must decide whether to adopt standardized or customized marketing programs across countries such as pricing and promotional programs.

New product development is a firm activity that frequently leads to conflicts between the marketing and production functions (Shapiro, 1977). On one hand the production function of a firm would prefer to be more cost efficient by producing large production volumes of standardized products. On the other hand, the marketing function tries to satisfy customers by offering a broader product line and in some cases customized products. As such, firms try to achieve a balance between product standardization and cost efficiency with product customization at higher costs. The pressure to find the golden section of the above problem between marketing and production and the intensity of competition, has forced firms to rethink and change their practices when developing new products (Krishnan and Ulrich, 2001; Ulrich and Eppinger, 2000). Changes have been made on the following two dimensions of new product development: (1) the process of new product development and (2) the architecture and underlying philosophy of the design of tangible products.

The main characteristic of the traditional new product development process is that decisions by each business function regarding the new product have been made in a serial fashion, one business function passing its output to the next function. This serial/relay approach to new product development has been proven to be time-consuming and too rigid to react to quick market changes and trends. As such, this recognition further highlighted the need that the business functions of marketing, manufacturing, and engineering that are involved in the development of new products must be coordinated in a more effective way. The traditional process to new product development, an outcome of the departmentalization of the firm into its distinct functional areas, is being replaced by concurrent engineering. Under the concurrent engineering philosophy, advocating the cross-functional team approach, members of the various functions and sub-functions work together with the customer (or with customer service in focus) on all phases of product/service development and introduction, as well as on all phases of process improvement along the product/service life cycle (introduction, growth, maturity and obsolescence/rebirth), sharing information and participating. Turino (1992) provides a detailed discussion on concurrent engineering, ways to implement it as well as the experiences of companies that have adopted the method. Lately, the quest for innovative ways to make the new product development process more effective, capture the needs of customers, and reduce the time from concept to delivering the actual product, has gone beyond concurrent engineering (where the customer is involved and consulted during the development of the product) into the phase where the customer is given the tools to design his/her product (Thomke and von Hippel, 2002)

The effort to achieve a balance between product standardization and cost efficiency with product customization has also brought changes on the architecture and underlying philosophy of the design of tangible products with the emergence of platform products and the concept of postponement. Platform products share components, processes, knowledge, and people, in order to develop differentiated products efficiently (Robertson and Ulrich, 1998). While platform products are designed and produced on the philosophy of sharing the above assets, they are targeted at specific market segments. Robertson and Ulrich (1998) describe the response of Kodak to Fuji's introduction of the QuickSnap, a 35mm single-use camera, in the U.S. market. Even though Kodak introduced its product later than Fuji, it captured 70% of the U.S. market. This is attributed to Kodak's strategy of developing many distinctively different models from a common platform. Kodak designed its products to share components and process stages, leading to faster and cheaper new product development.

Postponement is a management approach where the point in time of differentiating a product is delayed. Van Hoek (p. 161, 2001) defines postponement as the "organizational concept whereby some of the activities in the supply chain are not performed until customer orders are received." To apply postponement in a firm, it is implied that its products must share a common unfinished product that is further customized to create the final products. This unfinished common product is basically designed under the principles of platform products. Hewlett-Packard (HP) is a company that has successfully employed the principles of postponement to its DeskJet printers (Lee et al., 1993).

In order to facilitate the analysis and discussion in this chapter, henceforth, we refer to an *International Enterprise* (IE) as a company that has operations (production, sales, etc) in more than one country (Prasad and Babbar, 2000). Depending on the nature of the activities of the IE, researchers have proposed various typologies to classify IEs. Bartlett and Ghoshal (p.15, 1989) for example, identify three broad classes of companies with operations in more than one country, namely, multinational, global, and international companies. According to their typology, multinational companies "build strong local presence through sensitivity and responsiveness to national differences", global companies "build cost advantages through centralized global-scale operations", and finally international companies "exploit parent company knowledge and capabilities through world-wide diffusion and adaptation." However, Barteltt and Ghoshal (1989) argue that the companies of the above typology adopt a unidimensional strategic capability that emphasizes only efficiency, or responsiveness, or leveraging of parent company knowledge and competencies. They suggest that in order for companies to be successful, they must evolve into a transnational company that pursues all three goals at the same time.

The purpose of this chapter is to build a modeling framework for IEs that adopt the concept of platform products by combining factors from manufacturing, marketing, and product design. The model will be used to assess the profitability of IEs that adopt platform products as compared to the profitability of IEs that adopt other product policies. Manufacturing factors include factory location, inventory, and economies of scale. Marketing factors include product positioning (relative to consumer preferences) and pricing while design factors include the level of the product's attributes and postponement. For the sake of brevity, the modeling framework deals with IEs operating in two countries; however, it considers international factors such as the exchange rate, transportation cost and import taxes.

The combination of the above factors creates ten options that an IE can adopt. These options are summarized in Table 1. Under the first two options, termed *core product options,* the IE manufactures a "core product" in one of the two countries. In the first core product option, the facility in this country performs additional operations on the core product to develop custom-tailored products that meet the needs of the local and foreign markets. The final products are also shipped to the other country. Under the second core product option, the core product is shipped to the production facility in the second country to be customized. Customization of the core product can be additional machining, assembly processes, or simple attachment of components. Apparently, the above two core product options are an implementation of the platform product approach. Typically, these two options benefit from the fact that tariff rates are different for semi-finished goods (lower) and fully finished goods (higher) since governments provide incentives to firms to add value locally.

The other eight options are characterized by the following three dimensions: (1) the IE offers a standardized or a customized product design; (2) the IE customizes its

Table 1. Options for international enterprises defined by manufacturing, marketing, and product achitecture elements

Manufacturing-marketing options	Product	Price	Manufacturing
1	Core product-customized finished products	Customized	Centralized production of core product-centralized customization of finished products
2	Core product-customized finished products	Customized	Centralized production of core product-decentralized customization of finished products
3	Customized finished products	Customized	Decentralized production of finished products
4	Customized finished products	Customized	Centralized production of finished products
5	Standardized finished products	Customized	Centralized production of finished product
6	Standardized finished product	Customized	Decentralized production of finished product
7	Standardized finished product	Standardized	Centralized production of finished product
8	Standardized finished product	Standardized	Decentralized production of finished product
9 (not valid)	Customized finished products	Standardized	Decentralized production of finished products
10 (not valid)	Customized finished products	Standardized	Centralized production of finished products

pricing policy or uses a uniform pricing policy; (3) the IE centralizes its production to a single facility in one country or decentralizes its production to facilities located in each country. The two options that deal with centralized or decentralized production and customized products that are sold at uniform prices are not considered as feasible. This is justified by the fact that customized products have different features leading to different production costs and eventually different pricing policies.

The main result of the analysis is that under certain conditions, the core product options dominate the other six options. If these conditions are not met, four options provide the highest profits for an IE. These options are the two core product options and the two options that offer customized products at customized prices, adopting either centralized or decentralized production. We also show that options adopting customized product and pricing policies yield equal or higher profits than options adopting standardized product and pricing policies, regardless of the structure of the production function. These results show to product designers/managers that the core product options exploit the best of both worlds, customized pricing and product policies as well as savings derived from economies of scale from the centralized production of the core product. As such, the modeling framework proves that firms adopting platform products enjoy superior profitability.

The organization of this chapter is as follows. Section 2 provides a description of global product policies and establishes a linkage of these policies with platform products. Section 3 contains the modeling framework whose manufacturing, marketing, and product design elements serve as the basis of the analysis. Section 4 includes descriptions and model formulations of the eight options an IE may adopt. Section 5 includes profitability comparisons of the options, numerical examples, and discussion. Finally, Section 6 presents concluding remarks and directions for future research.

2. PRODUCT POLICIES AND PLATFORM PRODUCTS FOR GLOBAL MARKETS

An IE can adopt a number of international product policies (Kumar and Hadjinicola, 2000). Some of these policies are closely related to platform products. In this section we describe the various types of international policies and establish a link between these policies and platform products.

Takeuchi and Porter (1986) identified three basic types of international product policies: the *universal, the country-tailored, and the modified.* Two of the above three product policies, the universal and the country-tailored product policies, represent the polar positions in the product standardization versus customization debate. The third policy, modified product, deals with products that are relatively similar in overseas markets, but minor adaptations are made to these products to meet basic market needs or customer idiosyncracies.

UNIVERSAL PRODUCT POLICY. The first international product policy identified by Takeuchi and Porter (1986) is the universal product policy. A universal product, also referred to as standardized product or uniform product, is physically identical in all overseas markets in which it is sold, with the exception of such elements as labeling and the language used in the manuals. Studies by Sorenson and Wiechmann (1975), and Hill and Still (1984) provide evidence that product policy is an area where IEs have the highest propensity for standardization. Products such as basic materials, components, high technology products, industrial products, steel, chemicals, plastics, ceramic castings used in memory chips, aircraft turbine engines, and cameras can be considered as universal products. Jain (1989) states that industrial and high technology products are more suitable for standardization than consumer products.

The primary benefits derived from product standardization are cost savings. Five main reasons contribute to these cost savings. First, offering a standardized product across countries implies higher production volumes that enable the IE to reduce the unit cost due to economies of scale. Fixed costs such as wages, insurance costs on the plant and equipment, property taxes, and interest costs are distributed among a larger number of products causing the unit cost decrease. Such economies of scale are referred to as short-term economies of scale (Hayes and Wheelright, 1984). Second, cost savings can be derived from the design of a single product instead of a number of products. Third, the production of a standardized product enables the IE to increase the volume of the purchased raw materials and components. This assists the IE to increase its bargaining power over its suppliers and eventually enjoy better purchasing contracts. Fourth, the production of a standardized product may allow the IE to centralize the production of the product to few specialized facilities. The construction of fewer but bigger facilities results in the cost savings.Lastly, promoting a standardized product across countries results in economies of scale in marketing. Advertising the standardized product requires the same type of campaign in each country so as to maintain the global image of the product. The commonality in the advertising campaigns results in cost savings. In general, savings from standardization allow the company to position itself as a low-cost producer.

The higher production volume of the standardized product, as discussed above, leads to low production costs that eventually ensure lower prices for the customers. Nevertheless, this benefit may come at the risk that the standardized product may not meet the needs of customers in a specific country. The universal product approach can result in overstandardization of a product, and in general, marketing programs that may prove to be catastrophic for some companies. Kashani (1989) describes the negative experiences of overstandardization in the Danish toy-maker Lego. Lego was marketing around the world its educational toys in the same fashion. In the United States though Lego faced fierce competition after Tyco, its major competitor, started marketing its toys in plastic buckets that could store the toys after each play. Lego's universal approach to packaging was the elegant transparent boxes. This approach was not appealing to American parents who preferred the functional packaging approach of toys in buckets. Lego's eroding market position alarmed the U.S. management but headquarters refused the idea of introducing toys in buckets, claming that this could harm Lego's reputation for high quality. Nevertheless, massive losses forced Lego to design its own bucket for storing toys. The result was a reversal of its eroding position.

Walters and Toyne (1989) describe three methods that an IE can follow to develop standardized products. The first method is *product extension* under which the product developed for a single market, usually the home market of the IE, is sold to foreign countries. Companies that adopt this method have an ethnocentric orientation such as companies that are at the first stages of their internationalization and have started exporting. Textile and fitted carpet manufacturers are examples of such companies. Fuji Xerox designed and targeted its high performance plain paper copiers exclusively for the Japanese market. Subsequent expansion to foreign markets was simply limited to exporting the copiers designed for the Japanese market. The probability of success increases when the foreign markets are "similar" to the home market of the enterprise.

The second method is referred to as *premium prototype*. The universal product developed for this method meets the needs of the most demanding customers whether the group is overseas or in the home market. The product is also designed to withstand the most severe product condition use. Such a high quality product can be appealing to customers in foreign markets, but it may often include superfluous design features that are not necessary for the average foreign customer. The high quality design and, perhaps, high production cost due to the enhanced nature of the product, may result in high prices that discourage a group of potential buyers.

The third method, known as global *common denominators,* strives to identify a global segment of international customers with homogeneous preferences and conditions of product use. After identifying such a segment, a product is developed to meet the common needs of customers in that segment. The difficulty with this approach is the recognition of global segments. Douglas and Wind (1987), for example, argue that substantial heterogeneity exists not only among customers across countries but also within countries, making the existence of a global segment almost utopic.The new hotel chain Courtyard by Marriott introduced by Marriott is an example of the effort a company to identify a global segment. Marriott tried to identify the attributes that travellers value most in a hotel in order to design the hotel of their preference.

COUNTRY-TAILORED PRODUCT POLICY. A country-tailored product is designed and produced to meet the idiosyncratic needs of the country (or group of countries) for which it is intended for. Czinkota and Ronkainen (1993) present various factors that encourage the design of products that are exclusively targeted for certain countries. Such factors include: (1) different conditions of use depicted by different climatic conditions, (2) government regulations that have be to met in each country, (3) different consumer behavior patterns, and (4) an attempt to meet local competition when competing firms customize their products to the local market. The primary benefit obtained from designing products specifically for certain countries or regions, is that this customized product fully meets the needs of local consumers. This leads to higher demand, higher market share thus establishing a stronger position of the IE in the host country, and eventually higher profitability.

The case of Fuji Xerox and its high performance plain paper copiers is an example of a product designed exclusively for the Japanese market. Bausch & Lomb developed 25 new products in 1986 with only one being custom-tailored for foreign markets. Since then, Baush & Lomb has changed its strategy. In 1991 more than half of its products were developed for foreign markets. In Europe, Bausch & Lomb's Ray-Ban glasses are made flashier, more avant-garde, and are costlier, in order to accommodate the high fashioned European consumer. Ray-Ban glasses in Asia have been redesigned to better suit the Asian face with its flatter bridge and higher cheekbones (Fortune Magazine, 1992, p.76). Nokia, forced by flattening sales and the need to satisfy different customer segments, has reorganized its structure into nine profit-and-loss centers. One of this centers, the mobile phones unit, is responsible for high-end GSM phones and their successors where as a second unit, the mobile entry products unit, is responsible to produce and market cheap phones for developing countries such as China, India, and Russia (Reinhardt, 2002).

Consider the case of Mars producing its traditional chocolate bar in Europe and the United States. The two chocolate bars, even though they have the same brand name, have different tastes, texture, and packaging in order to meet the consumers' preferences in these two regions. Colgate toothpaste, offers a more spicy toothpaste which has been exclusively designed for the Middle East market. In fact, many consumer products, especially food and household products, are country-tailored products because of their strong cultural dependency. Furthermore, the production costs of consumer-packaged goods tend to be relatively lower than marketing costs. The low production costs provides a disincentive for IEs to standardize their products since the potential benefits from cost reduction are not very appealing.

Designing a customized product that meets the consumer preferences in a specific country or region has the drawback that the price of the product may be higher. The higher price is caused primarily by two reasons. First, the design of localized products significantly affects the new product identification and development costs. Costs increase because market research has to be conducted in each country or region and in addition, more funds must be allocated to design more than one product. As a result, the customized product designed for a specific country has to absorb all the development cost. If the product was to be marketed in all countries this cost would have been distributed among the higher volume demanded by the various countries. Second, since customized products are sold in each country, the smaller production volumes may prevent the company from exploiting benefits derived from economies of scale.

Another drawback of the country-tailored product policy is that the logistic operations for distributing the customized products may prove inefficient. Take for example the experience of Hewlett-Packard and its DeskJet printers. Initially Hewlett-Packard was producing and packaging printers for the United States and European countries. Customization involved the installation of the appropriate power supply, software, and addition of manuals in the specific language. Inaccurate sales predictions left the distribution centers with inventories of printers for some countries, and shortages for others. The distribution centers were forced to rework the packages and printers to create the customized products for the countries that demand was not satisfied.

MODIFIED PRODUCT POLICY. A modified product is substantially similar in each country in which it is sold, but minor alterations/adaptations are made on the basic product to conform to local needs and regulations. Modified products are considered products whose adaptations represent a "modest" percentage of the total cost (Takeuchi and Porter, 1986). The modified product policy enables an IE to reap the benefits of standardization and centralized production, and at the same time, respond to local needs. Modified products include industrial products such as CT scanners, broadcast video cameras, mainframe computers, copiers, precision testing equipment, and construction equipment. In consumer goods, modified products include cars, motorcycles, calculators, and microwave ovens.

Product adaptations are classified as involuntary or voluntary. Involuntary adaptations are made to meet legal requirements of the foreign government on packaging, labeling, local metric system and also alignment with local climatic and economic conditions. Examples include the use, by copier manufacturers, of different paper trays for copiers in Japan, the United States, and Europe since all three regions use papers of different size. Similarly, car manufacturers must modify the speedometer on cars sold in the United States to show the speed of the car in miles per hour where as car speedometers in Europe must comply with the European Union's standards and provide the reading in kilometers per hour. Manufacturers of electronic appliances must install the proper power supply module in order to comply with the country's voltage requirements.

Voluntary adaptations are made as part of the company's marketing strategy to meet local needs and consumer idiosyncracies such as color, the product's size, and its accessories. For example, the color of the package especially on consumer products is very important since different colors are perceived differently across countries. Packages in the Middle East tend to include the green color that is preferred by consumers in this region. McDonalds, besides the standard products that it serves, tries to adapt to local needs by selling wine in France, vodka in Russia, beer in Germany, and special dishes during the fasting period in Greece. Besides the world famous sandwiches that it serves, Subway prepares sandwiches that contain cheeses and ham made in the country of operation. Similarly, Pizza Hut prepares pizzas not only with the toppings that have traditionally been used, but also uses toppings that include cheese and processed meat found in each country. As such, these two companies modify their overall products to meet local tastes. Heineken, the Dutch beer manufacturer, offers its beers in bottles of various sizes depending on the country of operation. Finally, an example of product modification involving car accessories is the cup-holder which is almost a standard feature in cars in the United States but not so commonly found (until recently) in cars destined for the European countries.

Two approaches that allow an international enterprise to apply a modified product policy are the modular approach, and core-product approach. Adoption of either approach permits the firm to reap the benefits of standardization, and at the same time, respond to local needs and differences in conditions of product use across countries (Walters and Toyne, 1989).

Modular Approach. The notion of modular design was pioneered by Starr (p.137, 1965) who suggested that products should be made of interchangeable modules to supply "consumers with apparent variety even though the production output is based on the concepts of mass production." The modular approach entails the design and production of a range of standard parts that can be assembled in a variety of combinations yielding a series of products. These products can be assembled from the standard components to meet, as closely as possible, the preferences of local consumers in a particular foreign market. IKEA, the Swedish furniture manufacturer, is an IE that has adopted the modular approach. Specifically, IKEA is selling furniture that can be assembled from a standard collection of pieces.

Cost savings can be achieved when the components are mass-produced in a single facility, and the number of variant product combinations is limited when the number of components is limited. Nevertheless, under the modular approach, one must consider the tradeoffs between the number of product variants that enhance the attractiveness of the product to local markets (due to customization) and the number of components produced that promote economies of scale. Apparently, such a solution resides in the specifics of each firm.

Core-Product Approach. The core-product approach is based on the same principles as the modular approach. Under this approach, a uniform "central/core" product is designed that can accept a number of standard attachments, parts or components. The combination of attachments to the core product allows it to meet performance criteria and local consumer preferences in a country. The difference with the modular approach is that the core product is of higher value relative to the total value of the finished product where as the components in the modular approach are of lower value relative to the total value of the finished product. As such, the core product represents a significant proportion of the total value of the finished product. Apparently, the core-product approach is an implementation of the platform product philosophy

The production of a standard core product enables a firm to reap the benefits of economies of scale and thus enjoy lower production costs. On the other hand, the flexibility allowed by the attachment of components to the core product, helps a firm localize its product and increase its attractiveness. Yip (1989) argues that "in a global strategy, the ideal is a standardized core product that requires minimal local adaptation." An additional advantage of the core product approach is that it facilitates delayed product differentiation which enables marketers to react to market demand and manufacturing managers to reduce cost due to the reduced value added to the inventory at hand.

An example of an IE adopting the core-product approach is a manufacturer of agricultural machinery. Across countries, the IE was confronted with differences in climatic conditions, topography, and customer needs. To overcome these differences, the IE produced a robust and standard product that could operate in a variety of conditions. This core-product was also able to accept a number of attachments to additional equipment that gave the basic product, different performance characteristics meeting local requirements and conditions of use (Walters and Toyne, 1989). In a similar fashion, car manufacturers employ the core-product approach by using a basic product, e.g. the chassis of the car, and then attach to it features in order to customize it for specific countries and even consumer needs. Such attachments include bigger engines, air conditioning units, electric windows, and other accessories. The core product is also referred to as platform product in the automobile industry or as a generic product in the electronics industry (Lee et al., 1993). McGrath and Hoole (1992) state that each new product development should revolve around the design of a core product, since manufacturing globally implies the economic integration or rationalization of all the steps of the production process.

Honda recognizes the need for localized products since customers in different countries use them in different ways. To suit the needs of their customers and conform to the socio-economic conditions of the diverse marketplace, Honda uses its common basic technology to develop different types of cars for different regions. In fact, the core product concept reduced the cost of Honda's products, which "cheaply built its new car lines from the platforms of existing cars" (Fortune Magazine, p.32, 1996). Yip (1989) describes the product modification made by Boeing on its 737 in order to revive leveling sales. Boeing recognized that its current design (early 1970) was not appealing to developing countries because the shortness and softness of their runways, combined with the lower technical skills of their pilots, made the planes bounce. Bouncing resulted in brake failure. Boeing modified its 737 design by adding thrust to the engines, redesigning the wings and landing gear, and installing lower pressure tires.

Lee et al. (1993) describe Hewlett-Packard (HP)'s adoption of the concept of generic/core product. Specifically, HP is currently applying the principles of postponement in product differentiation on its DeskJet printers. Postponing the point of product differentiation means delaying, as much as possible, the stage at which value is added to the work-in-process inventory in order to obtain the final product. Under this strategy, a generic/core product is manufactured at Vancouver, Canada, and Singapore that is shipped to the distribution centers in the U.S., Far East, and Europe. The distribution centers perform the product customization by inserting into the packaging boxes the language specific manuals, power supply modules, and software. HP has even adopted the concept of packaging postponement under which final packaging is performed at the distribution center and not at the manufacturing facilities. The above concept requires the design of a generic/core product that can accommodate and facilitate the customization process. Lee et al. (1993) refer to this product design as design for localization or design for customization

3. MODELING FRAMEWORK

3.1 Model preliminaries

Products can be represented as a set of coordinates in an attribute space where each dimension in the attribute space designates a product characteristic. The quantifiable attributes should be actionable, that is, "indicate specific actions the manufacturer must take to build the product" (p.922, Shocker and Shrinivasan, 1974). Under this framework, each consumer is assumed to have a set of most preferred attributes termed the *ideal point*. The greater the proximity of the product in the attribute space to the consumers' ideal point, the greater the product *attraction* (appeal), and in general the propensity/probability of consumers purchasing the product. The probability of purchase is also affected by the product's price. Corstjens and Gautschi (1983) provide a review on consumer buying behavior as related to deterministic and probabilistic choice models.

Using the above, we model competition between IEs through multiplicative competitive interaction market share models, also known as attraction models (Gruca et al., 1992). These models are *logically consistent* market share models because they predict market shares between zero and one, and since the sum of the market shares equals to one (McGuire et al., 1968). As described by Cooper and Nakanishi (1988), logically consistent market share models use the relationship (us/(us + them)) to capture the market share of a firm in a competitive environment; "us" represents the marketing effort of the firm and "them" represents the sum of the marketing efforts of all competing firms. Bell et al. (1975) consider the market share of a firm to be the ratio of the *attraction* that consumers feel towards a particular brand to the sum of the attractions of the brands of all competing firms.Specifically, we employ the model:

$$M_{kj} = Attr_{kj} / \sum_{K=1}^{N} Attr_{kj}, \; k = 1, \ldots, N, \qquad j = 1,2, \tag{1}$$

where M_{kj} is the market share and $Attr_{kj}$ is the product attraction of IE k in country j. Attraction can be viewed as a measure of the "willingness" of consumers to purchase a product. Following the studies by Cooper and Nakanishi (1988) and Gruca et al. (1992), we define the product attraction for IE k in country j as:

$$\begin{aligned} Attr_{kj} &= f_{kj}\,(\mathbf{x_{kj}})g_{kj}(P_{kj}) = [1 - (\mathbf{x_j^*} - \mathbf{x_{kj}})'\,\mathbf{W_j}\,(\mathbf{x_j^*} - \mathbf{x_{kj}})/\mathbf{x_j^*}\mathbf{W_j}\,\mathbf{x_j^*}] \\ &\quad \times [1 - (\,b_j P_{kj}/\,Y_j)], \end{aligned} \tag{2}$$

where $\mathbf{x_{kj}}$ is an $L \times 1$ vector containing the attribute coordinates (x_{kij}, $i = 1, \ldots, L$) of the product offered by IE k in country j, $\mathbf{x_j^*}$ is an $L \times 1$ vector containing the attribute coordinates (x_{ij}^*) of country j's ideal point. $\mathbf{W_j}$ is an $L \times L$ diagonal matrix whose diagonal elements (w_{ij}) represent the weights consumers in country j place on attribute i. The functional form of f_{kj} ($\mathbf{x_{kj}}$) $= 1 - [\sum_{i=1}^{L} w_{ij} (x_{ij}^* - x_{kj})^2 / \sum_{i=1}^{L} w_{ij} x_{ij}^{*2}]$ indicates that the smaller the distance to the ideal point is, the larger the value of the function is. This function follows from Shocker and Srinivasan (1974) who modeled the distance between the product offering and the ideal point as a weighted Euclidean distance.

The variable P_{kj} represents the price of the product offered by IE k in country j, b_j is the price elasticity of demand in country j, and Y_j is the consumers' average income in country j. The function $g_{kj}(P_{kj})$ is a down-sloping function of price and follows by linearizing Gruca's et al.(1992) nonlinear function $g(P) = (Y - P)^b$. We assume that $P_{kj} \leq Y_j / b_j$, where Y_j / b_j is country j's reservation price. In this model, we neglect the effects of advertising and distribution in order to simplify the analysis. Note that we make the simplifying assumption that each country has one ideal point that may represent the median or mode of the distribution of preferences. In the product positioning literature, the optimal product position is usually derived after considering ideal points for each and every consumer (Albers and Brockhoff, 1977). The above assumption is a limitation of this study. Nevertheless, this study is not aimed at proposing a new positioning model but rather use some of the existing concepts to facilitate the analysis. Nevertheless, studies such as Sudharshan et al.(1988) consider one or more ideal points for each market segment. Eliashberg and Manrai (1992) presented a model for the positioning of a new product in a single market using a single ideal point that was the median value of consumers' ideal points.

3.2 Average cost

We define the average cost of IE k in country j producing a product with attribute vector $\mathbf{x_{kj}}$ as:

$$AC_{kj} = \boldsymbol{\beta}'_{kj} \cdot \mathbf{x_{kj}} - h(\mathbf{x_{kj}})\, K_{kj}\, M_{kj}\, Q_j, \tag{3}$$

where $\boldsymbol{\beta}_{kj}$ is an $L \times 1$ vector containing the costs (β_{kij}) for furnishing a product with one unit of attribute i by IE k in country j. K_{kj} are the unit savings derived from economies of scale by IE k in country j, and Q_j is the market sales potential of country j. The function $h(\mathbf{x_{kj}})$ is an indicator function assuming the value of zero if $\mathbf{x_{kj}} = \mathbf{0}$ and one otherwise.

In this formulation, we assume that the cost for IE k to furnish a product with its attributes ($\mathbf{x_{kj}}$) in country j is given by $\boldsymbol{\beta}'_{kj} \cdot \mathbf{x_{kj}}$. As such, we assume that the enhanced product features require more effort to be furnished resulting in higher production cost. Note that we assume a linear relationship between the cost of furnishing a product with its attributes given in $\mathbf{x_{kj}}$, and the magnitude of these attributes. Bachem and Simon (1981), Choi et al. (1990), and Hadjinicola (1999) also present product positioning models that use cost functions that increase linearly with increasing attribute levels. The average cost function also includes the term $h(\mathbf{x_{kj}})\ K_{kj}\ M_{kj}\ Q_j$ which represents the savings obtained from economies of scale by producing a volume of $M_{kj}\ Q_j$ units. In general, learning curves in production planning models have been assumed to be nonlinear (Kroll and Kumar, 1989). In this formulation, in order to simplify the analysis, we use the above linear relationship to indicate the savings derived from economies of scale. The indicator function $h(\mathbf{x_{kj}})$ is used in order to capture the case where no product is offered, i.e. for $\mathbf{x_{kj}} = \mathbf{0}$, $h(\mathbf{0}) = 0$ implying no savings from economies of scale.

3.3 Inventory holding cost

In the formulation of the profit functions of the eight MM options, we also consider the inventory holding cost I_k for IE k. Our formulation deals with a single period and the inventory holding cost is considered, as in all Economic Order Quantity (EOQ) models, to be a function of the average inventory. Specifically, we treat $I_{average} = (I_{max} + I_{min})/2$. We assume that $I_{min} = 0$ and that I_{max} equals to the total demand of the period. The inventory holding cost I_k is also assumed to be a percentage of the cost of the item (Chase et al., 2001). As such, we define z_{kj} to be the percentage of the cost of the product for IE k in country j associated with the inventory holding cost. For example, if demand in country j for IE k is equal to $M_{kj}Q_j$ and the cost of the product is equal to $\boldsymbol{\beta}'_{kj} \cdot \mathbf{x_{kj}}$, then the inventory holding cost associated with country j is given by $I_k = (M_{kj}\ Q_j\ /2)\ z_{kj}\ (\boldsymbol{\beta}'_{kj} \cdot \mathbf{x_{kj}})$.

4. OPTIONS FOR INTERNATIONAL ENTERPRISES

In this section we present the models of the profit functions of eight options that an IE may adopt. The modeling framework uses factors from manufacturing, marketing, product design, as well as international economic factors. To facilitate the analysis, the modeling framework for all eight options deals with a two-country scenario.

4.1 Option 1: Centralized production of core product-centralized customization (core-centr)

Under the core-product approach, a uniform "central" product is designed that can accept a number of standard attachments, parts or components. The combination of attachments to the core product allows it to meet performance criteria and preferences of local consumers. Adaptations on the core product can vary from simple assembly operations to more complex machining operations. The core product is usually centrally produced in order to exploit savings from economies of scale. Additional operations on the core product may be performed in the same facility that the core product is produced, or in production/distribution facilities located in foreign markets. The core product is also referred to as platform product in the automobile industry (Robertson and Ulrich, 1998; Muffatto, 1999) or as a generic product in the electronics industry (Lee et al. 1993).

Umax, Microtek, Primax are three of the major Taiwanese companies producing scanners for the global market. In 2001, Taiwanese manufacturers controlled 91% of the global production volume share in the scanner industry. They produced for the likes of global brand names such as Hewlett-Packard, Agfa, Kodak, IBM, Epson, Canon, Ricoh, Nikon, Matsushita etc. More than 75% of these are produced in China with the remaining in Taiwan. The core product here is the opto-electronics and mechanical engine of scanners, with various scan configurations specified in terms of dots per inch (dpi) resolution and bits color quality. The customization is in terms of housing configuration and quality, software and controls features as well as packaging. These products are custom produced, mainly in China, for these original equipment manufacturing (OEM) customers, and delivered globally to customer-desired distribution points (Chang and Sauve, 2000).

Without the loss of generality, we assume that under the first option, the core product is produced in country 1. The facility in country 1 is further responsible for adapting the core product into customized products that meet the needs of consumers in countries 1 and 2. Following the scenario described above, the average cost of IE k in country 1 is given by:

$$AC_{kl} = \boldsymbol{\beta}'_{kl} \cdot \mathbf{x}^{c}_{k} - h\,(\mathbf{x}^{c}_{k})K_{k1}(M_{k1}Q_1 + M_{k2}Q_2) + \boldsymbol{\beta}^{c\prime}_{k1}(\mathbf{x}_{kl} - \mathbf{x}^{c}_{k}) - h(\mathbf{x}_{kl} - \mathbf{x}^{c}_{k})K^{c}_{k1}M_{k1}Q_1,$$

where $\mathbf{x}^{c}_{k}$ is an L × 1 vector containing the attribute coordinates (x^{c}_{ki}) of the core product of IE k, $\boldsymbol{\beta}^{c}_{kj}$ is an L × 1 vector denoting the unit costs (β^{c}_{kij}) for adapting the core product of IE k on each attribute dimension i in country j, and K^{c}_{kj} are the unit savings derived from economies of scale from the adaptation of the core product in country j by IE k. The first two terms of the average cost represent the cost for producing the core product, where as the second two terms represent the cost for customizing the core product into the final product that meets the needs of country 1. The function $h(\mathbf{x}_{k1} - \mathbf{x}^{c}_{k})$ assumes the value of zero if $\mathbf{x}_{k1} - \mathbf{x}^{c}_{k} = \mathbf{0}$ and one otherwise. In a similar fashion, country's 2 average cost is given by:

$$AC_{k2} = \boldsymbol{\beta}'_{k1} \cdot \mathbf{x}^{c}_{k} - h\,(\mathbf{x}^{c}_{k})K_{k1}(M_{k1}Q_1 + M_{k2}Q_2) + \boldsymbol{\beta}^{c\prime}_{k1}(\mathbf{x}_{k2} - \mathbf{x}^{c}_{k}) - h(\mathbf{x}_{k2} - \mathbf{x}^{c}_{k})K^{c}_{k1}M_{k1}Q_2,$$

Note that in the above definition of average costs, the cost of furnishing the attributes of the core product ($\boldsymbol{\beta}_{k1}$) is allowed to be different from the cost of furnishing the core product with customized features ($\boldsymbol{\beta}^{c}_{k1}$). In addition, the unit savings derived from economies of scale in the production of the core product (K_{k1}) is allowed to be different from the unit savings of the core product adaptation (K^{c}_{k1}). The incorporation of the above parameters enables us to capture the cost of customization across countries. To facilitate our modeling framework, we define the exchange rate between the two countries by r and the transportation cost and import taxes by c.

The profit function of IE k adopting the *Core-Centr* option is given by:

$$\begin{aligned}\Pi_k^{\text{Core-Centr}} = & \{P_{k1} - [\boldsymbol{\beta}'_{k1} \cdot \mathbf{x}_k^c - h(\mathbf{x}_k^c) K_{k1}(M_{k1}Q_1 + M_{k2}Q_2) \\ & + \boldsymbol{\beta}_{k1}^c \cdot (\mathbf{x}_{k1} - \mathbf{x}_k^c) - h(\mathbf{x}_{k1} - \mathbf{x}_k^c) K_{k1}^c M_{k1} Q_1]\} M_{k1} Q_1 \\ & + \{rP_{k2} - [\boldsymbol{\beta}'_{k1} \cdot \mathbf{x}_k^c - h(\mathbf{x}_k^c) K_{k1}(M_{k1}Q_1 + M_{k2}Q_2) \\ & + \boldsymbol{\beta}_{k1}^{c'} \cdot (\mathbf{x}_{k2} - \mathbf{x}_k^c) - h(\mathbf{x}_{k2} - \mathbf{x}_k^c) K_{k1}^c M_{k2} Q_2] - c\} M_{k2} Q_2 - I_k\end{aligned} \tag{4}$$

The exchange rate affects only the revenues obtained from country 2, since production of the core product and its adaptation to the customized product for country 2 is performed in country 1. Also note that since the final customized product for country 2 is shipped from country 1, the transportation cost and import taxes are subtracted from the profit margin for country 2. The inventory holding cost for IE k adopting the *Core-Centr* option is given by:

$$\begin{aligned}I_k = & [(M_{k1}Q_1 + M_{k2}Q_2)/2]\ z_{k1}\alpha_k(\boldsymbol{\beta}'_{k1} \cdot \mathbf{x}_k^c) \\ & + [(M_{k1}Q_1/2]\ [z_{k1}(\boldsymbol{\beta}'_{k1} \cdot \mathbf{x}_k^c + \boldsymbol{\beta}_{k1}^c \cdot (\mathbf{x}_{k1} - \mathbf{x}_k^c))](1 - \alpha_k h(\mathbf{x}_k^c)) \\ & + [(M_{k2}Q_2/2]\ [z_{k2}(\boldsymbol{\beta}'_{k1} \cdot \mathbf{x}_k^c + \boldsymbol{\beta}'_{k1} \cdot (\mathbf{x}_{k2} - \mathbf{x}_k^c))](1 - \alpha_k h(\mathbf{x}_k^c))\end{aligned} \tag{5}$$

The parameter α_k in equation (5) represents the proportion of the single time period that the core product is kept at the central facility where it is produced, before additional operations are performed for its customization into the final products (i.e. more value added to it). At this α_k proportion of the time period, we also assume that the customized final product is shipped to country 2. The first term of equation (5) represents the inventory holding cost for keeping an α_k proportion of the period an average inventory of $[(M_{k1}Q_1 + M_{k2}Q_2)/2]$ core products, whose unit cost is equal to $\boldsymbol{\beta}'_{k1} \cdot \mathbf{x}_k^c$. The second and third terms represent the inventory holding cost of the final products being kept in country 1 and 2, $(1 - \alpha_k)$ proportion of the time period. The term $[1 - \alpha_k h(\mathbf{x}_k^c)]$ in the inventory holding cost accounts for the case where no core product is produced.

Given the above, the profit maximization program of an IE adopting the *Core-Centr* option is given by max P_{k1}, P_{k2}, $\mathbf{x}_{k1}$, $\mathbf{x}_{k2}$, $\mathbf{x}_k^c$ $\Pi_k^{\text{Core-Centr}}$ s.t. $\mathbf{x}_k^c \leq \mathbf{x}_{kj}$, $j = 1,2$. The constraint in the program ensures that the attribute levels of the core product do not exceed those of the customized products in the two countries. While this enforces the core product to be a subset of the final customized product implying value-addition, it is quite possible to visualize value-deletion as a potential option, i.e. superfluous design, and then limit functionality as done in some electronics industries.

4.2 Option 2: Centralized production of core product-decentralized customization (core-decentr)

Under the *Core-Decentr* option, the core product is produced in country 1. The facility in country 1 is further responsible for adapting the core product into a customized product that meets the needs of consumers in country 1. In addition, the core product is shipped to the production facility of country 2 which adapts the core product into a customized product that meets the needs of consumers in this country. This option is adopted by HP for its Deskjets (Lee et al., 1993; Lee and Billington, 1994; Feitzinger and Lee, 1997). "By delaying product differentiation one delays as much as possible that moment when different product versions assume their unique identities thereby gaining the greatest possible flexibility to customer demands. Flexibility can improve the cost-effectiveness of the supply chain, because inventories are stocked in their predifferentiation form" (Lee and Billington, p.55, 1995).

First International Computers (FIC), with revenues of US$2.2 billion, is a major Taiwanese company that supports desktop production to order for Hewlett-Packard/Compaq. FIC supplies motherboards and assembles both business and consumer desktops for HP. Its shipments of Pavilion, Presario and Evo desktops will focus on the European and US markets. Pavilion and Presario are focused on the home and home-office markets while Evo is focused on the corporate segment. FIC has an R&D Center in Taiwan, an Engineering and Contract Manufacturing Center in Guangzhou, China, 4 supplier hubs and 8 final assembly sites all over the world to support its customers (Hua-hsin, 2002). Another company adopting this option example is Intel where semiconductor wafers for the Pentium products are produced mainly in the US and Europe but packaging and testing are done in cheaper places, including China.

Given the above description, the profit function of IE k adopting the *Core-Decentr* option is given by:

$$\begin{aligned}\Pi_{\mathrm{k}}^{\text{Core-Decentr}} = {} & \{P_{k1} - [\boldsymbol{\beta}'_{k1} \cdot \mathbf{x}_{\mathrm{k}}^{\mathrm{c}} - h(\mathbf{x}_{\mathrm{k}}^{\mathrm{c}})K_{k1}(M_{k1}Q_1 + M_{k2}Q_2) \\ & + \boldsymbol{\beta}_{k1}^{c\prime} \cdot (\mathbf{x}_{k1} - \mathbf{x}_{\mathrm{k}}^{\mathrm{c}}) - h(\mathbf{x}_{\mathrm{k1}} - \mathbf{x}_{\mathrm{k}}^{\mathrm{c}})K_{k1}^{c}M_{k1}Q_1]\}\ M_{k1}Q_1 \\ & + \{\mathrm{r}P_{\mathrm{k2}} - [\boldsymbol{\beta}'_{k1} \cdot \mathbf{x}_{\mathrm{k}}^{\mathrm{c}} - h(\mathbf{x}_{\mathrm{k}}^{\mathrm{c}})K_{k1}(M_{k1}Q_1 + M_{k2}Q_2) \\ & + \mathrm{r}(\boldsymbol{\beta}_{k2}^{c\prime} \cdot (\mathbf{x}_{\mathrm{k2}} - \mathbf{x}_{\mathrm{k}}^{\mathrm{c}}) - h(\mathbf{x}_{\mathrm{k2}} - \mathbf{x}_{\mathrm{k}}^{\mathrm{c}})K_{k2}^{c}M_{k2}Q_2]\} - h(\mathbf{x}_{\mathrm{k}}^{\mathrm{c}})\mathrm{c}\}\ M_{k2}Q_2 - \mathrm{I}_{\mathrm{k}}\end{aligned} \tag{6}$$

The exchange rate affects the revenues obtained from country 2 as well as the cost of furnishing the core product with additional features to obtain the customized product that meets the needs of consumers in country 2. Furthermore, the transportation cost and import taxes are multiplied by the function $h(\mathbf{x}_{\mathrm{k}}^{\mathrm{c}})$ in order to capture the case where no core product is shipped from country 1 to 2. This is the case where the facility in country 2 undertakes the complete production of the customized product for country 2.

Using the same concepts as in option 1, the inventory holding cost for option 2 is given by:

$$\begin{aligned}I_k = {} & [(M_{k1}Q_1 + M_{k2}Q_2)/2]\ z_{k1}\alpha_{k}(\boldsymbol{\beta}'_{k1} \cdot \mathbf{x}_{\mathrm{k}}^{\mathrm{c}}) \\ & + [(M_{k1}Q_1/2]\ [z_{k1}(\boldsymbol{\beta}'_{k1} \cdot \mathbf{x}_{\mathrm{k}}^{\mathrm{c}} + \boldsymbol{\beta}_{k1}^{c\prime} \cdot (\mathbf{x}_{\mathrm{k1}} - \mathbf{x}_{\mathrm{k}}^{\mathrm{c}}))](1 - \alpha_k h(\mathbf{x}_{\mathrm{k}}^{\mathrm{c}})) \\ & + [(M_{k2}Q_2/2]\ [z_{k2}(\boldsymbol{\beta}'_{k1} \cdot \mathbf{x}_{\mathrm{k}}^{\mathrm{c}} + r\,\boldsymbol{\beta}_{k2}^{c\prime} \cdot (\mathbf{x}_{\mathrm{k2}} - \mathbf{x}_{\mathrm{k}}^{\mathrm{c}}))](1 - \alpha_k h(\mathbf{x}_{\mathrm{k}}^{\mathrm{c}}))\end{aligned} \tag{7}$$

Note that the customization of the core product is now performed in the facilities in countries 1 and 2, after α_k proportion of the single time period has elapsed. Since customization takes place in the facility in country 2, the contribution to the inventory from country 2 is affected by the exchange rate as evident in the third term of equation (7). Given the above, the profit maximization program of IE *k* adopting the *Core-Decentr* option is given by $\max P_{k1}, P_{k2}, \mathbf{x}_{k1}, \mathbf{x}_{k2}, \mathbf{x}_k^c\ \Pi_k^{\text{Core-Decentr}}$ s.t. $\mathbf{x}_k^c \leq \mathbf{x}_{kj}$, j = 1,2.

4.3 Option 3: Decentralized production-customized product-customized prices (decentr-cust-cust)

The subsidiaries of an IE adopting this option are solely responsible for the production and development of the marketing programs for the country they serve. As such, products are customized to the needs of local consumers and their production takes place in facilities located in each country. Prices in each country are customized and are dependent on the product features and the income of the consumers in each country. This option is synonymous to a multinational corporation which follows a decentralized policy where local subsidiaries are given the autonomy to design and produce their own products as well as determine their pricing policies (Bartlett and Ghoshal, 1989). Following the characteristics of a multinational corporation as described by Hout et al.(1982) and Yip (1989), we assume that the responses of one subsidiary in its local market are independent of the responses of the other subsidiaries. Therefore, an IE *k* adopting the *Decentr-Cust-Cust* option has the following profit function:

$$\Pi_k^{\text{Decentr-Cust-Cust}} = [P_{k1} - (\boldsymbol{\beta}'_{k1} \cdot \mathbf{x}_{k1} - h(\mathbf{x}_{k1})K_{k1}(M_{k1}Q_1)]\ M_{k1}Q_1 + [rP_{k2} - [\boldsymbol{\beta}'_{k2} \cdot \mathbf{x}_{k2} - h(\mathbf{x}_{k2})K_{k2}M_{k2}Q_2)]\ M_{k2}Q_2 - I_k, \quad (8)$$

where

$$I_k = (M_{k1}Q_1/2)\ z_{k1}(\boldsymbol{\beta}'_{k1} \cdot \mathbf{x}_{k1}) + r(M_{k2}Q_2/2)\ z_{k2}(\boldsymbol{\beta}'_{k2} \cdot \mathbf{x}_{k2}). \quad (9)$$

Since production takes place in country 2 as well, the profits generated by the subsidiary in country 2 are affected by the exchange rate. The first and second terms of equation (9) represent the inventory holding cost associated with each country. Note that the inventory holding cost in country 2 is affected by the exchange rate since the customized product is produced in country 2. The profit maximization program of an IE *k* adopting the *Decentr-Cust-Cust* option is given by $\max P_{k1}, P_{k2}, \mathbf{x}_{k1}, \mathbf{x}_{k2}, \Pi_k^{\text{Decentr-Cust-Cust}}$. Due to the decentralized nature of the *Decentr-Cust-Cust* option, from equation (8) we see that the profit maximization program of an IE adopting this option degenerates to the profit maximization programs of each subsidiary.

A company that adopts this option is Emerson Electric Copeland Compressors. The company produces its scroll technology compressors in manufacturing plants in eleven countries in North America, Europe and Asia. Because of high volumes, the company is able to focus each of its production facilities on specific markets, achieving maximum productivity and cost effectiveness. The plants produce over 10 million units per year for both the air conditioning and commercial refrigeration market. The presence of the plants in major markets allows the firm to reduce its delivery times and satisfy specific local needs and requirements (e.g. voltage). Data Card is also a company adopting this option. The company produces credit cards for various financial institutions in plants located in North America, Europe, and Asia. Each plant serves customers in a specific region and all cards manufactured are customized to suit each customer's design.

4.4 Option 4: Centralized production-pustomized product-customized prices (centr-cust-cust)

Under this option, the IE manufactures customized products for both countries it serves in a central facility located in country 1. The product targeted for country 2 is shipped from the facility in country 1 to country 2. The IE also follows a price customization policy. The profit function of IE k adopting the *Centr-Cust-Cust* option is given by:

$$\Pi_k^{Centr\text{-}Cust\text{-}Cust} = [P_{k1} - (\boldsymbol{\beta}'_{k1} \cdot \mathbf{x_{k1}} - h(\mathbf{x_{k1}})K_{k1}(M_{k1}Q_1)]\ M_{k1}Q_1 + [rP_{k2} - [\boldsymbol{\beta}'_{k1} \cdot \mathbf{x_{k2}} - h(\mathbf{x_{k2}})K_{k2}(M_{k2}Q_2)]\ M_{k2}Q_2 - I_k, \quad (10)$$

where

$$I_k = (M_{k1}Q_1/2)\ z_{k1}(\boldsymbol{\beta}'_{k1} \cdot \boldsymbol{x_{k1}}) + (M_{k2}Q_2/2)(\boldsymbol{\beta}'_{k1} \cdot \boldsymbol{x_{k2}})[z_{k1}\alpha_k + z_{k2}(1 - \alpha_k)] \quad (11)$$

Note in equation (10), that under the *Centr-Cust-Cust* option, the exchange rate affects only the revenues obtained from country 2 since production of the customized products takes place in country 1. In addition, the profit margin for country 2 is decreased by the transportation cost and import taxes. The first term of equation (11) represents the inventory holding cost incurred by keeping an average inventory of $M_{k1}Q_1/2$ products of value $\boldsymbol{\beta}'_{k1} \cdot \boldsymbol{x_{k1}}$ in country 1. The second term in equation (11) represents the inventory holding cost incurred by keeping the customized product for country 2, α_k proportion of the time period in the facility in country 1 and $1 - \alpha_k$ proportion of the period in the facility in country 2. The profit maximization program of IE k adopting this option is given by $\max P_{k1}, P_{k2}, \mathbf{x_{k1}}, \mathbf{x_{k2}}, \Pi_k^{Centr\text{-}Cust\text{-}Cust}$.

Sony is an example of a company adopting this option. The company produces videorecorders centrally in Japan which are customized to meet various country specifications such as the need for NTSC or the PAL system, power supply etc. However, companies have started manufacturing superfluous products, that can function in various environments (NTSC or PAL, various voltage supplies) in order to minimize the number of different components required for the production of the products and standardize the production processes. Pohang Steel, a major Korean producer of standard and specialty steel products, also adopts this option. The company maintains centralized manufacturing plants in Pohang, Korea, and produces customized products that are sold at customized prices.

4.5 Option 5: Centralized production-standardized product-customized prices (centr-stand-cust)

Under the *Centr-Stand-Cust* option, IE k offers a standardized product across the two countries, denoted by $\boldsymbol{x}_k$ (its elements are denoted by x_{ki}). The standardized product is produced in a facility located in country 1 and then shipped to country 2. This option requires centralized decision-making since the IE needs to design and produce a product with such features that yield the maximum total profits. Centralized decision making may be viewed as a measure of coordination and tighter control. Prices of the standardized product are customized for each country. The profit function of IE k adopting this option is given by:

$$\begin{aligned}\Pi_k^{Centr\text{-}Stand\text{-}Cust} = {} & [P_{k1} - (\boldsymbol{\beta}'_{k1} \cdot \mathbf{x_k} - h(\mathbf{x_k})K_{k1}(M_{k1}Q_1 + M_{k2}Q_2))]\ M_{k1}Q_1 \\ & + [rP_{k2} - [\boldsymbol{\beta}'_{k2} \cdot \mathbf{x_k} - h(\mathbf{x_k})K_{k1}\ (M_{k1}Q_1 \\ & + M_{k2}Q_2)) - c]\ M_{k2}Q_2 - I_k,\end{aligned} \tag{12}$$

where

$$I_k = (M_{k1}Q_1/2)\ z_{k1}(\boldsymbol{\beta}'_{k1} \cdot \boldsymbol{x}_k) + (M_{k2}Q_2/2)(\boldsymbol{\beta}'_{k1} \cdot \boldsymbol{x}_k)[\ z_{k1}\alpha_k + z_{k2}(1 - \alpha_k)] \tag{13}$$

Note that by replacing $\mathbf{x_{kj}}$ with $\mathbf{x_k}$ in the inventory holding cost of the *Centr-Cust-Cust* option in equation (11), we obtain the inventory holding cost of the *Centr-Stand-Cust* option in equation (13). Since the standardized product is centrally produced in country 1 and then shipped to country 2, the average cost will be a function of the total volume produced for both countries. Since the product is manufactured in country 1, it follows that the parameters $\boldsymbol{\beta}'_{k1}$ and K_{k1} will determine the average cost of the standardized product. The average cost for country 2 is also burdened by the transportation cost and import taxes. In addition, the revenues obtained from country 2 are affected by the exchange rate. The profit maximization program of IE k adopting this option is given by $\max P_{k1}, P_{k2}, \mathbf{x_k}\ \Pi_k^{\text{Centr-Stand-Cust}}$

A firm that adopts the above option is Compal, a major Taiwanese company that makes laptops to stock for Dell. With factories in China and Taiwan, Compal supports Dell's 15″ notebook line with volumes of around 150,000-175,000 per month, estimated for 2002-3. Compal will produce these products at both its Taiwan plant but mainly at its expanding plant in Jiangsu Province, China. Interestingly, Dell intends to market the new model to both the consumer and corporate markets, contrary to its custom of differentiating PC offerings for the two markets. However, industry experts consider it a smart move, given the cost savings in design and other benefits from the economies of scale. (Tien and Teng, 2002; Tzeng and Hua-hsin, 2002).

4.6 Option 6: Decentralized production-standardized product-customized prices (decentr-stand-cust)

Under this option the IE offers a standardized product to the consumers of the two countries. The standardized product is manufactured in production facilities located in both countries. As a result, each production facility supplies the local market it serves with the standardized product. Prices of the standardized product are customized for each country. The profit function of IE k adopting the *Decentr-Stand-Cust* option is given by :

$$\Pi_k^{Decentr\text{-}Stand\text{-}Cust} = [P_{k1} - (\boldsymbol{\beta}'_{k1} \cdot \mathbf{x_k} - h(\mathbf{x_k})K_{k1}(M_{k1}Q_1)]\ M_{k1}Q_1 + [rP_{k2} - [\boldsymbol{\beta}'_{k2} \cdot \mathbf{x_k} - h(\mathbf{x_k})K_{k2}M_{k2}Q_2)]\ M_{k2}Q_2 - I_k, \quad (14)$$

where

$$I_k = (M_{k1}Q_1/2)\ z_{k1}(\boldsymbol{\beta}'_{k1} \cdot \boldsymbol{x_k}) + r(M_{k2}Q_2/2)\ z_{k2}\ (\boldsymbol{\beta}'_{k2} \cdot \boldsymbol{x_k}) \quad (15)$$

By replacing $\boldsymbol{x_{kj}}$ with $\boldsymbol{x_k}$ in the inventory holding cost of the *Decentr-Cust-Cust* option in equation (9), we obtain the inventory holding cost of the *Decentr-Stand-Cust* option in equation (15). From equation (14) we also see that the exchange rate affects the profits obtained in country 2. The profit maximization program of IE k adopting this option is given by $\max P_{k1}, P_{k2}, \mathbf{x_k}\ \Pi_k^{\text{Decentr-Stand-Cust}}$. Gillette, the manufacturer of the Mach 3 razors and Papermate pens, adopts such an option. The company maintains manufacturing facilities around the world that produce these standardized products. The products are sold at customized prices for each country/local area. Similarly, Oral B produces its standard toothbrushes in decentralized facilities and prices these products in a customized way.

4.7 Option 7: Centralized production-standardized product-standardized prices (centr-stand-stand)

One of the issues that arises in cases where the IE offers a standardized product is the standardization of prices across countries. Standardization of prices in the case where the IE offers dissimilar products across countries is not pursued since differences in product features result in differences in production costs and eventually prices. Kasa (p.1, 1992) states that "one of the cornerstones on international economic modelling is the notion of the "law of one price", which states that the same good selling in two different countries should sell for the same price when expressed in common currency units." In fact, equalization of prices across countries is aimed at preventing arbitrage opportunities. Sorenson and Wiechmann (p.42, 1975) report that companies "while realizing that actual retail prices for their products have to vary because of differences in local manufacturing and market situations, try to standardize at least the relative price from country to country."

The *Centr-Stand-Cust* option, previously described, can adopt the standardized pricing policy. If product prices in both countries are equal when translated in a common currency, it is implied that $P_{k1} = rP_{k2} = P_k$. Using equation (12), the profit function of IE k adopting the *Centr-Stand-Stand* option is given by:

$$\Pi_k^{Centr\text{-}Stand\text{-}Stand} = [P_k - (\boldsymbol{\beta}'_{k1} \cdot \mathbf{x_k} - h(\mathbf{x_k})K_{k1}(M_{k1}Q_1 + M_{k2}Q_2))]\ M_{k1}Q_1 + [P_{k2} - [\boldsymbol{\beta}'_{k2} \cdot \mathbf{x_k} - h(\mathbf{x_k})K_{k1}\ (M_{k1}Q_1 + M_{k2}Q_2)) - c]\ M_{k2}Q_2 - I_k. \quad (16)$$

The inventory holding cost of the *Centr-Stand-Stand* option has the same functional form as the inventory holding cost of the *Centr-Stand-Cust* option given in equation (13). The profit maximization program of IE *k* adopting the *Decentr-Stand-Stand* option is given by $\max P_k, \mathbf{x_k}\ \Pi_k^{\text{Centr-Stand-Stand}}$.

IBM is a company that follows the option of centralized production of a standardized product that is sold at standardized prices for its Z-series mainframe/supercomputers. These products are almost monopoly products, with standardized specifications resulting in standardized prices. The small production volumes force IBM to produce these products in a centralized facility. Rolex can also be considered as a company adopting this option; it produces its standardized products in centralized facilities and provides suggested retail prices to all its authorized dealers.

4.8 Option 8: Decentralized production-standardized product-standardized prices (decentr-stand-stand)

An IE which adopts the *Decentr-Stand-Cust* option may also employ a standardized pricing policy by setting $P_{k1} = rP_{k2} = P_k$ in equation (14). Therefore, the profit function of IE *k* adopting the *Decentr-Stand-Stand* option is given by:

$$\begin{aligned}\Pi_k^{\textit{Decentr-Stand-Stand}} = {} & [P_k - (\boldsymbol{\beta}'_{k1} \cdot \mathbf{x_k} - h(\mathbf{x_k})K_{k1}(M_{k1}Q_1)]\ M_{k1}Q_1 \\ & + [P_k - r[\boldsymbol{\beta}'_{k2} \cdot \mathbf{x}_k - h(\mathbf{x}_k)K_{k2}M_{k2}Q_2)]\ M_{k2}Q_2 - I_k\end{aligned} \tag{17}$$

The inventory holding cost of the *Decentr-Stand-Stand* option has the same functional form as the inventory holding cost of the *Decentr-Stand-Cust* option given in equation (15). The profit maximization program of IE *k* adopting the *Decentr-Stand-Stand* option is given by $\max P_k, x_k\ \Pi_k^{\text{Decentr-Stand-Stand}}$.

Examples of companies following this option are companies dealing with commodities. Oil and cement companies are classic examples since their product is fairly standardized, with the prohibitive transportation cost forcing the companies to locate facilities around the globe. The price of these commodity-type products is also standardized since their price is affected by international supply and demand.

5. ANALYSIS AND DISCUSSION

5.1 Profitability comparison of the international options

This section focuses on the profitability comparison of the eight options presented in this paper. This comparison can help an IE to select the optimal option. The proposition that follows provides insights as to which option an IE should adopt to obtain maximum profits.

PROPOSITION 1 *The following hold true for any IE k:*

(1) $\Pi_k^{\text{Core-Decentr}} \geq \Pi_k^{\text{Centr-Stand-Cust}} \geq \Pi_k^{\text{Centr-Stand-Stand}}$,

(2.1) $\Pi_k^{\text{Core-Decentr}} \geq \Pi_k^{\text{Decentr-Cust-Cust}}$, if $\beta_{k1} = \beta^c_{k1}$, $\beta_{k2} = \beta^c_{k2}$, $K_{k1} = K^c_{k1}$, $K_{k2} = K^c_{k2}$

(2.2) $\Pi_k^{\text{Decentr-Cust-Cust}} \geq \Pi_k^{\text{Decentr- Stand-Cust}} \geq \Pi_k^{\text{Decentr-Stand-Stand}}$,

(3) $\Pi_k^{\text{Core-Centr}} \geq \Pi_k^{\text{Centr-Cust-Cust}}$, if $\beta_{k1} = \beta^c_{k1}$, $K_{k1} = K^c_{k1}$, $z_{k1} = z_{k2}$

Proof of Proposition 1: See Appendix.

From parts (1) and (2.2) of Proposition 1,we see that $\Pi_k^{\text{Centr-Stand-Cust}} \geq \Pi_k^{\text{Centr-Stand-Stand}}$ and $\Pi_k^{\text{Decentr-Stand-Cust}} \geq \Pi_k^{\text{Decentr- Stand- Stand}}$. This shows that if an IE adopts the option of uniform pricing its standardized product, regardless of its production configuration, it would generate equal or less profits than when customizing the price of the standardized product across the two countries. The intuitive reason for this phenomenon is that price customisation enables the IE to exploit differences in the reservation prices and price sensitivity of consumers across countries. As such, the IE increases its profitability by exploiting local consumers' buying capability to the maximum possible degree.

From part (2.2) of the proposition we also see that $\Pi_k^{\text{Decentr-Cust-Cust}} \geq \Pi_k^{\text{Decentr- Stand-Cust}}$. The explanation for this result lies in two facts. First, decentralized production of a standardized product does not yield the maximum cost savings from economies of scale. Second, customized products, in any type of production environment, enhance the propensity of consumers to purchase the product, which further results in improved revenues. As a result, under the *Decentr-Cust-Cust*, the IE produces customized products in facilities located in the two countries. If the benefits derived from the improved sales of the customized products do not exceed the costs of customization, then the IE will be forced to decentrally produce a standardized product, thus adopting the *Decentr-Stand-Cust* option. Note that the above comparison of the two options does not take into consideration the costs of designing the products. For example, options offering a standardized product may benefit from economies of design and thus allow for lower product prices. On the other hand, options offering customized products may be burdened by the cost of designing more products and the amortization of such cost may result in higher product prices.

Part (1) of the proposition shows that $\Pi_k^{\text{Core-Decentr}} \geq \Pi_k^{\text{Centr-Stand-Cust}}$. The key to understanding why the *Core-Decentr* option may yield equal or more profits than the *Centr-Stand-Cust* option, is the fact that the *Centr-Stand-Cust* option can be seen as a special case of the *Core-Decentr* option. We know that the core product's attribute levels are less than or equal to those of the customized products offered in the two countries. At the extreme case, the core product can be the product offered in one of the countries. Assume, without the loss of generality, that the core product is the same as the product offered in country 1, the country where it is manufactured. Then, this core product is shipped to country 2 to be custom-tailored. If the benefits derived from customizing the product in country 2 are less than the cost of customization, then the optimal product policy is to market the core product itself in country 2 which is equivalent to the *Centr-Stand-Cust* option. In other words, the minimum profits that the *Core-Decentr* option can generate are those of the *Centr-Stand-Cust* option.

From part (3) of the proposition we see that $\Pi_k^{\text{Core-Centr}} \geq \Pi_k^{\text{Centr-Cust-Cust}}$. The intuition behind this result lies in the fact that the differentiating factor among the two options is the production of the core product since both options offer customized products at customized prices. Both options customize their products in country 1 and subsequently, the customized product for country 2 is shipped to this country from country 1. Now, under the *Core-Centr* option, if the benefits derived from the production of the core product are not significant enough due to high production costs, the IE will not produce the core product but it will simply produce centrally the two customized products. This of course is equivalent to adopting the *Centr-Cust-Cust* option. The parametric conditions under which the above result holds, simply state that the production related parameters for producing the customized products in country 1 and the respective parameters for producing the core product must be the same. We argue that this condition is highly likely to be valid since producing the core product or customized products in a single plant will dictate the use of the same technology for all production processes and hence the cost conditions indicated in part (3). If the parametric conditions are not met, then cases exist where the savings from centrally producing the core product will be outperformed from the direct production of customized products.

Part (2.1) shows that $\Pi_k^{\text{Core-Decentr}} \geq \Pi_k^{\text{Decentr-Cust-Cust}}$ under certain production-related conditions. The intuition behind this result is that at the extreme case, the *Core-Decentr* option will not ship from country 1 to country 2 any form of core product for further customization, i.e., the core product is vacuous. This may occur in environments where the savings derived from the centralized production of the core product are not sufficient to compensate for high transportation costs and import taxes of the core product. In this case, the subsidiary in country 2 will have to produce the customized product for country 2 from scratch since there is no core product. This is equivalent to the *Decentr-Cust-Cust* option. In other words, the minimum profits that the *Core-Decentr* option can generate are those of the *Decentr-Cust-Cust* option. Similar arguments for the parametric conditions can be used as above.

The results of Proposition 1 have significant implications on the choice of options adopted by an IE. The proposition for example, indicates that, under certain production-related conditions, two of the eight options considered generate higher or equal profits than the other six options, regardless of the number and nature of competitors. If the conditions are not met, the proposition indicates that the IE will have to structure itself in accordance to one of the following four options: *Core-Centr, Core-Decentr, Decentr-Cust-Cust, Centr-Cust-Cust.* Note that the above four "dominant" options deal with customized product and pricing policies. The choice will depend on the profitability comparison of the above four options in the specific environment that the IE operates.

In the analysis presented in this paper, we have assumed that the transportation cost and import taxes are the same whether the IE exports the core product or finished products. This may not be true since governments provide incentives to IEs to manufacture their products locally by reducing import taxes on raw materials and semi-finished goods. As a result, in such environments, the core product strategies may yield higher profits. In addition, the issue of the cost of product development is critical for an IE's profitability and must be taken into consideration when adopting an option. For example, customized product policies may require higher research and development costs than standardized product policies.

5.2 Numerical examples

To demonstrate the results of the comparison of the options, we present numerical examples with two competing IEs. Since the profitability comparison of the options is performed in a competitive environment, we need to compute the Nash equilibrium in prices and product positions of the various competing IEs. The computation of the Nash equilibrium in prices and product positions was achieved through the solution of a system of nonlinear equations composed of the first order conditions of the profit functions of the two IEs. The solution was obtained through a Fortran program that used a subroutine from the International Mathematics and Statistics Library (IMSL) employing the finite differences method by Levenberg-Marquadt which numerically computes the Jacobian of a system of equations.

Proposition 1 pruned the number of the options that IEs can adopt to two (when the conditions are met) and four options (when the conditions are not met). We illustrate the computation of the Nash equilibrium for both the two-option case and the four-option case when there are two competing IEs which have the same firm-specific parameters. When the competing IEs have the same parameters, the Nash equilibrium point in option, prices, and product positions will correspond to the case where both IEs adopt the same option. Given the discrete choice of the options, the game is presented in matrix form. Furthermore, in the numerical example that follows, the two IEs offer products with a single attribute.

Consider first the case where the parametric conditions presented in Proposition 1 hold and the two competing IEs have exactly the same parameter values. More specifically, we use the following set of parameter values:

$c_1 = 1$	$Q_1 = 300$	$Y_1 = 100$	$b_1 = 1.6$	$w_{11} = 1$	$x_1^* = 5.5$	$\alpha_1 = 0.5$
$c_2 = 3$	$Q_2 = 250$	$Y_2 = 110$	$b_2 = 1.5$	$w_{12} = 1$	$x_2^* = 6$	$\alpha_2 = 0.5$
$K_{11} = 0.02$	$K_{11}^c = 0.02$	$K_{12} = 0.02$	$K_{12}^c = 0.02$	$K_{21} = 0.03$	$K_{21}^c = 0.03$	$K_{22} = 0.03$
$\beta_{11} = 2$	$\beta_{11}^c = 2$	$\beta_{12} = 3$	$\beta_{12}^c = 3$	$\beta_{21} = 2$	$\beta_{21}^c = 2$	$\beta_{22} = 3.0$
$K_{22}^c = 0.03$	$\beta_{22}^c = 2.9$	$z_{11} = 0.1$	$z_{21} = 0.1$	$z_{12} = 0.1$	$z_{22} = 0.1$	$r = 1$

Table 2 shows the profits of the two competing IEs when both adopt one of the following options: *Core-Centr, Core-Decentr, Decentr-Cust-Cust,* and *Centr-Cust-Cust,* for two values of the transportation cost and import taxes ($c_k = 1$ and $c_k = 7$). For $c_k = 1$, the IEs generate the highest profits when they both adopt the *Core-Decentr* option. Therefore, for this set of parameter values, the adoption of the *Core-Decentr* option by both IEs represents the Nash equilibrium in option, prices, and product positions. If the IEs select any other option, this would lead to lower profits. For $c_k = 1$ we also observe that the profits generated by the *Core-Decentr* option are higher than the case where the IEs adopt the *Decentr-Cust-Cust* option, a direct result of Proposition 1 when the parametric conditions hold. Furthermore, for $c_k = 1$ we observe that the *Core-Centr* option generates the same profits as the *Centr-Cust-Cust* option. This demonstrates the fact in Proposition 1 that if the parametric conditions hold, the minimum profits that the *Core-Centr* option yields are those of the *Centr-Cust-Cust* option.

Table 2. Profits of four manufacturing–marketing options, including the two dominant core product options, for the case where the parametric conditions hold. The two international enterprises have the same firm-specific parameters. Values not underlined represent the case where $\mathbf{c_k = 1}$ and underlined values the case where $\mathbf{c_k = 7}$.

		IE2			
		Option 1 (*Core-Centr*)	**Option 2** (*Core-Decentr*)	**Option 3** (*Decentr-Cust-Cust*)	**Option 4** (*Centr-Cust-Cust*)
IE 1	**Option 1** (*Core-Centr*)	$\Pi_1 = \Pi_2 = 17863.53$* $\underline{\Pi 1 = \Pi 2 = 17556.21}$			
	Option 2 (*Core-Decentr*)		$\Pi_1 = \Pi_2 = 18223.05$ $\underline{\Pi_1 = \Pi_2 = 17775.56}$**		
	Option 3 (*Decentr-Cust-Cust*)			$\Pi_1 = \Pi_2 = 17556.56$ $\underline{\Pi_1 = \Pi_2 = 17556.56}$	
	Option 4 (*Centr-Cust-Cust*)				$\Pi_1 = \Pi_2 = 17863.53$ $\underline{\Pi_1 = \Pi_2 = 17384.62}$

*The *Core-Centr* Option degenerates to the *Centr-Cust-Cust* Option

**The *Core-Decentr* Option degenerates to the *Decentr-Cust-Cust* Option

From Table 2 we see that for $c_k = 7$, implying an environment of higher transportation cost and import taxes, the profits of the *Core-Decentr* option are reduced to such levels that the option degenerates to the *Decentr-Cust-Cust* option, a direct result of Proposition 1. The high transportation cost and import taxes force the IEs to stop shipping the core product from country 1 to country 2 and start producing customized products in each country. As such, for $c_k = 7$, the Nash equilibrium for the two IEs is for both to adopt either the *Core-Decentr* option of the *Decentr-Cust-Cust* option. Note also that for $c_k = 7$ the profits of the *Core-Centr* option are higher than the profits generated by the *Centr-Cust-Cust* option, where as for $c_k = 1$ the two options generated the same profits. Apparently, the high transportation cost and import taxes force the IEs to sustain their profitability by exploiting the benefits from centrally producing the core product. In a related study, Hadjinicola and Kumar (1997) show that an IE should change its production configuration from decentralized to centralized in environments of high tariffs and its product policies from standardized policies to customized policies when the savings from economies of scale are high and the exchange rate increases.

Consider now the case where the parametric conditions in Proposition 1 do not hold and the two competing IEs have again the same parameter values. Specifically we use the market parameter values presented earlier, except for $K_{k1} = K_{k2} = K^c_{k1} = K^c_{k2} = 0.02$, $\beta_{k1} = 2$, $\beta^c_{k1} = 2.1$, $\beta_{k2} = 3$, and $\beta^c_{k2} = 3.1$. Since the conditions in Proposition 1 do not hold, all four of the above options need to be considered. Table 3 includes the profits of two competing IEs when both adopt one of the four options described above, for $c_k = 1$ and $c_k = 7$. We can clearly see that for $c_k = 1$, both IEs are better off adopting the *Core-Decentr* option, whereas for $c_k = 7$ are better off adopting the *Decentr-Cust-Cust* option.

Finally, to examine the impact of postponement on the profits and the inventory holding cost, we perform a sensitivity analysis on the parameter α_k. Specifically, all parameters assume the values presented at the beginning of this section except α_1 which is allowed to vary. Figure 1 presents the profits and inventory holding cost of two IEs adopting the *Core-Centr* option. The figure shows that as the value of α_1 increases, implying that the first IE holds the core product for a longer period of time without performing the customization, its inventory holding cost decreases. This result demonstrates the fact that for this set of parameter values, postponement can reduce the inventory holding cost. It is interesting to note that as the first IE delays the customisation of the core product, the competing IE also reduces its inventory holding cost. This reduction stems from the fact that the competing IE tries to remain cost competitive by reducing the attribute level of its product, thus its inventory holding cost, since the latter is a function of the cost of the product. Profits for the IE, which postpones the customization, increase where as profits of the competing IE decrease. Similar results are obtained when both IEs adopt the *Core-Decentr* option. Postponement is closely related to parts commonality. According to the square root law (see Chopra and Meindl, 2001 for an extensive discussion), when the number of distinct components used for the production of the final product is reduced by a factor of n, the safety stock required to achieve the desired service level reduces by a factor equal

Table 3. Profits of four Profits of four manufacturing–marketing options for the case where the parametric conditions do not hold. The two international enterprises have the same firm-specific parameters. Values not underlined represent the case where $\mathbf{c_k = 1}$ and underlined values the case where $\mathbf{c_k = 7}$.

		IE 2			
		Option 1 *(Core-Centr)*	**Option 2** *(Core-Centr)*	**Option 3** *(Decentr-Cust-Cust)*	**Option 4** *(Centr-Cust-Cust)*
IE 1	**Option 1** *(Core-Centr)*	$\mathbf{\Pi_1 = \Pi_2 = 17057.95}$ $\underline{\mathbf{\Pi_1 = \Pi_2 = 17575.61}}$			
	Option 2 *(Core-Decentr)*		$\mathbf{\Pi_1 = \Pi_2 = 18241.66}$ $\underline{\mathbf{\Pi_1 = \Pi_2 = 17762.82}}$		
	Option 3 *(Decentr-Cust-Cust)*			$\mathbf{\Pi_1 = \Pi_2 = 17775.56}$ $\underline{\mathbf{\Pi_1 = \Pi_2 = 17775.56}}$	
	Option 4 *(Centr-Cust-Cust)*				$\mathbf{\Pi_1 = \Pi_2 = 17863.53}$ $\underline{\mathbf{\Pi_1 = \Pi_2 = 17384.62}}$

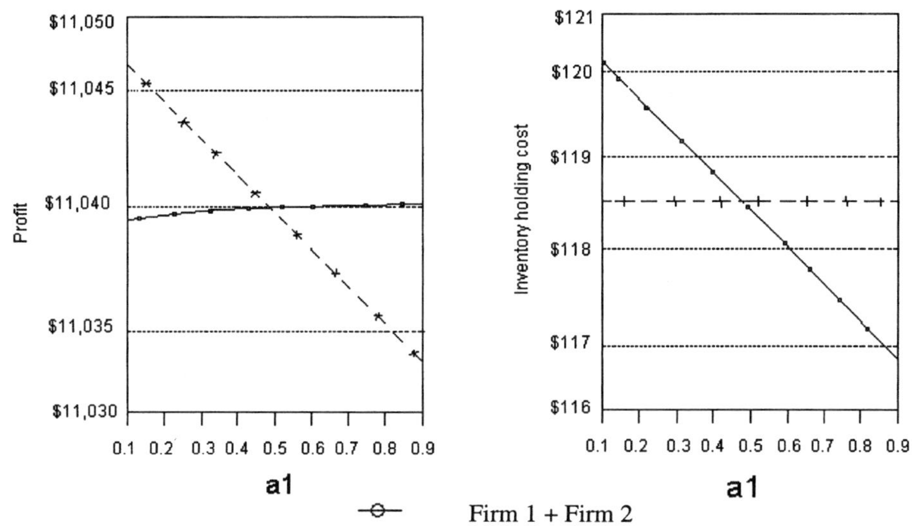

Figure 1. Sensitivity analysis on α_1, the time that the first **IE** performs the customisation on its core product. Both IEs adopt the *Core-Centr* option.

to the square root of n. This, of course, results in the reduction of the inventory holding cost.

6. CONCLUSION

Companies with operations in more than one country, termed International Enterprises, have to make a number of decisions concerning their manufacturing and marketing functions. Manufacturing and product design-related decisions include factory location, inventory, economies of scale, product design, and postponement. Marketing-related decisions include whether the enterprise should customize or standardize its product and pricing policies. In addition, the international enterprise needs to consider such international factors as exchange rate, transportation cost and import taxes.

In this chapter, we combine all of the above factors in a single model in order to examine the profitability of eight options an international enterprise can adopt. The analysis is performed in a competitive environment for a two-country scenario. We identify the options that if adopted by IEs will result to higher profitability. In summary, options that adopt core or platform products as well as options adopting customized products lead to higher profitability.

Research can be extended in several dimensions. First, the analysis can be extended to account for economies of product design. Second, the framework can be extended from a two-country scenario to a multi-country scenario. Another direction for future research is to examine the effect of parameter changes such as exchange rate and savings from economies of scale, in order to assess the impact of these changes on the product and pricing policies as well as the profitability of the international enterprises.

APPENDIX

Proof of Proposition 1

$\Pi_k^{\text{Centr-Stand-Cust}} \geq \Pi_k^{\text{Centr-Stand-Stand}}$ and $\Pi_k^{\text{Decentr-Stand-Cust}} \geq \Pi_k^{\text{Decentr-Stand-Stand}}$.

If we impose the constraint $P_{k1} = rP_{k2} = P_k$ on the profit functions (12) and (14) of the *Centr-Stand-Cust* and *Decentr-Stand-Cust* options, respectively, for any value $Attr_{kj}$ and $Attr_{mj}$, $m \neq k$, we obtain the following relationships:

$$\Pi_k^{\text{Centr-Stand-Cust}} \geq \Pi_k^{\text{Centr-Stand-Cust}}\Big|_{P_{k1} = rP_{k2} = P_k} = \Pi_k^{\text{Centr-Stand-Stand}},$$

$$\Pi_k^{\text{Decentr-Stand-Cust}} \geq \Pi_k^{\text{Decentr-Stand-Cust}}\Big|_{P_{k1} = rP_{k2} = P_k} = \Pi_k^{\text{Decentr-Stand-Stand}}$$

Since the above relationship holds true for any value of $Attr_{kj}$ and $Attr_{mj}$, $m \neq k$ then it will hold true at the Nash equilibrium point as well.

$\Pi_k^{\text{Decentr-Cust-Cust}} \geq \Pi_k^{\text{Decentr-Stand-Cust}}$.

If we impose the constraint $\mathbf{x}_{k1} = \mathbf{x}_{k2} = \mathbf{x}_k$ on the profit function (8) of the *Decentr-Cust-Cust* option, we obtain the profit function in (14) of the *Decentr-Stand-Cust* option, for any value of $Attr_{kj}$ and $Attr_{mj}$, $m \neq k$. As a result, we obtain the following relationship:

$$\Pi_k^{\text{Decentr-Cust}} \geq \Pi_k^{\text{Decentr-Cust}}\Big|_{\mathbf{x}_{k1} = \mathbf{x}_{k2} = \mathbf{x}_k} = \Pi_k^{\text{Decentr-Stand}}$$

Since the above relationship holds true for any value of $Attr_{kj}$ and $Attr_{mj}$, $m \neq k$ then it will hold true at the Nash equilibrium point as well.

$\Pi_k^{\text{Core-Decentr}} \geq \Pi_k^{\text{Centr-Stand-Cust}}$.

If we impose the constraint $\mathbf{x}_{k1} = \mathbf{x}_{k2} = \mathbf{x}_k^c = \mathbf{x}_k$ on the profit function (6) of the *Core-Decentr* option and take into account the inventory holding cost in (7) and (13), we obtain the profit function in (12) of the *Centr-Stand-Cust* option, for any value of $Attr_{kj}$ and $Attr_{mj}$, $m \neq k$. If the above constraints are imposed, the terms $\boldsymbol{\beta}_{k1}^{c\prime} \cdot (\mathbf{x}_{k1} - \mathbf{x}_k^c) - h(\mathbf{x}_{k1} - \mathbf{x}_k^c)K_{k1}^c M_{k1} Q_1$ and $\boldsymbol{\beta}_{k2}^{c\prime} .(\mathbf{x}_{k2} - \mathbf{x}_k^c) - h(\mathbf{x}_{k2} - \mathbf{x}_k^c)K_{k2}^c M_{k2} Q2$ drop to zero, since $h(\mathbf{x}_{k1} - \mathbf{x}_k^c) = 0$ and $h(\mathbf{x}_{k2} - \mathbf{x}_k^c) = 0$. This implies that no cost in incurred from customizing the core product and thus no savings from economies of scale are derived. As a result, we obtain the following relationship:

$$\Pi_k^{\text{Core-Decentr}} \geq \Pi_k^{\text{Core-Decentr}}\Big|_{\mathbf{x}_{k1} = \mathbf{x}_{k2} = \mathbf{x}_k^c = \mathbf{x}_k} = \Pi_k^{\text{Centr-Stand-Cust}}.$$

Since the above relationship holds true for any value of $Attr_{kj}$ and $Attr_{mj}$, $m \neq k$, then it will hold true at the Nash equilibrium point as well.

$\Pi_k^{\text{Core-Centr}} \geq \Pi_k^{\text{Centr-Cust-Cust}}$.

If we set $\mathbf{x}_k^c = \mathbf{0}$, $K_{k1} = K_{k1}^c$, $\boldsymbol{\beta}_{k1} = \boldsymbol{\beta}_{k1}^c$ and $z_{k1} = z_{k2}$ in the profit function in (4) of the *Core-Centr* option and take into account the inventory holding cost in (5) and

(11), we obtain the profit function in (10) of the *Centr-Cust-Cust* option, for any value of $Attr_{kj}$ and $Attr_{mj}$, $m \neq k$. Note that if $\mathbf{x}_k^c = \mathbf{0}$ then $h(\mathbf{x}_k^c) = \mathbf{0}$. This implies that at the above extreme case of $\mathbf{x}_k^c = \mathbf{0}$ the *Core-Centr* option will yield profits equal to those of the *Centr-Cust-Cust* option. This further implies that at the optimum, the least profits that the *Core-Centr* option can generate are those of the *Centr-Cust-Cust* option, i.e.,

$$\Pi_k^{\text{Core-Centr}} \geq \Pi_k^{\text{Core-Centr}} \Big|_{\mathbf{x}_k^c = 0,\, k_{k1} = k_{k1}^c,\, \beta_{\mathbf{k1}} = \beta_{\mathbf{k1}}^c,\, z_{k1} = z_{k2}} = \Pi_k^{\text{Centr-Cust}}$$

Since the above relationship holds true for any value of $Attr_{kj}$ and $Attr_{mj}$, $m \neq k$, then it will hold true at the Nash equilibrium point as well.

$\Pi_k^{\text{Core-Decentr}} \geq \Pi_k^{\text{Decentr-Cust-Cust}}$.

If we set $\mathbf{x}_k^c = 0$, $\boldsymbol{\beta}_{\mathbf{k1}} = \boldsymbol{\beta}_{\mathbf{k1}}^c$, $\boldsymbol{\beta}_{\mathbf{k2}} = \boldsymbol{\beta}_{\mathbf{k2}}^c$, $K_{k1} = K_{k1}^c$ and $K_{k2} = K_{k2}^c$ in the profit function in (6) of the *Core-Decentr* option and take into account the inventory holding cost in (7) and (9), we obtain the profit function in (8) of the *Decentr-Cust-Cust* option, for any value of $Attr_{kj}$ and $Attr_{mj}$, $m \neq k$. Note that if $\mathbf{x}_k^c = \mathbf{0}$ then $h(\mathbf{x}_k^c) = \mathbf{0}$. This implies that at the above extreme case of $\mathbf{x}_k^c = 0$ the *Core-Decentr* option will yield profits equal those of the *Decentr-Cust-Cust* option. This further implies that at the optimum, the least profits that the *Core-Decentr* option can generate, are those of the *Decentr-Cust-Cust* option, i.e.,

$$\Pi_k^{\text{Core-Decentr}} \geq \Pi_k^{\text{Core-Decentr}} \Big|_{\mathbf{x}_\mathbf{k}^c = 0,\, \beta_{\mathbf{k}1} = \beta_{\mathbf{k}1}^c,\, \beta_{\mathbf{k}2} = \beta_{\mathbf{k}2}^c,\, k_{\mathbf{k}1} = k_{k1}^c,\, k_{k2} = k_{k2}^c} = \Pi_k^{\text{Decentr-Cust}}$$

Since the above relationship holds true for any value of $Attr_{kj}$ and $Attr_{mj}$, $m \neq k$, then it will hold true at the Nash equilibrium point as well.

REFERENCES

Albers, S., Brockhoff, K. (1977), "A procedure for new product positioning in an attribute space," *European Journal of Operational Research,* 1(4), 230–238.

Bachem, A., Simon, H. (1981), "A product positioning model with costs and prices," *European Journal of Operational Research,* 7(4), 362–370.

Bartlett, C.A., Ghoshal, S. (1989), *Managing Across Borders: The Transnational Solution,* Harvard Business School Press, Boston.

Bell, D.E., Keeney, R.L., Little, J.D.C. (1975), "A market share theorem," *Journal of Marketing Research,* 12(2), 136–141.

Boddewyn, J.J., Soehl, R., Picard, J. (1986), "Standardization in international marketing: Is Ted Levitt in fact right?," *Business Horizons,* 29(6), 69–75.

Chang, O., Sauve, N. (2000), "Microtek to reach target, Umax looks to ODM order in 2001," *DigiTimes.com,* November 13.

Chase, R.B., Aquilano, N.J., Jacobs, F.R. (2001), *Production and Operations Management,* IRWIN, Chicago.

Choi, S.C., Desabro, W.S., Harker, P.T. (1990), "Product positioning under price competition," *Management Science*, 36(2), 175–199.

Chopra, S., Meindl, P. (2001), *Supply Chain Management*, Prentice Hall, Upper Saddle River, New Jersey.

Cooper, L.G., Nakanishi, M. (1988), *Market Share Analysis,* Kluwer Academic Press, Boston.

Corstjens, M.L., Gautschi, D.A. (1983), "Formal choice models in marketing," *Marketing Science,* 2(1), 19–56.

Czinkota, M.R., Ronkainen, I.A (1993), *International Marketing,* Dryden Press, Fort Worth.

Douglas, S.P., Wind, Y. (1987), "The myth of globalization," *Columbia Journal of World Business,* 22(4), 19–29.
Eliashberg, J., Manrai, A.K. (1992), "Optimal positioning of new–product concepts: Some analytical implications and empirical results," *European Journal of Operational Research,* 63(3), 376–397.
Feitzinger, E., Lee, H.L. (1997), "Mass customization at Hewlett–Packard: The power of Postponement," *Harvard Business Review,* 75(1), 116–121.
Ferdows, K. (1989), *Managing International Manufacturing,* North–Holland, Amsterdam.
Gruca, T.S., Kumar, K.R., Sudharshan, D. (1992), "An equilibrium analysis of defensive responses to entry using a coupled response function model," *Marketing Science,* 11(4), 348–358.
Hadjinicola, G.C. (1999) "Product positioning under production cost considerations," *Decision Sciences,* 30(3), 849–864.
Hadjinicola, G.C., Kumar, K.R. (1997),"Factors affecting international product design," *Journal of the Operational Research* Society, 48(11), 1131–1143.
Hayes, R.H.,Wheelwright, S.C. (1984), *Restoring Our Competitive Edge: Competing Through Manufacturing,* John Wiley & Sons, New York.
Hill, J.S., Still, R.R. (1984), "Adapting products to LDC tastes," *Harvard Business Review,* March–April, 92–101.
Hout, T., Porter, M.E., Rudden, E. (1982), "How global companies win out," *Harvard Business Review,* 60(5), 98–108.
Hua-hsin, C. (2002), "FIC, Foxconn, Asustek and Tatung to become HP's desktop suppliers in 2003," DigiTimes.com, June 26.
Huchzermeier, A., Cohen, M.A. (1996), "Valuing operational flexibility under exchange rate risk," *Operations Research,* 44(1), 100–113.
Jain, S.J. (1989). "Standardization of international marketing strategy: Some research hypotheses," *Journal of Marketing,* 53, 70–79.
Kasa, K. (1992), "Adjustment costs and pricing-to-market: Theory and evidence," *Journal of International Economics,* 32(1), 1–30.
Kashani, K. (1989). "Beware of the pitfalls of global marketing." *Harvard Business Review,* September-October, 91–98.
Kogut, B. (1985), "Designing global strategies: Comparative and competitive value-added chains," *Sloan Management Review,* 26(4), 15–28.
Kogut, B., Kulatikala, N. (1994), "Operating flexibility, global manufacturing, and the option value of a multinational network," *Management Science,* 40(1), 123–139.
Krishnan, V., Ulrich, K.T. (2001), "Product development decisions: A review of the literature," *Management Science,* 47(1), 1–21.
Kroll, D.E., Kumar, K.R. (1989), "The incorporation of learning in production planning models," *Annals of Operations Research,* 17, 291–304.
Kumar, K.R., Hadjinicola, G.C. (2000), "Product design for global markets," *Encyclopedia of Production and Manufacturing Management,* P. Swamidass (Ed.), Kluwer Academic Publishers, Boston, 559–563.
Lee, H.L. Billington, C., Carter, B. (1993), "Hewlett-Packard gains control of inventory and service through design for localization," *Interfaces,* 23(4), 1–11.
Lee, H.L., Billington, C. (1994), "Designing products and processes for postponement," *Management of Design: Engineering and Manufacturing Perspectives,* Dasu, S., and Eastman C. (Eds.),. Kluwer Academic Press, Boston, 105–122.
Lee, H.L., Billington, C. (1995), "The evolution of supply–chain management models at Hewlett–Packard," *Interfaces,* 25(5), 42–63.
Levitt, T. (1983), "The globalization of markets," *Harvard Business Review,* 61(3), 92–102.
McGrath, M.E., Hoole, R.W. (1992). "Manufacturing's new economies of scale." *Harvard Business Review,* May–June, 94–102.
McGuire, T.W., Farley, J.W., Lucas Jr., R.E., Ring, L.W. (1968), "Estimation and inference for linear models in which subsets of the dependent variables are constrained," *Journal of the American Statistical Association,* 63, 1201–1213.
Muffatto, M. (1999), "Introducing a platform strategy in product development," *International Journal of Production Economics,* 60, 145–153.
Pine, B.J. (1993), *Mass Customization: The New Frontier in Business Competition,* Harvard Business School Press, Boston.
Pine, B.J., Victor, B., Boynton, A.C., "Making Mass Customization Work," Harvard Business Review, 71 September-October, 108–111.
Porter, M.E. (1990), *The Competitive Advantage of Nations,* The Free Press, New York.

Prasad, S., Babbar, S. (2000), "International operations management research," *Journal of Operations Management,* 18(2), 209–247.
Reinhardt, A. (2002), "Nokia's next act," *Business Week,* July 1, 18–22.
Robertson, K., Ulrich, K. (1998), "Planning for product platforms," *Sloan Management Review,* 19–31.
Shapiro, B.S. (1977), "Can marketing and manufacturing coexist?," *Harvard Business Review,* September-October, 104–114.
Shi, Y., Gregory, M. (1998), "International manufacturing networks to develop global competitive capabilities," *Journal of Operations Management,* 16, 195–214.
Shocker, A.D., Shrinivasan, V. (1974), "A consumer-based methodology for the identification of new product ideas," *Management Science,* 20(6), 921–937.
Sorenson, R.Z.,Wiechmann, E.E. (1975), "How multinationals view marketing standardization," *Harvard Business Review,* 53(3), 38+.
Starr, M.K. (1965), "Modular production—A new concept," *Harvard Business Review,* 137–145.
Stalk, G., Hout, T.M. (1990), *Competing Against Time,* Free Press, New York.
Sudharshan, D., May, J.H., Gruca, T. (1988), "DIFFSTRAT: An analytical procedure for generating optimal new product concepts for differentiated-type strategy," *European Journal of Operational Research,* 36(1), 50–65.
Takeuchi, H., Porter, M.E. (1986), "Three roles on international marketing in global strategy," *Competition in Global Industries,* Porter, M.E. (Ed.), Harvard Business School Press, Boston, 111–146.
Thomke, S., Von Hippel, E. (2002), "Customers as innovators: a new way to create value," *Harvard Business Review,* 80(4), 74–81.
Tien, H.K., Teng, W. (2002), "Compal, Arima step up notebook production in China," *DigiTimes.com,* March 18.
Turino, J. (1992), *Managing Concurrent Engineering,* Van Nostrand Reinhold, New York.
Tzeng, D., Hua-hsin, C. (2002), "Compal rumoured to land Dell contract for 2.5 million notebooks," *DigiTimes.com,* May 24.
Ulrich, K.T., Eppinger, S.D. (2000), *Product Design and Development,* Irwin-McGraw-Hill, Bostson
Van Hoek, R.I. (2001), "The rediscovery of postponement, a literature review and directions of research," *Journal of Operations Management,* 19(2), 161–184.
Walters, P.G., Toyne, B. (1989), "Product modification and standardization in international markets: Strategic options and facilitating policies," *Columbia Journal of World Business,* 24(4), 37–44.
Yip, G.S. (1989), "Global strategy...in a world of nations?," *Sloan Management Review,* 31(1), 29–41.

11. THE MARKETING/MANUFACTURING INTERFACE: STRATEGIC ISSUES

UDAY S. KARMARKAR and MILIND M. LELE

ABSTRACT

Most firms separate the marketing and manufacturing functions into distinct organizational groups without fully considering the interactions and conflicts between the two functions. These interactions can have very significant costs for the firm, if they are not recognized. We discuss situations in which strategy and policy in the two areas are not synchronized. They include issues such as capacity decisions, inventory deployment and location in manufacturing; and price-promotion policy, forecasting and market intelligence in marketing. These topics are discussed in the context of case examples, which illustrate problems that can arise when interactions are not recognized.

1. INTRODUCTION

In most manufacturing firms, the marketing (advertising, pricing, sales and distribution) and manufacturing functions are organizationally separate. There are several underlying reasons for this separation. One has to do with the desire to achieve economies of scale with centralized manufacturing in few locations, while marketing and sales, by their very nature, are likely to be more concerned with the geographical distribution of customers. Another might be that manufacturing is typically associated with cost minimization, while marketing is concerned with revenue maximization. Certainly, it is true that rather different skills and different knowledge bases are required for the two areas, and the organizational forms, incentives and procedures are likely to be very different. At any rate, the end result is a functional separation, which despite the reasons adduced may lead to certain systematic problems.

A typical approach to separation of the functions is to designate the manufacturing function as a cost center, required to produce the required output at minimum cost. Marketing, on the other hand, might be required to maximize revenue net of marketing and sales expenses. Marketing thus sets prices and advertising policy; the market determines the quantities demanded and manufacturing produces the quantities demanded at minimum unit cost. On the face of it, this appears to be a reasonable decomposition of the goal of profit maximization. However, there are interactions between manufacturing and marketing that are not taken into account in such a decomposition, which results in conflicts between the functions due to inconsistent objectives or externalities imposed by one function on the other.

In the following, we focus on examples of such interactions. These examples are drawn from experience with several case studies including:

- A manufacturer of laboratory and medical products, sold worldwide through distributors,
- A fine chemicals manufacturer with direct and catalog sales,
- A manufacturer of electronic equipment sold internationally, directly and through distributors and
- A consumer products company with dedicated wholesalers.

We first discuss the effect of annual promotional cycles on the demand that is placed on manufacturing. We then describe the effect of capacity and location decisions in manufacturing on responsiveness to marketing situations. Finally, some aspects of forecasting and market intelligence on inventory costs are discussed.

2. THE PROMOTIONAL ROLLER-COASTER

In many companies, price promotions are instituted on a cyclical basis every year. For example, promotional drives may be instituted on a quarterly basis, which have the result of driving up sales at the end of each quarter. In many cases, it appears that these promotions primarily have the effect of shifting demand from one week to another to create artificial seasonalities. The effect on manufacturing is to create a seasonal demand pattern that leads to a varying load on capacity. One result is a need to hold seasonal inventory accumulations, which are built up in periods of low demand and then drawn down in periods of high demand, while leaving production levels fairly constant. If the seasonal variations become large enough, there may even be reason to add capacity, which starts to be driven to some extent by the seasonal peaks. Alternatively, instead of adding hard capacity (i.e., equipment), there may be a need to add labor capacity at peak demand levels or at least to incur overtime.

The reasons for these promotions are varied. A simple one is that sales quotas are reviewed on a quarterly basis and that as the end of the quarter approaches, any shortfall starts to become visible. This leads to a push to make the sales to meet the quota. On occasion, the push may arise because of quarterly internal reporting requirements. One company division that manufactures testing equipment sold through distributors is required to make quarterly reports to corporate headquarters. Due to the way in

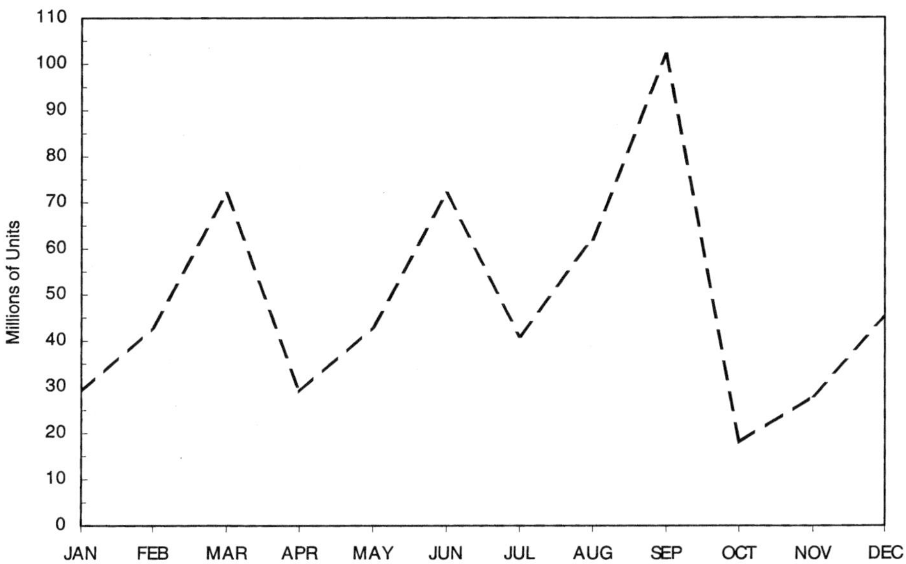

Figure 1. Seasonal demand pattern for an industrial product with quarterly peaks induced by promotion activity.

which profits for the division are computed, the cost-of-goods-sold (COGS) calculation looks more favorable if inventories are high and worse if they are low. This tends to lead to an end of quarter push either to make sales targets or at least to build to inventory, so that the report looks good. Another firm, which sells industrial products through distributors, has argued that promotions help the firm capture distributor's dollars and shelf space by periodic (quarterly) promotions. This is certainly a plausible idea, but the costs imposed by the policy on manufacturing and on inventory levels, may not always be apparent to the group setting the promotion and price policy.

Figure 1 shows the actual demand pattern observed for one particular product line of the manufacturer of industrial equipment. There is a pronounced quarterly pattern of demand, with peaks occurring every three months, starting in December and culminating in September. The pattern is apparently largely due to promotional activity. One way to determine this is to look at the end demand at the distribution level for the products under consideration. In this case (and with the testing equipment manufacturer as well), the products have no intrinsic seasonality in their end demand patterns. Furthermore, the peaks in demand as seen by manufacturing are highly correlated to the timing of promotions and price discounts. Figure 2 shows the production pattern for the product line and the seasonal inventory accumulation, as determined by a standard linear programming (LP) seasonal planning model. The production plan is actually nearly level, once an allowance is made for the varying number of days per month. The only variation is due to shutdown and vacation periods for the plant

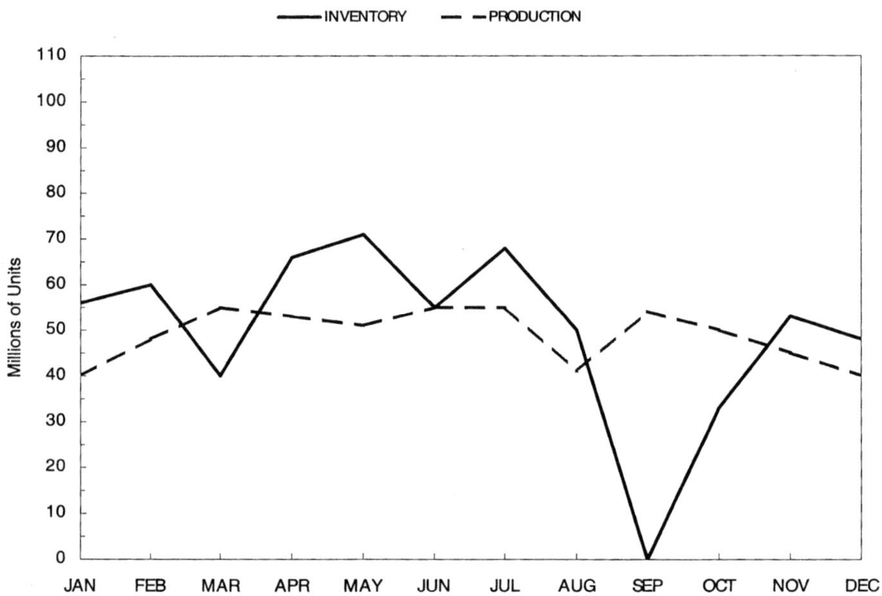

Figure 2. Aggregate production plan and seasonal inventory accumulation to meet the seasonal demand pattern with limited production capacity.

in August, December and January. Note the quarterly inventory accumulation pattern is approximately counter-cyclical to the demand pattern. We conducted an analysis of what would happen to inventory accumulation if demand was level. The resulting inventory pattern is shown in Figure 3, in which the inventories are only held to cover the shut down and vacation periods. The difference in inventories and holding costs from the previous case is obviously substantial.

In this particular case, it was a strongly held view that the promotional policy was an important tool in the competition for distribution assets. However, note that commitment by the distributor to purchase a large enough quantity is enough to capture market share. It is not really necessary to deliver the purchased material immediately; in fact, it only occupies the distributor's space and requires more storage capacity. One possibility is to get a commitment from the distributor to purchase, and then to commit to delivery of the material only as it is needed (just-in-time). Seasonality of demand at the plant level is thus effectively removed. This would also reduce the distributor's inventories and space requirements, and decrease the need for advance forecasting of demand on the distributor's part. The match between production and demand can be closer with less need for safety stocks. At the same time, the purpose of the promotional activity would be served.

In the case of the testing equipment manufacturing division with quarterly reporting requirements, it should be noted that the COGS calculation requires con-

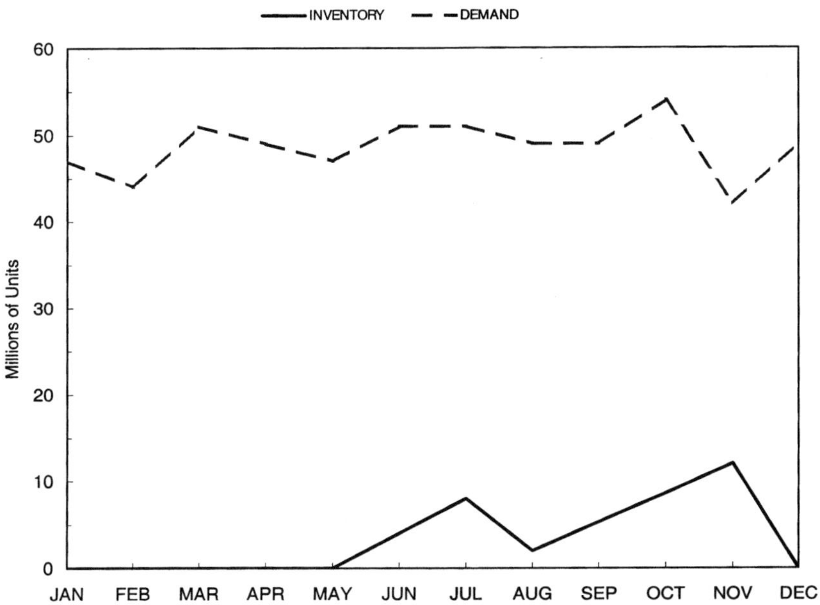

Figure 3. Inventory accumulation under deseasonalized (level) demand conditions.

tinually increasing inventory levels to be manipulated. Obviously, this cannot be kept up indefinitely with level demand. In particular, turn measures start to deteriorate even though quarterly "profitability" may look good due to a poor measurement scheme. If it were recognized that increased inventory level is not an asset, and if the COGS calculation is changed (or the effect on the cost of inventory holding is assessed correctly) the incentive for the sales push would largely go away.

3. PLANNING MODELS FOR EVALUATING CAPACITY STRATEGIES AND DEMAND VARIATIONS

The problem of aggregate capacity allocation and seasonal inventory planning is often formulated as a linear program (Hanssman and Hess, 1960; Hax, 1978; Saad, 1982). Two of our case companies used such models for seasonal planning of manufacturing facilities. This familiarity allowed us to use similar LP models to evaluate both the implications of promotional activity for capacity planning and the effect of large demand surges. For illustration purposes, we will assume that all products for the plant in question are aggregated into a single product. This is reasonable in situations where the products are similar, and have similar demand patterns and costs. Such was the case in one of our case examples. In the other, a multi-product analysis was necessary. Time periods are indexed by *t,* and the model minimizes variable production costs and inventory holding costs.

Define:

X_t: Production in period t (units)
I_t: Inventory at end of period t
c_t: Variable cost of production ($/unit)
h_t: Cost of holding inventory for one period ($/unit)
P_t: Production capacity in period t (units)
D_t: Demand in period t (units)

The model may be written as:

$$\text{Min } \Sigma_t c_t X_t + \Sigma_t h_t I_t$$

$$\begin{aligned} \text{s.t.} \quad I_{(t-1)} + X_t - I_t &= D_t, \quad \text{each } t = 1, \ldots, T \\ X_t &\leq P_t, \quad \text{each } t = 1, \ldots, T \\ X_t, I_t &\geq 0, \quad \text{each } t = 1, \ldots, T \end{aligned}$$

In addition, the starting and ending inventory levels may either be given or be set equal to simulate a repeating cycle. The production level in each period may not be equal due to differences in the number of days in each period and to holidays and plant closedowns. Similarly, the demand, though actually quite level, may also vary by time period due to these reasons. It is straightforward to include extensions such as multiple products, lost sales, backorders, overtime and extra shifts, which are often important concerns in planning plant capacity over a budgeting cycle.

In order to evaluate the effect of promotion policy, the model is run with demand based on actual historical pattern, or the forecast demand pattern using a seasonal forecasting model. The result for one of our case examples gives the pattern shown in Figure 1. This is then compared with the level demand case, under the assumption that the production output of the plant will closely match end market demand. Of course, some allowance has to be made for transportation times, as well as pipeline filling in the distribution channel. The result for our example case is shown in Figure 2. This type of planning model can also be used to evaluate capacity and inventory strategies to address sudden surges in demand. We discuss this topic in the next section.

4. MANUFACTURING CAPACITY AND DEMAND SURGES

The cost minimization role given to manufacturing leads to capacity decisions aimed at maximum utilization of capacity. Unit costs are typically calculated as the total volume of production divided by the total costs of equipment (amortization) and operation. Naturally, the higher the capacity utilization, the lower the unit cost of production. Consequently, there is a tendency to have very little excess capacity at any time. This implies level production so that seasonal inventory accumulation is required if demand is seasonal. In addition, if an unpredictable demand surge occurs,

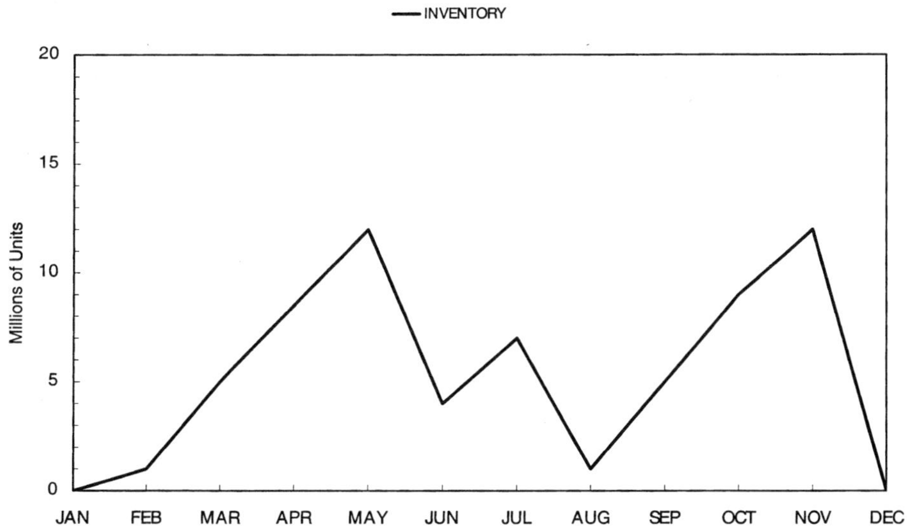

Figure 4. Inventory accumulation pattern for level demand with a 10% demand surge in june.

the time taken to produce the necessary material can be very long, because there is no volume flexibility left in the system.

The industrial products company mentioned earlier experienced significant demand surges due to large military and government contracts that it would sometimes win, often from countries in Europe or Southern Asia. When such a contract was won, the resulting surge would clean out inventories of certain products. There appear to be only two ways of handling these surges. One is to hold buffer inventories and the other is to hold excess capacity. Holding inventories is not a viable policy, because the inventories have to be built up before the surge occurs. If the contract is not won, the inventories are in excess and production is left "holding the bag." Furthermore, the exact composition of the inventory required is not always easy to predict. As a result, the protective inventory held as a substitute for capacity will have to be larger than the capacity cushion that is required. In particular, the tighter the capacity constraint, the longer the lead time to build up inventories, and the larger the safety stock component due to uncertainty in demand. In many cases, bidding and contracting cycles limit the lead time available for building safety buffers. Finally, an effective inventory policy often requires more cooperation between marketing and production than is usually forthcoming. Furthermore, such surge inventories should logically be under the ownership of marketing and sales, since they are caused by bid activity; however, manufacturing is often asked to hold the inventories.

One way to show the time taken to work off the effects of a demand surge is to consider how long it would take to produce the order, if advance notice of the surge was available. Figure 4 shows the results for a specific case of a demand surge equal to about 10% of annual demand in the month of June. It has been assumed in this example that demand does not have promotional seasonality, but that shutdown and vacations periods are in effect. An LP model is used to determine the production plan. It can be seen that it takes about six months to prepare for the surge. This is a conservative analysis; in the presence of seasonality and if the surge occurred with little warning, the situation would be worse. Incidentally, we found managers to be surprised that the surge took more than a month's time to compensate for by preparation or reaction. There appeared to be the fallacious impression that a 10% surge should take 10% of the year to fix. However, the real issue is the surge volume relative to excess capacity, not to total demand. Furthermore, if the surge coincides with the approach of a seasonal peak, capacity may be even tighter than on average.

The development of a suitable capacity and inventory plan to handle surges is quite complex. It depends upon many factors including the timing of the bid cycle, the information available about the contract requirements, the frequency of occurrence of such opportunities, the probability of winning the contract and the delivery requirements of the contract. Such uncertainties are likely to exist in any situation where demand surges occur. However, the LP model can be used as a tool to analyze the effect of a surge. In particular, the time taken to work off a surge can be used to make a rough estimate of the inventory level needed to buffer against the surge (it is actually very conservative since it assumes that the size and timing of the surge is known). The costs of such inventory buffers can be compared with the cost of having excess capacity available. A complication here is that the excess capacity can also be used to reduce seasonal stocks at times when there are no surges, and the two issues have to be considered together. Our experience is that there is often a very strong argument to be made to deliberately hold excess capacity as a strategic buffer for markets with combinations of uncertain, seasonal and surge demands.

5. MANUFACTURING CAPACITY, LEAD TIMES AND DOWNSTREAM INVENTORY

A direct effect of high capacity utilization for regular operations is that lead times in manufacturing increase. These increased lead times mean that larger safety stocks are required in finished inventory to protect against the longer replenishment times. This effect is quite distinct from the seasonal inventory and surge response issues dealt with above, although there is an underlying similarity.

The mechanism behind the increased manufacturing lead times is as follows: as capacity utilization is increased, there is less time available for setups and changeovers from product to product. As a result, run lengths (batch sizes) have to be larger. Consequently, the length of the production cycle is increased (i.e., the time between successive runs of a given product increases). This cycle time is, roughly speaking, proportional to the lead time for a product, since it determines the waiting time for an order to be filled in having to wait for the next run of the product. As this lead time increases, so does the safety stock that has to be held, since that depends on the standard deviation of product demand over the lead time. An early discussion of this problem can be found in Magee (1956).

Figure 5 shows cycle time in days for a particular synchronous production facility

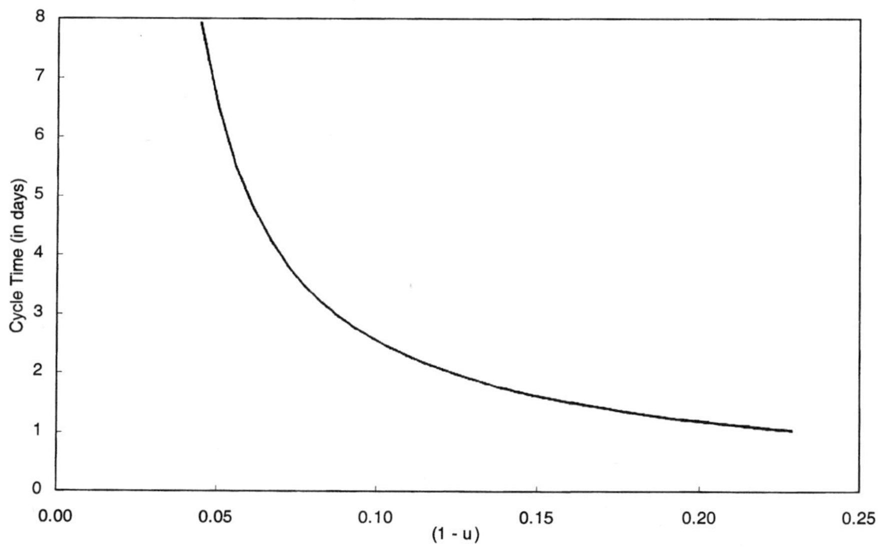

Figure 5. Cycle time (in days) as a function of excess capacity (1-u).

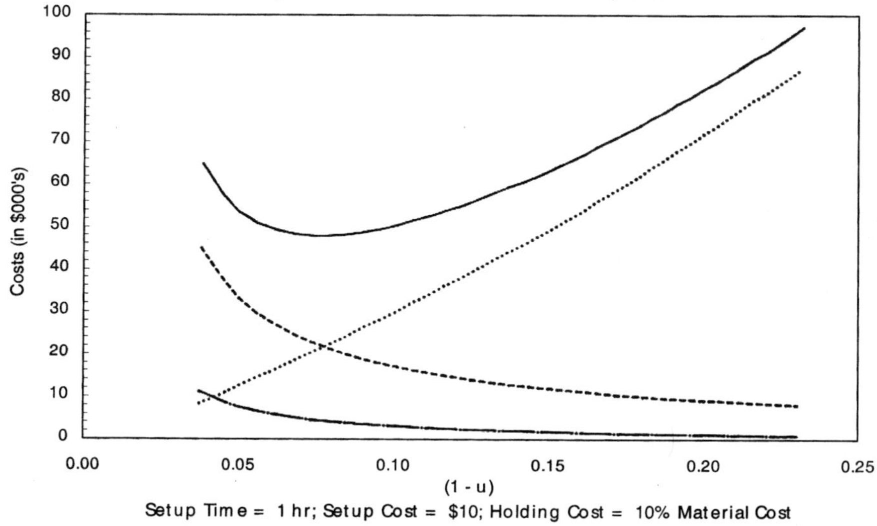

Figure 6. Inventory holding costs (safety and cycle stocks) and setup costs as a function of excess capacity (1-u).

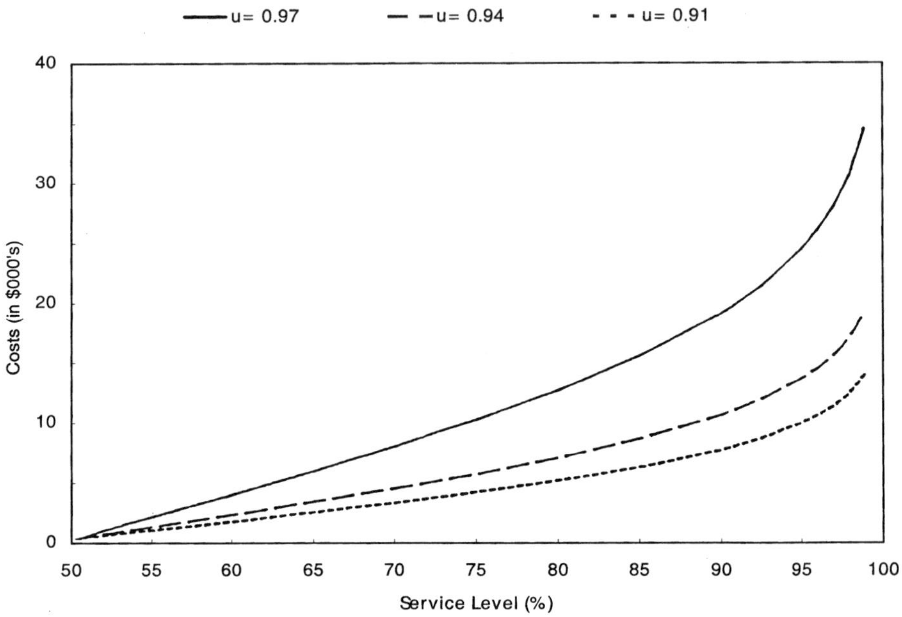

Figure 7. Safety stock costs as a function of service level for different levels of capacity utilization.

as a function of the fraction of excess capacity. In other words, 0.1 on the x-axis indicates 10% excess production capacity in terms of time availability. Note that the cycle time rises sharply as utilization is driven to 100%, and as excess capacity is driven to zero. The model used to determine the cycle time is described in the next section. The synchronous or paced facility is actually a very good situation in terms of the lead time implications; the problem is much worse for asynchronous facilities such as dynamic batch shops, where there are severe congestion and queuing problems. It appears that in the latter case, the effect of high capacity utilization can be worse by an order of magnitude (Karmarkar, Kekre, Kekre and Freeman, 1985).

The effect of capacity utilization on safety stocks, on cycle stocks and on setup costs (if present) is shown in Figure 6 for a particular case. Note that safety stocks dominate cycle stocks, which in our experience is often, if not always, the case. In fact, this calculation assumes that there is just one finished goods stocking location; the situation would be worse for stock levels in an extended multi-echelon distribution system. Furthermore, the safety stock costs are larger as service levels are increased. Figure 7 shows the relationship between safety stock costs and service levels for different capacity utilization levels.

In many cases, the inventory implications of capacity utilization are so significant that they can justify the deliberate addition of excess capacity. Note that the excess capacity has to be considered in terms of its impact on lead times. In the case under

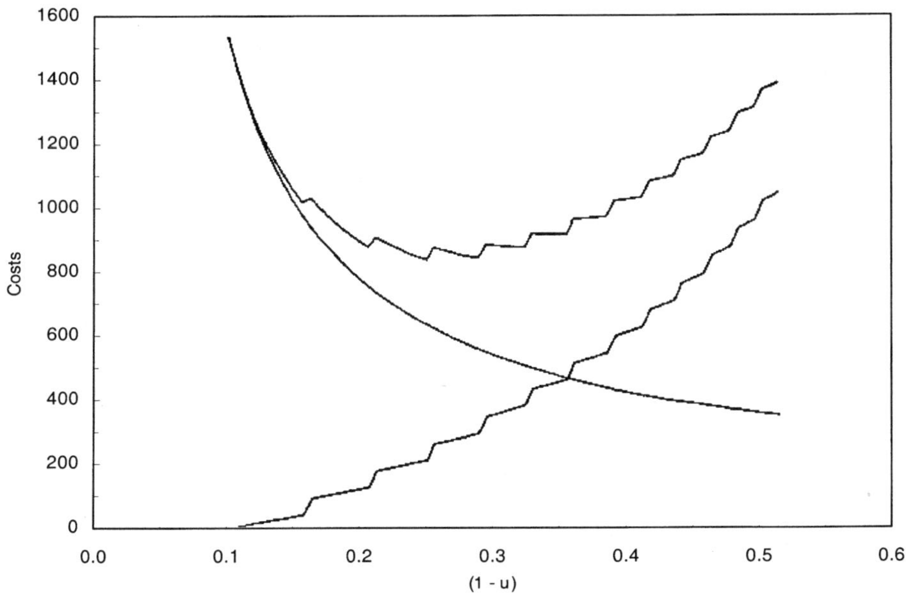

Figure 8. Inventory costs (cycle and safety stock holding) and capacity costs (overtime and added shifts) relative to excess capacity (1-u).

consideration, excess capacity already existed in terms of actual equipment. The capacity addition considered was in terms of overtime and additional shifts on certain facilities. Figure 8 shows the effect of adding overtime and shift capacity on inventory and setup costs. The stepped curve reflects that overtime can only be tolerated up to a certain average per period, beyond which an extra shift has to be added. For the particular numbers that are being considered the total cost curve reaches its minimum at the point where two shifts are added and overtime is used to the maximum allowable level on available equipment. At this point, there is about 13% excess capacity deliberately being created to permit shorter production cycles.

6. MODELING EXCESS CAPACITY, LEAD TIME AND INVENTORY

The relationship between capacity, manufacturing lead times and inventory consequences of lead times depends on the nature of the manufacturing facility, as stated in the previous section. In our case example, since production was undertaken in a high volume, synchronous or paced facility, the relation between capacity and lead time could be modeled using a conventional "rotation cycle" model for batching. This model is appropriate when all the products (or families) produced in the facility are produced in each cycle because of similar demand levels, setup times and inventory costs. More generally, products may have differential characteristics, and may then be produced with different frequencies. In those cases, a more general ELS (Economic

Lot Scheduling) model may have to be employed to determine production patterns (Elmaghraby, 1988; Dobson, 1989). For the rotation cycle analysis, let i index products, and define:

T: Length of a production cycle (time)
s_i: Setup time for the ith product (time)
D_i: Demand rate for the ith product (units/time)
P_i: Production rate for the ith product (units/time)

The production rate increases, in terms of units produced per week, month or year, as machines, overtime, capacity or shifts are added. It is well known that the cycle time T must satisfy:

$$T \geq \Sigma_i s_i \,/\, (1 - \Sigma_i(D_i \,/\, P_i))$$

If there are substantial setup costs, then the cycle time may be larger than the right-hand side of this inequality, which also implies that there will be idle time during the cycle. In the latter case, capacity may be decreased, if possible, to eliminate the idle time. In actual practice with synchronous production, no idle time is seen and the relation holds as an equality. Note that the denominator on the right-hand side is precisely excess capacity.

The cycle time T may at a first approximation be taken as manufacturing lead time. When a production batch of a particular item is completed, there is an interval of approximately T until production is undertaken again. Hence, the safety stock maintained in the system has to cover the appropriate fractile of demand over the cycle time T, so as to provide the desired level of service and product availability. Of course, this model can be refined further, but for illustrative purposes, this simple approximation suffices. The lead time, or the time for the rotation cycle, determines both the "cycle stock" due to batch size as well as the level of safety inventories. The cycle stock is simply $D_iT/2$ for the ith product. The safety stock may be approximated by $kT\beta\sigma_i$ for the ith product, where k depends on the service level to be maintained, β is a factor between ½ and 1 depending on the degree of correlation in the demand, and σ_i is the standard deviation of demand per period for the ith product.

7. MANUFACTURING PRESENCE

The economics of manufacturing, considered in isolation, often favor large, centralized plants exploiting economies of scale. Again, this is a direct consequence of performance measures based on minimizing unit cost. It has begun to be recognized that there are diseconomies of scale even at the plant level that are often overlooked. Even more significant are some of the diseconomies of centralization that arise from the interactions with distribution, marketing and sales. One of these has to do with manufacturing "presence." Plant location is often the result of historical accident; at least, it is frequently the outcome of decisions made under circumstances that no longer exist. Even when the decision is current, market factors are sometimes only weakly reflected in the tradeoffs made. Typically, transportation costs are the only factor considered; and, given economies of scale arguments, they may only cause small shifts in the location of one or two plants.

There is a set of issues that becomes particularly important in the context of global markets. In these markets, it is often the case that service tradeoffs and product design requirements are quite different from the "home" market in which the plant is situated. For example, in the case of a Far East division of the industrial equipment company, the service requirements for fill rate from inventory were simply stated to us as being "100%." The concept of a tradeoff between inventory level and service level was not seen as being acceptable, and the inventory to provide 100% service was not perceived as being infinite or even too costly. As one of the managers commented, "If a distributor does not fill an order to a major customer, he would have to go with his senior managers as a group to the customer and, standing in a row before his counterpart, apologize profusely for the missed order." This might be hyperbole; but there is no doubt that the service requirements were much more stringent than the home market. At the very least, missing a shipment was likely to mean a customer permanently lost. Unfortunately, this issue was simply not perceived with enough seriousness at the central manufacturing plant, which then appeared to be indifferent to the division's concerns. In many cases, if there was a shortage of available production, the tendency was for it to be allocated to orders from the home market rather than to divisions abroad. This was a natural consequence of the home marketing divisions having more "clout" with the plant, partly due to their geographical (and cultural) proximity. The idea of establishing a "Far East Desk" at the plant site was seriously proposed as a way to circumvent the communication gap.

Similar issues arose with respect to product design. In Europe, minor product improvements made by local competitors were not being matched, since to a U.S. plant they simply meant added manufacturing costs with no visible benefits. Similarly, products designed to U.S. or Western ergonometric standards, did not work well in Eastern markets. For example, certain kinds of equipment proved to be difficult to operate for someone with a smaller-sized hand. Again, the response from manufacturing was listless.

These problems are directly traceable first to communication and transactions costs, and second to the fact that the objectives for manufacturing do not include any incentives for responding to foreign marketing and sales groups. The organizational route to exerting pressure on manufacturing runs from the foreign marketing division up to the division's leadership, across to the corresponding senior level of the home division and down to the manufacturing function. This is not a smooth communication channel and is unlikely to be used to expedite a given shipment or to reallocate inventories. Since organizational form is so heavily driven by geography, there is no obvious reporting or incentive mechanism for inducing a high level of responsibility towards the offshore marketing groups. Although these concerns are intangible and hard to quantify, they provide real reasons for considering a more decentralized manufacturing strategy. Of course, this does not mean that it is necessary to have every product manufactured in every market area. Just the presence of some basic manufacturing capabilities and a local branch of the manufacturing function improves communication and provides more organizational leverage. It is in fact useful to route communication from the manufacturing function to the manufacturing function across geographic divisions rather than the more circuitous route described earlier, which is forced to drive decisions higher up the corporate structure.

An intangible benefit to locating manufacturing closer to customers is that the firm has more credibility. This is likely to be much more important in the case of sales to OEM's and in situations where the design of the product is of concern to the user. However, it may also be true of consumer goods. A rather subtle effect of manufacturing "presence" is that the firm is taken more seriously because the investment in facilities implies a commitment to the market. In a sense, the commitment to facilities is viewed as a bonding mechanism. Incidentally, this is true not only for customers, but also for the firm itself; it acts as an internal signal that the firm is serious about competing in the market.

While many of these effects are hard to quantify, they can have substantial implications for marketing success. When the tangible tradeoffs are close, or are a small proportion of the total costs, these intangible issues can become the decisive ones.

8. BEACH-HEAD INVENTORIES AND NORMAL INVENTORIES

When a new product is introduced or when new markets are penetrated, normal inventory control methods are inappropriate. First, the usual forecasting methods used for stable conditions are inapplicable. Second, the tradeoff between service and inventory costs must be slanted heavily in favor of service, since the opportunity costs of not meeting demand in a new market are likely to be very high. Notice that both of these issues are such that manufacturing can have no control over them; indeed, it is unrealistic to require manufacturing to make the necessary assessments. Yet, it is very common to find that in practice, the responsibility for setting and maintaining finished product inventory levels falls on manufacturing practically by default.

In addition, the demand pattern with new products or new geographic markets is likely to be much more variable even past the early startup stage. This demand pattern, or more accurately its forecast, cannot be controlled or improved by manufacturing. Furthermore, the variability inherent in the demand is likely to create uncertainties in production quantities.

With new products and markets, inventories play a strategic role in the process of establishing a beachhead. The information necessary to set these inventories lies with the marketing and sales function, and the benefits also flow in large part to that function. The responsibility for determining these levels and the ownership of the material should both be placed with the marketing/sales function. The role of manufacturing is to provide the rapid response necessary to support the mission. However, the performance that is expected has to be based on what is possible. For example, excess capacity may be the best way to provide good responsiveness to new market needs. If so, manufacturing cannot be penalized because their unit production cost measure of performance is then calculated to include the cost of strategic capacity. Alternatively, if inventories are to be accumulated, then advance planning is required to build these up. As the surge example earlier showed, the time to build up adequate inventories with highly utilized facilities could be very long.

Even under normal circumstances, the amount of finished goods inventory held depends partly on the lead time in manufacturing and partly on forecast accuracy. Specifically, service inventories are proportional to the standard deviation of demand forecast over lead time, as was discussed earlier in the section on capacity and lead times. One scheme, which assigns responsibilities for these interactions in a reasonable way, is to have the marketing/sales function be responsible for inventories, while manufacturing is responsible for lead times. Groenevelt and Karmarkar (1987) discuss the application of such an approach in a pull system.

9. FORECASTS, PLANS AND MARKET INTELLIGENCE

For any product line that experiences seasonal demands, stochastic demand variations, mix changes and volume changes due to competitive action, forecasting is a critical input for production and sales activity management. It is typically the case that the primary responsibility for forecasting demands falls on the sales and marketing function. This is as it should be, since it represents the location of the best information about the marketplace. Furthermore, sales and promotion activity can directly affect demand. It is very difficult for manufacturing to develop any real sense of the marketplace since the function is typically buffered from the market and has no direct contact with customers. Marketing and sales, therefore, typically provide forecasts for long-term planning, as well as short-term inventory and production management.

A very common problem with this arrangement arises first from confusion between forecasts and plans, and secondly from the very different incentives associated with forecasts in different functions. The former issue is relatively straightforward in the manufacturing function: care should be taken to distinguish between the best expectation, or forecast, of demand and the level of production that is planned to meet this demand. The issue is less clear in sales and marketing. The point is that the actual demand that is seen is very heavily dependent on pricing, promotions and sales call activity. It is very difficult to distinguish an underlying market demand from the demand that will be seen as a result of planned sales activity. The earlier discussion on artificial demand seasonality engendered by promotion policies emphasizes this point.

A second issue of incentives is related to the confusion between plans and forecasts in sales and marketing. Sales managers asked to make "forecasts" of sales in their regions have an incentive to present an aggressive and optimistic sales plan. Furthermore, the most important numbers in this plan are aggregate sales dollar figures, and there is less emphasis on the specific mix that will be seen. Ex post too, if the observed mix does not match "forecasts," then there are no significant repercussions in sales as long as overall revenue and contribution expectations are realized. Obviously, the situation is very different in production and inventory management. Here, inventories are held at the item or SKU level; if the actual mix seen varies from the forecast mix, the result is simultaneously high inventory levels and poor service. Similarly, in production scheduling and purchasing, capacity choices, production schedules and purchasing activity are very dependent upon the actual mix to be produced.

As a result of these differences, it is typical to have the production function second-guess forecasts obtained from marketing. Unfortunately, this does not always work well simply because of the lack of market intelligence in the production function. If a product or item has relatively stable demand patterns, then using conventional forecasting techniques based on booked orders can work. However, if demand is significantly dependent on competitive action or is influenced strongly by sales and marketing activity, then automated forecasting at the inventory or production level can be a futile and misleading exercise.

The issue of forecasts is closely related to that of inventory ownership, discussed at the end of the preceding section. In the case of forecasts, their quality affects inventory levels directly. One way to give the sales function incentives to make accurate forecasts is to let them decide on the level of finished inventories to be maintained—accumulations, as well as buffers—and to then give them ownership of the inventories. Then, the responsibility of production is to keep inventories at the desired level. In some firms, the desired inventory levels are set by agreement in the course of an annual planning exercise. Such a planning discussion also allows commitments to aggregate capacity to be made by production, based on aggregate sales forecasts. A large manufacturer of electronic equipment uses this scheme. Implementation of the planning agreements is done through a two-tier production control system in which a "push" scheme controls the planned inventories, while a "pull" scheme allows rapid reaction to short-term changes in volume (limited) and mix.

10. REFERENCE NOTES ON CROSS-FUNCTIONAL INTEGRATION

This paper emphasizes observations collected over many years of industrial project work. Of course, the topic of cross-functional integration has not escaped other researchers and practitioners and there are many other sources in the research and applications literature that address the area. On our own part, two overviews of the subject can be found in Karmarkar (1993, 1996). We have also dealt with the implementation realities of an integrated and strategic solution in the area of post-sales product support, where a number of cross-functional interactions arise (Lele and Karmarkar, 1985).

One of the earliest general discussions of the topic is by Shapiro (1977), who addresses the organizational conflicts that can arise between manufacturing and marketing. Subsequently, there have been many articles that have emphasized that competitive success requires attention to cross-functional issues such as quality and time. An example is the paper by Stalk (1988). There have also been a number of research efforts studying the marketing-manufacturing interface from a variety of perspectives. Among the contributors to this literature are Chakravarty et al. (1992, 1996, 1997), Dobson and Kalish (1988, 1993), Eliashberg and Steinberg (1987, 1993), Hausman and Montgomery (1993) and Karmarkar and Pitbladdo (1995, 1997). We do not presume to give a complete survey of this growing field here.

11. SUMMARY

The conventional separation between the marketing/sales and the manufacturing functions has been to regard manufacturing as a supplier to sales. Manufacturing responsibility is seen as supplying products at the required times in the required amounts, while minimizing costs. The marketing and sales functions then carry the responsibility for revenue generation, including pricing and promotion decisions. This conventional approach can be inadequate when rapidly changing market conditions or increased competition require a finer degree of coordination across the interface between the functions. In this paper, we have described some strategic issues related to this interface. In many of these cases, the specific problems raised can be solved on a case basis. However, a basic and interesting question remains unanswered: is there a general scheme for coordinating these functions that will remove these issues? It may be that despite the increased costs, the only answer is close coordination, increased communication across the interface and perhaps even organizational integration.

REFERENCES

Chakravarty, Amiya (1992). "Supporting Manufacturing-Marketing Interface Management with a Partially Automated Knowledge Base—A Precursor to Computer Integrated Business Enterprise." *Journal of Intelligent Manufacturing*. 3: 347–362.

Chakravarty, Amiya and N. Balakrishnan (1996). "Managing Engineering Change: Market Opportunities and Manufacturing Costs." *Production and Operations Management*. 5(4): 335–356.

Chakravarty, Amiya, N. Balakrishnan and S. Ghose (1997). "Role of Design Philosophy in Interfacing Manufacturing with Marketing." *European Journal of Operational Research*. 103: 453–469.

Dobson, G. (1987). "The Economic Lot-Scheduling Problem: Achieving Feasibility Using Time-Varying Lot Sizes." *Operations Research*. 35: 764–771.

Dobson, G. and S. Kalish (1988). "Positioning and Pricing a Product Line." *Marketing Science*. 7(2): 107–125.

Dobson, G. and S. Kalish (1993). "Heuristics for Pricing and Positioning a Product Line Using Conjoint and Cost Data." *Management Science*. 39: 160–175.

Eliashberg, J. and R. Steinberg (1987). "Marketing-Production Decisions in an Industrial Channel of Distribution." *Management Science*. 33(8): 981–1000.

Eliashberg, J. and R. Steinberg (1993). "Marketing-Production Joint Decision-Making." *Handbooks in OR & MS*. Eliashberg, J. and G.L. Lilien (eds.). Elsevier Science Publishers. 3: Chapter 18.

Elmaghraby, S.E. (1978). "The Economic Lot Scheduling Problem (ELSP): Review and Extensions." *Management Science*. 24: 587–598.

Groenevelt, H. and U.S. Karmarkar (1988). "A Dynamic Kanban System Case Study." *Production and Inventory Management Journal*. 29: 46–50.

Hanssman, F. and F.W. Hess (1960). "A LP Approach to Production and Employment Scheduling." *Management Technology*. 1: 46–51.

Hausman, Warren and D. Montgomery (1993). "The Manufacturing/Marketing Interface: Critical Strategic and Tactical Linkages." *Perspectives in Operations Management: Essays in Honor of Elwood S. Buffa*. R. Sarin (ed.). Kluwer Academic Publishers.

Hax, A.C. (1978). "Aggregate Production Planning." *Handbook of Operation Research*. Van Nostrand, Reinhold. J. Moder and S. E. Elmaghraby (eds.). 2: Chapter 5.

Karmarkar, U.S. (1993). "Research in Manufacturing Strategy: A Cross-Functional Perspective." *Perspectives in Operations Management: Essays in Honor of Elwood S. Buffa*. R. Sarin (ed.). Kluwer Academic Publishers.

Karmarkar, U.S. (1996). "Integrative Research in Marketing and Operations." *Journal of Marketing Research*. 33: 125–133.

Karmarkar, U.S., S. Kekre, S. Kekre and S. Freeman (1985). "Lot Sizing and Lead Time Performance in a Manufacturing Cell." *Interfaces*. 15: 1–9.

Karmarkar, U.S. and M. Lele (1983). "Good Product Support is Smart Marketing." *Harvard Business Review*. 61: 124–132. Reprinted in *Marketing Renaissance*: J. Wiley and Sons (1987). Reprinted in *Keeping Customers*, Sviokla, J. and B. Shapiro (eds.): Harvard Business Press (1993).

Karmarkar, U.S. and R.C. Pitbladdo (1995). "Service Markets and Competition." *Journal of Operations Management*. 12: 397–411.

Karmarkar, U.S. and R.C. Pitbladdo (1997). "Quality, Class and Competition." *Management Science*, 43, 27–39

Magee, J.F. (1956). "Guides to Inventory Policy: Part II." *Harvard Business Review.* 34: 103–116.
Saad, G.H. (1982). "An Overview of Production Planning Models: Structural Classification and Empirical Assessment." *International Journal of Production Research.* 20(1): 105–114.
Shapiro, B.P. (1977). "Can Marketing and Manufacturing Co-Exist?" *Harvard Business Review.* 55:104–114.
Stalk, G. Jr. (1988). "Time - The Next Source of Competitive Advantage." *Harvard Business Review.* 66(4): 41–51.

SUBJECT INDEX